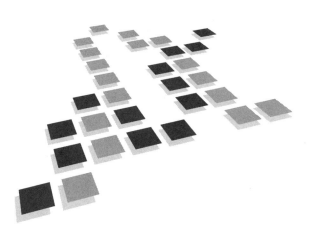

An SPSS Companion to Political Analysis

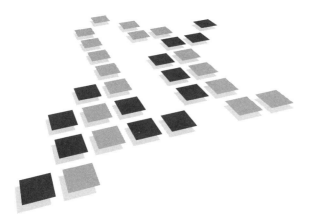

An SPSS Companion to Political Analysis

Third Edition

Philip H. Pollock III
University of Central Florida

CQ PRESS

A Division of SAGE
Washington, D.C.

CQ Press
2300 N Street, NW, Suite 800
Washington, DC 20037

Phone: 202-729-1900; toll-free, 1-866-4CQ-PRESS (1-866-427-7737)

Web: www.cqpress.com

Credits: Datasets on the CD: **GSS2006, GSS2006A_Student, and GSS2006B_Student:** selected variables obtained from General Social Surveys, 1972–2006 (machine-readable data file). Principal investigator, James A. Davis; director and co-principal investigator, Tom W. Smith; co-principal investigator, Peter V. Marsden, NORC ed. Chicago: National Opinion Research Center (producer), 2007; Storrs, Conn.: Roper Center for Public Opinion Research, University of Connecticut/Ann Arbor, Mich.: Inter-university Consortium for Political and Social Research (distributors). **NES2004, NES2004A_Student, and NES2004B_Student:** selected variables obtained from the American National Election Study, 2004: Pre- and Post-Election Survey (computer file). ICPSR04245-v1. Ann Arbor, Mich.: University of Michigan, Center for Political Studies, American National Election Study (producer), 2004. Ann Arbor, Mich.: Inter-university Consortium for Political and Social Research (distributor). **World:** all variables obtained from Pippa Norris, John F. Kennedy School of Government, Harvard University, www.pippanorris.com, Global Indicators Shared Dataset V2.0 (updated Fall 2005 [September 13, 2005]), http://ksghome .harvard.edu/~pnorris/Data/Data.htm. **States:** selected variables obtained from Gerald C. Wright, Indiana University. See Robert S. Erikson, Gerald C. Wright, and John P. McIver, *Statehouse Democracy* (Cambridge, U.K.: Cambridge University Press, 1993). The screenshot on page 227, "Americans for Democratic Action (ADA) 2003 Senate Voting Records," is reprinted with the permission of Americans for Democratic Action.

Cover design by Auburn Associates Inc., Baltimore, Md.
Typesetting by BMWW, Baltimore, Md.

♾ The paper used in this publication exceeds the requirements of the American National Standard for Information Sciences—Permanence of Paper for Printed Library Materials, ANSI Z39.48-1992.

Printed and bound in the United States of America

12   11   10   09   3   4   5

ISBN 978-0-87289-607-9

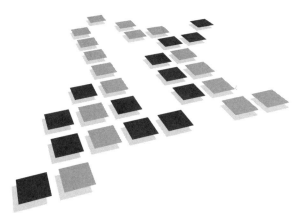

Contents

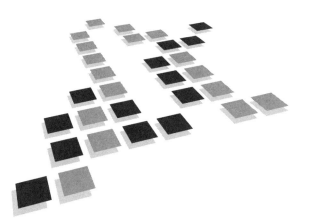

Tables and Figures

Preface

Ever since I began teaching research methods some 30 years ago, I have tried to introduce my students to the joys of doing their own political analysis. After all, that's the best way to learn it. Students who appreciate the practical side of research are better prepared to contribute to class discussions of methodological concepts and problems. Moreover, students often realize the salutary side benefit of developing a usable skill that they can hone as they continue their academic careers or pursue employment opportunities. To achieve these pedagogic and career-building outcomes, students need a solid foundation in basic political analysis techniques. They need to learn how to manipulate variables, explore patterns, and graph relationships. And they need a working knowledge of powerful yet easy-to-learn software, such as SPSS. This book instructs students in using SPSS for Windows to construct meaningful descriptions of variables and to perform substantive analysis of political relationships. The chapters cover all major topics in data analysis, from descriptive statistics to logistic regression. A final chapter describes several doable research projects, shows how to collect and code data, and lays out a framework for a well-organized research paper.

In its essential features—multiple datasets, guided examples, screenshots, graphics instruction, and end-of-chapter exercises—this book continues in the tradition of previous editions. Yet there are some noteworthy changes in emphasis, content, and sequence, more than a few of which were inspired by adopters' suggestions.

DATASETS AND EXERCISES

The SPSS datasets have been thoroughly revised and updated. The CD that accompanies this book contains four SPSS data files: selected variables from the 2006 General Social Survey (dataset GSS2006) and the 2004 National Election Study (NES2004), as well as datasets on the 50 states (States) and 191 countries of the world (World). An ASCII file containing information on the 2003 U.S. Senate (senate2003), used in Chapter 11 to demonstrate how to code and read data into SPSS, is the only dataset that I carried over from the second edition. As before, each chapter is written as a step-by-step tutorial, taking students through a series of guided examples and providing many annotated screenshots. Because of the revised and updated datasets, all of the examples are fresh and all of the screenshots are new.

This book contains 55 end-of-chapter exercises, 15 more than the second edition. All of the exercises are designed to give students opportunities to apply their new skills, and all engage students in discovering the meaning of their findings and learning to interpret them. In this edition I have included more exercises that reflect current scholarly debates in American political behavior and issues in comparative politics. The exercises in this volume test a fuller range of competency, and most chapters include at least one more-challenging exercise. Even so, I continue to assume that students using this workbook have never heard of SPSS and have never used a computer to analyze data. After completing this book, students will have become competent SPSS users, and they will have learned a fair amount about substantive political science, too.

Any student who has access to SPSS—the full version or the student version—can use this book. The States and World datasets (and the senate2003 raw data file) are fully compatible with the student version.

Because NES2004 meets one student version limitation (number of cases) but not another (number of variables), I split the dataset into two student version-compatible sets. Student version users will analyze NES2004A_Student or NES2004B_Student. SPSS output from these sets is identical to the output from the full-version file, NES2004. Compatible versions of GSS2006 (GSS2006A_Student and GSS2006B_Student) also are included. To stay within the student version's 1,500-case maximum, I created GSS2006A_Student and GSS2006B_Student from a random sample drawn from GSS2006. SPSS output from these sets will be similar to, but not the same as, output from the full-version GSS2006 dataset. To perform logistic regression, covered in Chapter 10, students will need access to the full version of SPSS.

DIFFERENT RELEASES OF SPSS

SPSS 16 is featured here, but anyone running release 12.0 or later can profitably use this book. There are many commonalities across releases 12 through 16, including the graphic dialogs and the Chart Editor. Although SPSS now uses one editor for all graphics output, there are three ways to obtain unedited charts: the Legacy Dialogs, Interactive graph, and (for release 14 or later) Chart Builder. Interactive graph is showing its age—SPSS no longer supports a dedicated editor for these charts—and so it is not covered in this book. Chart Builder was meant to be a one-stop, easy-to-use vehicle for constructing any chart, from the simple to the complex. However, the Legacy Dialogs offer superior flexibility and intuitiveness. Thus, for creating unedited charts this book uses the Legacy Dialogs, not Chart Builder. (In Chapter 4, endnote 1, I provide a fuller justification for this choice.) In any event, this edition carries forward the emphasis on elegant graphic display to complement and clarify empirical results. I have sought to instruct students in using the Chart Editor to emulate the techniques advocated by Edward R. Tufte and other experts on the visual display of data.

CHAPTER ORGANIZATION

With one exception, the chapters are organized as before. The "Getting Started" introduction describes the datasets, alerts students to differences between the full and student versions, and describes how to install the student version software. Chapter 1 introduces the SPSS Data Editor, discusses the output Viewer, and illustrates the print procedure. Chapter 2 covers central tendency and dispersion and guides students in using the Frequencies routine. I have added coverage of Case Summaries, which can be quite useful for providing insights into small datasets, such as States and World. Chapter 2 also shows how a frequency distribution, examined in conjunction with a bar chart or histogram, can enrich the description of a variable. Chapter 3, which appeared as Chapter 4 in earlier editions, describes the main SPSS data transformation procedures, Recode and Compute. I have added a discussion of Visual Binning (labeled Visual Banding through release 15), a powerful and efficient alternative to Recode, especially for collapsing interval variables into ordinal categories of roughly equal size. In Chapter 4, which covers Crosstabs and Compare Means, students learn bivariate analysis. Chapter 4 introduces line charts and bar charts, and students are given an initial tour of the Chart Editor. In Chapter 5 students use Crosstabs and Compare Means to obtain and interpret controlled comparisons. The chapter also discusses graphic support for controlled relationships. Chapter 6 uses One-Sample T Test and Independent-Samples T Test to demonstrate statistical significance for interval-level dependent variables. I have expanded the discussion of one-tailed versus two-tailed tests of significance—a persistent point of confusion for students. Chapter 7 covers chi-square and measures of association for nominal and ordinal variables. In Chapter 8 students work through an extended guided example to learn Correlate (Bivariate) and Regression (Linear). Chapter 8 discusses advanced editing using the Chart Editor. Chapter 9 shows how to create dummy variables (essentially an application of Recode, which students learned in Chapter 3), perform dummy variable regression analysis, and model interaction in multiple regression. Chapter 10 covers binary logistic regression, including a discussion of how to present logistic regression results in terms of probabilities. Chapter 11 guides students as they collect, code, and analyze their own data. With the help of the text data file included with this book (the senate2003 dataset), students learn to use the Read Text Data function. This experience greatly enhances their ability to create and manipulate their own SPSS datasets.

These chapters are organized in the way that I normally teach my methods courses. I prefer to cover the logic of description and hypothesis testing before introducing inferential statistics and statistical significance. However, with a little rearranging of the chapters, this book will prove useful for instructors who do things differently. For example, after covering basic data transformations (Chapter 3) and discussing cross-tabulation analysis (Chapter 4), an instructor could assign Chapter 7, which covers chi-square and bivariate measures of association for categorical variables. Instructors who prefer using the regression approach to evaluating the statistical significance of mean differences might decide to skip Chapter 6 and move on to Chapters 8 and 9.

ACCOMPANYING CORE TEXT

Instructors will find that this book makes an effective supplement to any of a variety of methods textbooks. However, it is a particularly suitable companion to my own core text, *The Essentials of Political Analysis*. The textbook's substantive chapters cover basic and intermediate methodological issues and ideas: measurement, explanations and hypotheses, univariate statistics and bivariate analysis, controlled relationships, sampling and inference, statistical significance, correlation and linear regression, and logistic regression.

Each chapter also includes end-of-chapter exercises. Students can read the textbook chapters, do the exercises, and then work through the guided examples and exercises in *An SPSS Companion to Political Analysis*. The idea is to get students in front of the computer, experiencing political research firsthand, fairly early in the academic term. An instructor's solutions manual, available for download online and free to adopters, provides solutions for all the textbook and workbook exercises.

ACKNOWLEDGMENTS

I received more than a few friendly e-mails suggesting ways to improve this book. I am grateful for this advice. Many thanks, as well, to the following reviewers for pointing me in the right direction: Jason Kehrberg, University of Kentucky; Thad Kousser, University of California, San Diego; Brian Vargus, Indiana University–Purdue University Indianapolis; and Julian Westerhout, Illinois State University. I thank my University of Central Florida colleagues Bruce Wilson and Kerstin Hamann for helping me with ideas for exercises on comparative politics. I thank Kerri Milita for her excellent technical review, and I am grateful to Erin Suzanne Greene for her editorial insights. I offer special thanks to Joe Simons-Rudolph of North Carolina State University for suggesting that I aim the exercises at a wider range of competencies and ground them more firmly in current scholarly debate. Thanks to Pete Furia of Wake Forest University and Bill Claggett of Florida State University for sharing their SPSS know-how with me. It was Claggett who reminded me of an ancient flaw in Compute: Multiply 0 times missing, and SPSS interprets the product as 0, not as missing. This quirk—and how to avoid it—is discussed in Chapter 9. Many encouraging people have helped me make this a better book. Any remaining errors, however, are mine.

I gratefully acknowledge the encouragement and professionalism of everyone associated with the College Division of CQ Press: Charisse Kiino, acquisitions editor; Steve Pazdan, managing editor; Amy Marks, copy editor; Gwenda Larsen, senior production editor; and Allie McKay, editorial assistant.

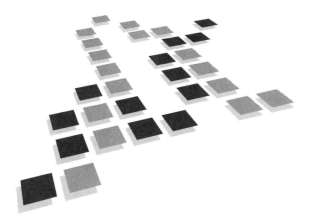

An SPSS Companion
to Political Analysis

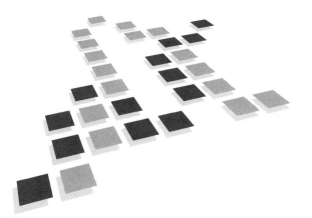

Getting Started

To get started with this book you will need
- Access to a Microsoft Windows–based computer
- The CD-ROM that accompanies this workbook
- A USB flash drive or other portable media

As you have learned about political research and explored techniques of political analysis, you have studied many examples of other people's work. You may have read textbook chapters that present frequency distributions, or you may have pondered research articles that use cross-tabulation, correlation, or regression analysis to investigate interesting relationships between variables. As valuable as these learning experiences are, they can be enhanced greatly by performing political analysis firsthand—handling and modifying social science datasets, learning to use data analysis software, obtaining your own descriptive statistics for variables, setting up the appropriate analysis for interesting relationships, and running the analysis and interpreting your results.

This book is designed to guide you as you learn these valuable practical skills. In this volume you will gain a working knowledge of SPSS, a data analysis package used widely in academic institutions and business environments. SPSS has been in use for many years (it first appeared in 1968), and it contains a great variety of statistical analysis routines—from basic descriptive statistics to sophisticated predictive modeling.[1] It is extraordinarily user friendly. In fact, although this book assumes that you have practical knowledge of the Windows operating system and that you know how to perform elemental file-handling tasks, it also assumes that you have never heard of SPSS and that you have never used a computer to analyze data of any kind. By the time you complete the guided examples and the exercises in this book, you will be well on your way to becoming an SPSS aficionado. The skills you learn will be durable, and they will serve you well as you continue your educational career or enter the business world.

This book's chapters are written in tutorial fashion. Each chapter contains several guided examples, and each includes exercises at the end. You will read each chapter while sitting in front of a computer, doing the analyses described in the guided examples, and analyzing the datasets that accompany this text. Each data analysis procedure is described in step-by-step fashion, and the book has many figures that show you what your computer screen should look like as you perform the procedures. Thus the guided examples allow you to develop your skills and to become comfortable with SPSS. The end-of-chapter exercises allow you to apply your new skills to different substantive problems.

This book will provide you with a solid foundation in data analysis. You will learn to obtain and interpret descriptive statistics (Chapter 2), to collapse and combine variables (Chapter 3), to perform cross-tabulation and mean analysis (Chapter 4), and to control for other factors that might be affecting your results (Chapter 5). Techniques of statistical inference (Chapters 6 and 7) are covered too. On the somewhat more advanced side, this book introduces correlation and linear regression (Chapter 8), and it teaches you how to use dummy variables and how to model interaction effects in regression analysis (Chapter 9). If you are running the full version of SPSS, Chapter 10 provides an introduction to logistic regression, an analytic technique that has gained wide currency in recent years. Chapter 11 shows you how to code raw data and read it into SPSS, and it provides guidance on writing up your results.

DATASET CD

The CD accompanying this book contains five datasets that you will analyze using SPSS. Insert the CD into your computer's drive and let's take a preliminary look at the datasets (Figure I-1). Ignoring (for the moment) the "Student" datasets—these are discussed below—five different datasets are available on the CD.

1. GSS2006 (For Student Version users: GSS2006A_Student and GSS2006B_Student). This dataset has selected variables from the 2006 General Social Survey, a random sample of 4,510 adults aged 18 years or older, conducted by the National Opinion Research Center and made available through the Inter-university Consortium for Political and Social Research (ICPSR) at the University of Michigan.[2] Some of the scales in GSS2006 were constructed by the author. These constructed variables are described in the appendix (Table A-1).

2. NES2004 (NES2004A_Student and NES2004B_Student). This dataset includes selected variables from the 2004 National Election Study, a random sample of 1,212 citizens of voting age, conducted by the University of Michigan's Institute for Social Research and made available through ICPSR.[3] See the appendix (Table A-4).

3. States. This dataset includes variables on each of the 50 states. Most of these variables were compiled by the author from various sources. A complete description of States is found in the appendix (Table A-2).

4. World. This dataset includes variables on 191 countries of the world. These variables are based on data compiled by Pippa Norris, John F. Kennedy School of Government, Harvard University, and made available to the scholarly community through her Internet site.[4] A complete description of World appears in the appendix (Table A-3).

5. senate2003 This dataset, compiled by the author, includes information on each member of the 2003 U.S. Senate. You won't use senate2003 until Chapter 11.

Figure I–1 Datasets on the CD–ROM

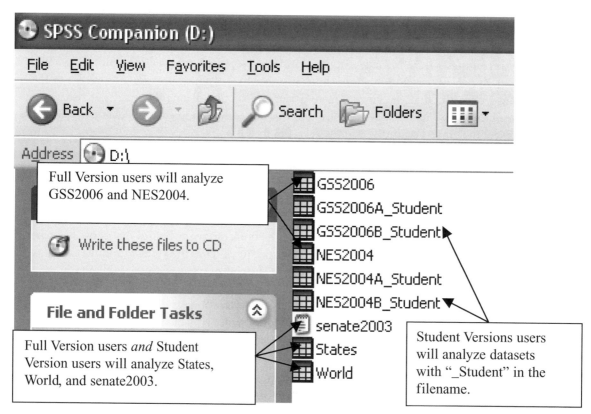

As you work your way through this book, you will modify these datasets—recoding some variables, computing new variables, and otherwise tailoring the datasets to suit your purposes. You will need to make personal copies of the datasets and store them on a removable drive, such as a USB flash drive.

 DO THIS NOW: Before proceeding, copy the datasets from the accompanying CD to a personal USB drive or other portable media.[5]

When you begin each chapter's guided examples, or when you do the exercises, you will want to insert your personal media into the appropriate computer drive. SPSS will read the data from the drive. (Chapter 1 covers this operation in detail.) If you make any changes to a dataset, you can save the newly modified dataset directly to your drive. Alternatively, your lab's administrator may permit you to work on datasets that have been copied to the lab computer's desktop or to a folder designated for such a purpose. In any case, if you have modified a dataset during a data analysis session, it is important that you copy the dataset to your personal drive and take the datasets with you! (A comforting thought: The original datasets are still stored safely on the CD-ROM.)

FULL VERSION SPSS AND STUDENT VERSION SPSS: WHAT IS THE DIFFERENCE?

The book you purchased contains only the dataset CD. The CD provides the data, but your campus computer lab provides the software. Alternatively, you may have purchased SPSS Student Version software along with the workbook. After you install the Student Version (covered below), you can run SPSS from your PC or laptop.

In terms of the guided examples and exercises in this book, how does Student Version compare with Full Version SPSS? Here are four facts worth knowing:

1. When a guided example or exercise calls for the States dataset or the World dataset (and, in Chapter 11, the senate2003 data), Student Version output will be identical in every respect to Full Version output. As shown in Figure I-1, everyone will analyze these datasets.
2. When a guided example or exercise calls for NES2004, Student Version users will analyze NES2004A_Student (Chapter 1 through Chapter 6) or NES2004B_Student (Chapter 7 through Chapter 9). Student Version output will be identical to Full Version output in every detail.
3. When a guided example or exercise calls for GSS2006, Student Version users will analyze GSS2006A_student (Chapter 1 through Chapter 8) or GSS2006B_Student (Chapter 9). Student Version output will not be the same as Full Version output. They will be close, but they won't be the same.
4. SPSS Student Version will not perform logistic regression, covered in Chapter 10. If your instructor plans to cover logistic regression, and you are running Student Version, then you will need to use your computer lab's Full Version installation to perform the analyses described in Chapter 10.[6]

These similarities and differences are summarized in the table that follows:

Version of SPSS				
Full Version		Student Version		
Dataset	Chapters	Dataset	Chapters	Same output as Full Version?
GSS2006	All	GSS2006A_Student GSS2006B_Student	1 through 8 9	No No
NES2004	All	NES2004A_Student NES2004B_Student	1 through 6 7 through 9	Yes Yes
States	All	States	All	Yes
World	All	World	All	Yes
senate2003	11	senate2003	11	Yes

INSTALLING STUDENT VERSION

If you purchased the Student Version, *now* would be a good time to install it on your PC or laptop. This software follows a standard installation protocol. Relevant instructions follow. (You will also find these instructions in the file Installation Instructions.pdf on the Student Version CD.[7])

1. Insert the Student Version CD-ROM into your CD-ROM drive. The AutoPlay feature presents a menu.
2. On the AutoPlay menu, click Install SPSS for Windows Integrated Student Version, and then follow the instructions that appear on the screen. (Note: If you already have another version of SPSS installed, then install the new version in a separate directory.)

You can also run the Setup program manually by following these steps:

1. From the Windows Start menu, choose Run.
2. In the Run dialog box, type D:\setup. (This assumes that drive D is your CD-ROM drive. If your CD-ROM resides on a different drive, then substitute that drive's letter for D.) Follow the instructions that appear on the screen.

NOTES

1. Originally, SPSS was an acronym for Statistical Package for the Social Sciences. SPSS developed many applications beyond the social sciences—for example, market research and data mining—and is now known simply as SPSS.
2. GSS2006 was created from the General Social Surveys 1972–2006 Cumulative Data File. James A. Davis, Tom W. Smith, and Peter V. Marsden. General Social Surveys, 1972-2006 [Cumulative File] [Computer file]. ICPSR04697-v2. Chicago: National Opinion Research Center [producer], 2007. Storrs, CT: Roper Center for Public Opinion Research, University of Connecticut/Ann Arbor, MI: Inter-university Consortium for Political and Social Research [distributors].
3. University of Michigan, Center for Political Studies, American National Election Study. American National Election Study, 2004: Pre- and Post-Election Survey [Computer file]. ICPSR04245-v1. Ann Arbor, MI: University of Michigan, Center for Political Studies, American National Election Study [producer], 2004. Ann Arbor, MI: Inter-university Consortium for Political and Social Research [distributor].
4. See www.pippanorris.com.
5. In the open CD window, click Edit → Select All. This selects all the datasets on the CD. Click Edit → Copy. Open your removable media. Click Edit → Paste. After the datasets have been copied, remove your portable media and the CD-ROM and put them in a safe place.
6. Student Version will not handle datasets containing more than 50 variables or having more than 1,500 cases. The States and World datasets fall comfortably below these limits. NES2004 passes muster with the number-of-cases limitation ($N = 1,212$), but it has more than 50 variables.

   To get around this limitation, the author split NES2004 into two sets (NES2004A_Student and NES2004B_Student), each of which contains fewer than 50 variables. Because all 1,212 cases are included in both the "A" and "B" datasets, Student Version users will obtain the same SPSS output as Full Version users.

   GSS2006 has many more than 50 variables, and it also violates the limitation on number of cases ($N = 4,510$). Therefore, the author took a random sample ($N = 1,500$) from the full dataset. This random sample set was divided into GSS2006A_Student and GSS2006B_Student, each of which contains fewer than 50 variables. Because the "A" and "B" datasets do not have all 4,510 cases, Student Version output will not look the same as the output generated with Full Version SPSS.
7. SPSS for Windows Integrated Student Version installation instructions.

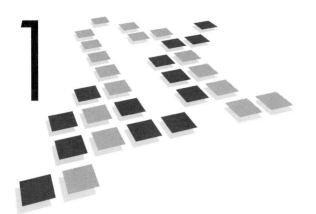

Introduction to SPSS

Suppose you were hired by a telephone-polling firm to interview a large number of respondents. Your job is to find out and record three characteristics of each person you interview: the region of the country where the respondent lives, how liberal or conservative the respondent claims to be, and the respondent's age. The natural human tendency would be to record these attributes in words. For example, you might describe a respondent this way: "Respondent #1 lives in the North Central United States, is politically 'moderate,' and is 50 years old." This would be a good thumbnail description, easily interpreted by another person. To SPSS, though, it would make no sense at all. Whereas people excel at recognizing and manipulating words, SPSS excels at recognizing and manipulating numbers. This is why researchers devise a *coding system,* a set of numeric identifiers for the different values of a variable. For one of the above variables, age, a coding scheme would be straightforward: Simply record the respondent's age in number of years, 50. In recording information about region and liberalism-conservatism, however, a different set of rules is needed. For example, the National Election Study (NES) applies these codes for region and liberalism-conservatism:

Variable	Response	Code
Region	Northeast	1
	North Central	2
	South	3
	West	4
Liberalism-Conservatism	Extremely liberal	1
	Liberal	2
	Slightly liberal	3
	Moderate	4
	Slightly conservative	5
	Conservative	6
	Extremely conservative	7

Thus the narrative profile "lives in the North Central United States, is politically 'moderate,' and is 50 years old" becomes "2 4 50" to SPSS. SPSS doesn't really care what the numbers stand for. As long as SPSS has numeric data, it will crunch the numbers—telling you the mode of the region codes, for example, or the median of liberalism-conservatism codes, or the mean age of all respondents. It is important, therefore, to provide SPSS with labels for each code so that the software's analytic work makes sense to the user. Accordingly, the SPSS Data Editor has two "views." The Data View shows the codes that SPSS recognizes and analyzes. The Variable View, among other useful features, shows the word labels that the researcher has assigned to the numeric codes.

THE DATA EDITOR

Let's open the 2004 National Election Study dataset, NES2004, and see how this works. (If you are running SPSS Student Version, open NES2004A_Student. You will use NES2004A_Student through Chapter 6.) Insert into the computer's appropriate drive (for example, the USB port) the storage device containing your personal copies of the datasets. You created these copies in the Getting Started chapter. After it recognizes the drive, Windows will probably ask you whether you want to open its files. If this offer is not forthcoming, double-click the My Computer icon and then double-click the icon representing your computer's drive where the datasets are stored. Double-click the NES2004 icon. (Alternatively, you may want to open NES2004 after you have copied it onto the desktop. SPSS will run faster if the dataset is located on the computer's hard drive.)

SPSS opens the data file and displays the Data Editor (Figure 1-1). Notice the two tabs at the bottom of the window: Data View and Variable View. Let's look at the Data View first. (Make sure the Data View tab is clicked.) This shows how all the cases are organized for analysis. Information for each case occupies a separate row. The variables, given brief yet descriptive names, appear along the columns of the editor. Because you are familiar with a few of the codes, you can tell that the first case—coded 1 on V040001, the 2004 NES preelection wave case identification number—is 50 years old. If you were to scroll to the right, you would also find that this respondent is a self-described "moderate" (coded 4 on the variable named libcon7_r) and resides in the North Central region (coded 2 on the variable region). To paint a more complete word-portrait of this respondent, however, you need to see how all the variables are coded. To reveal this information, click the Variable View tab (Figure 1-2). This view shows complete information on the meaning and

Figure 1–1 SPSS Data Editor: Data View

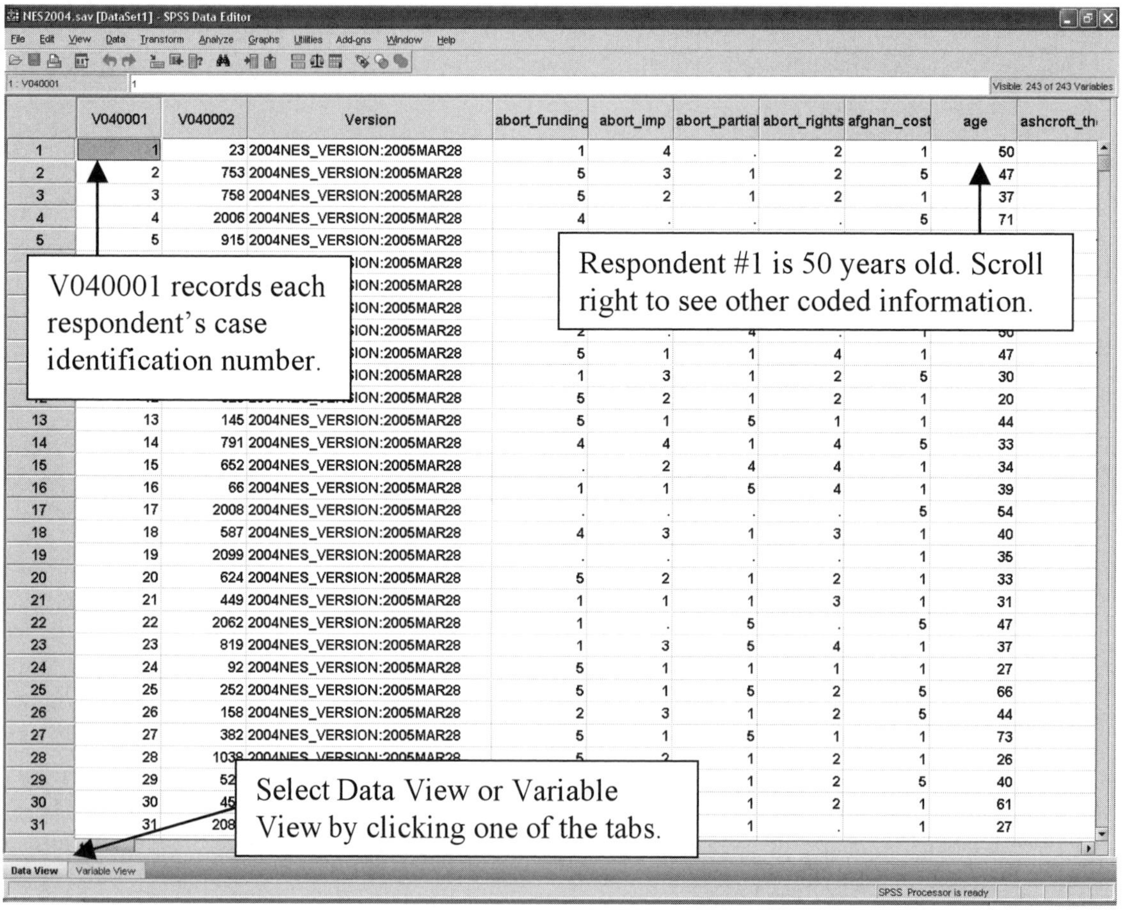

Figure 1–2 SPSS Data Editor: Variable View

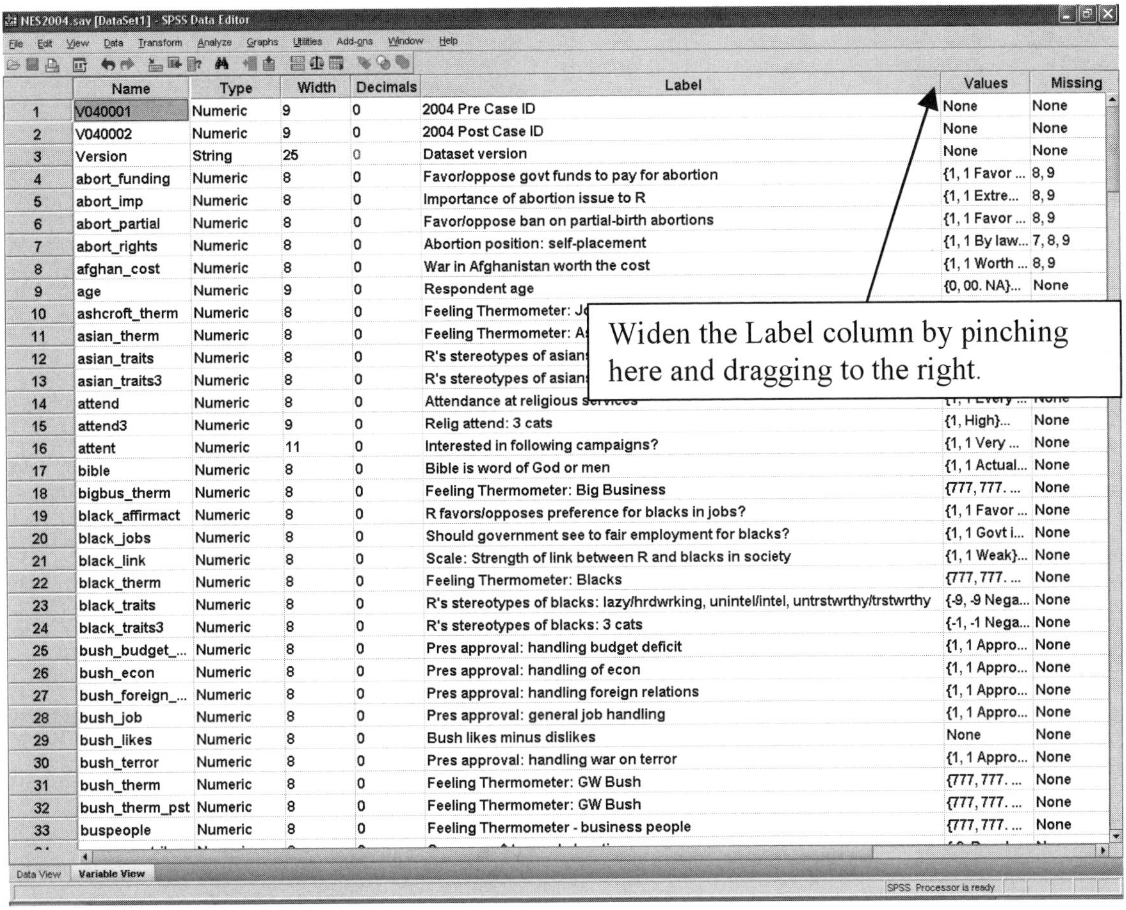

	Name	Type	Width	Decimals	Label	Values	Missing
1	V040001	Numeric	9	0	2004 Pre Case ID	None	None
2	V040002	Numeric	9	0	2004 Post Case ID	None	None
3	Version	String	25	0	Dataset version	None	None
4	abort_funding	Numeric	8	0	Favor/oppose govt funds to pay for abortion	{1, 1 Favor ...	8, 9
5	abort_imp	Numeric	8	0	Importance of abortion issue to R	{1, 1 Extre...	8, 9
6	abort_partial	Numeric	8	0	Favor/oppose ban on partial-birth abortions	{1, 1 Favor ...	8, 9
7	abort_rights	Numeric	8	0	Abortion position: self-placement	{1, 1 By law...	7, 8, 9
8	afghan_cost	Numeric	8	0	War in Afghanistan worth the cost	{1, 1 Worth ...	8, 9
9	age	Numeric	9	0	Respondent age	{0, 00. NA}...	None
10	ashcroft_therm	Numeric	8	0	Feeling Thermometer: J		
11	asian_therm	Numeric	8	0	Feeling Thermometer: A		
12	asian_traits	Numeric	8	0	R's stereotypes of asian		
13	asian_traits3	Numeric	8	0	R's stereotypes of asian		
14	attend	Numeric	8	0	Attendance at religious services	{1, 1 Every ...	None
15	attend3	Numeric	9	0	Relig attend: 3 cats	{1, High}...	None
16	attent	Numeric	11	0	Interested in following campaigns?	{1, 1 Very ...	None
17	bible	Numeric	8	0	Bible is word of God or men	{1, 1 Actual...	None
18	bigbus_therm	Numeric	8	0	Feeling Thermometer: Big Business	{777, 777. ...	None
19	black_affirmact	Numeric	8	0	R favors/opposes preference for blacks in jobs?	{1, 1 Favor ...	None
20	black_jobs	Numeric	8	0	Should government see to fair employment for blacks?	{1, 1 Govt i...	None
21	black_link	Numeric	8	0	Scale: Strength of link between R and blacks in society	{1, 1 Weak}...	None
22	black_therm	Numeric	8	0	Feeling Thermometer: Blacks	{777, 777. ...	None
23	black_traits	Numeric	8	0	R's stereotypes of blacks: lazy/hrdwrking, unintel/intel, untrstwrthy/trstwrthy	{-9, -9 Nega...	None
24	black_traits3	Numeric	8	0	R's stereotypes of blacks: 3 cats	{-1, -1 Nega...	None
25	bush_budget_...	Numeric	8	0	Pres approval: handling budget deficit	{1, 1 Appro...	None
26	bush_econ	Numeric	8	0	Pres approval: handling of econ	{1, 1 Appro...	None
27	bush_foreign_...	Numeric	8	0	Pres approval: handling foreign relations	{1, 1 Appro...	None
28	bush_job	Numeric	8	0	Pres approval: general job handling	{1, 1 Appro...	None
29	bush_likes	Numeric	8	0	Bush likes minus dislikes	None	None
30	bush_terror	Numeric	8	0	Pres approval: handling war on terror	{1, 1 Appro...	None
31	bush_therm	Numeric	8	0	Feeling Thermometer: GW Bush	{777, 777. ...	None
32	bush_therm_pst	Numeric	8	0	Feeling Thermometer: GW Bush	{777, 777. ...	None
33	buspeople	Numeric	8	0	Feeling Thermometer - business people	{777, 777. ...	None

measurement of each variable in the dataset. (You can adjust the width of a column by clicking, holding, and dragging the column border.)

The most frequently used variable information is contained in Name, Label, Values, and Missing. Name is the brief descriptor recognized by SPSS when it does analysis. Names can be up to 64 characters in length, although they need to begin with a letter (not a number). Plus, names must not contain any special characters, such as dashes or commas, although underscores are okay. You are encouraged to make good use of Label, a long descriptor (up to 256 characters are allowed), for each variable name. For example, when SPSS analyzes the variable attent, it will look in the Variable View for a label. If it finds one, then it will label the results of its analysis by using Label instead of Name. So attent shows up as "Interested in following campaigns?"—much more user friendly than simply "attent." Just as Label permits a wordier description for Name, Values attaches word labels to the numeric variable codes. Consider abort_funding, which, according to Label, gauges whether the respondent said they "Favor/oppose govt funds to pay for abortion." Click the mouse anywhere in the Values cell and then click the gray button that appears. A Value Labels window pops up, revealing the labels that SPSS will attach to the numeric codes of abort_funding (Figure 1-3). Unless you instruct it to do otherwise, SPSS will apply these labels to its analysis of abort_funding. (Click the Cancel button in the Value Labels window to return to the Variable View.)

Finally, a word about Missing: Sometimes a dataset does not have complete information for some variables on a number of cases. In coding the data, researchers typically give a special numeric code to these missing values. In coding age, for example, the NES coders entered a value of "0" for respondents who did not reveal their ages. To ensure that SPSS does not treat "0" as a valid code for age, coders defined "0" as missing, to be excluded from any analysis of age. Similarly, people who did not know who Hillary Clinton is (scroll down to the variable named hillary_therm and labeled "Feeling Thermometer: Hillary Clinton") or who did not know where to rate her were coded 777 or 888 on this variable. These two codes were then defined as missing.[1]

Figure 1–3 Value Labels Box

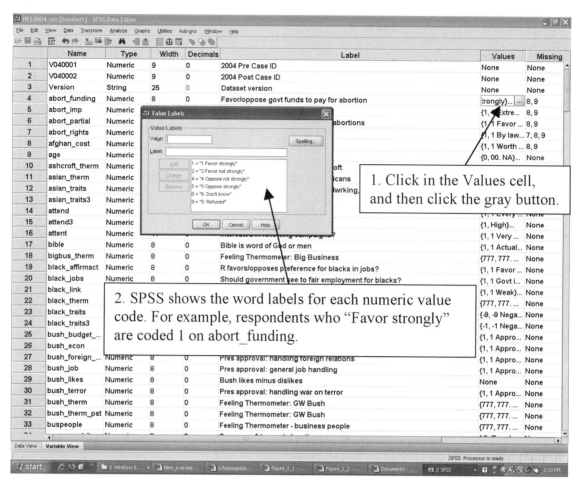

A MUST-DO: SETTING OPTIONS FOR VARIABLE LISTS

Now you have a feel for the number-oriented side and the word-oriented side of SPSS. Before looking at how SPSS produces and handles output, you must do one more thing. To ensure that all the examples in this workbook correspond to what you see on your screen, you will need to follow the steps given in this section when you open each dataset for the first time.

⚠️ **DO THIS NOW:** In the main menu bar of the Data Editor, click Edit → Options. Make sure that the General tab is clicked. (See Figure 1-4.) If the radio button Display names *and* the radio button Alphabetical were already selected when you opened the Options menu, you are set to go. Click Cancel. If, however, Display names and/or Alphabetical were not already selected when you opened the Options menu, select them (as in Figure 1-4). Click Apply. Click OK, returning to the Data Editor. Again, you will need to do this for each dataset. When you open a new set for the first time, go to Edit → Options and ensure that Display names / Alphabetical are selected and applied.

THE VIEWER

Let's run through a quick analysis and see how SPSS handles variables and output. On the main menu bar, click Analyze → Descriptive Statistics → Frequencies. The Frequencies window appears (Figure 1-5). There are two panels. On the right is the (currently empty) Variable(s) panel. This is the panel where you enter the variables you want to analyze. On the left you see the names of all the variables in NES2004 in alphabetical order, just as you specified in the Options menu. Although the names are not terribly informative, complete coding information is just a (right) mouse click away. Put the mouse pointer on the variable, abort_rights, and right-click. Then click on Variable Information. As shown in Figure 1-6, SPSS retrieves and displays the

Figure 1–4 Setting Options for Variable Lists

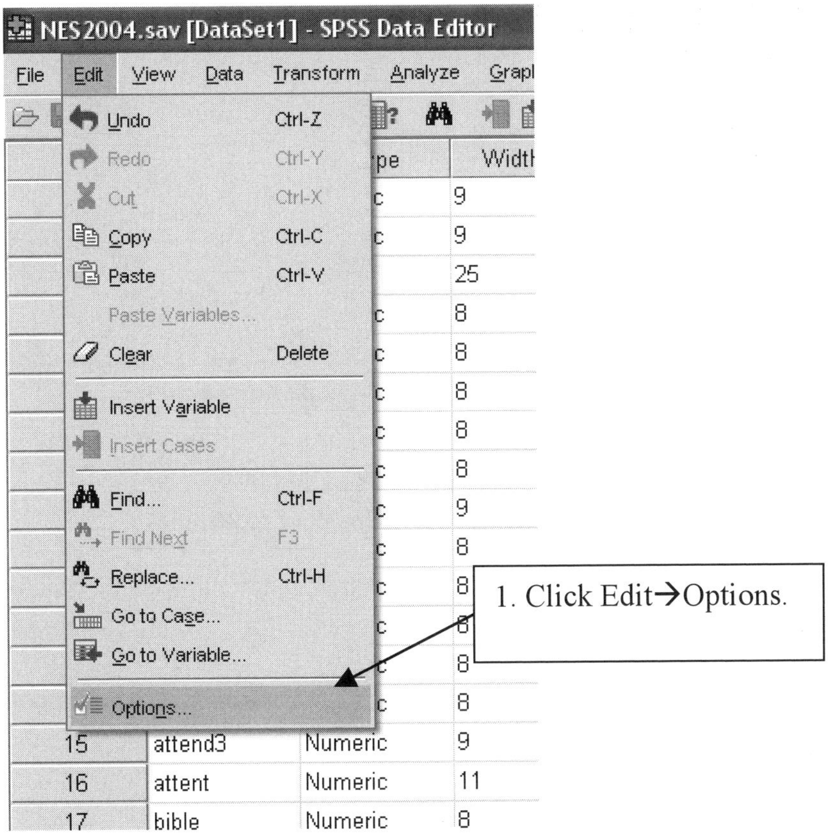

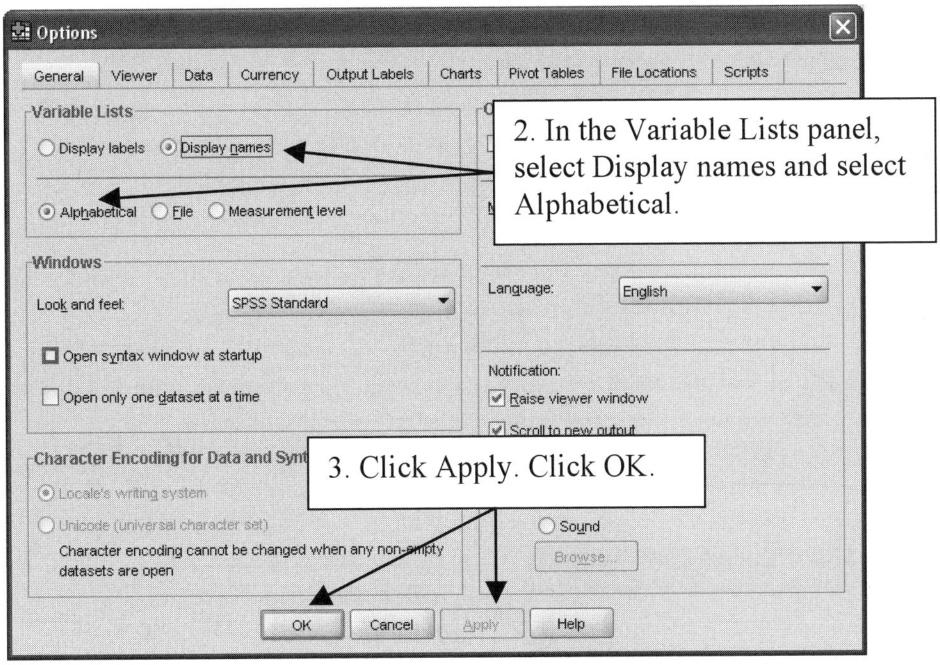

label (Abortion position: self-placement), name (abort_rights), and, most usefully, the value labels for the numeric codes. (To see all the codes, click the drop-down arrow in the Value Labels box.) Respondents who believe abortion should never be permitted are coded 1, those who think abortion should be permitted under specific circumstances (rape or incest) are coded 2, and so on.

Return the mouse to the Frequencies window and click abort_rights into the Variable(s) panel. (Click on abort_rights and then click the arrow between the panels.) Click OK. SPSS runs the analysis and displays

Figure 1–5 Requesting Frequencies

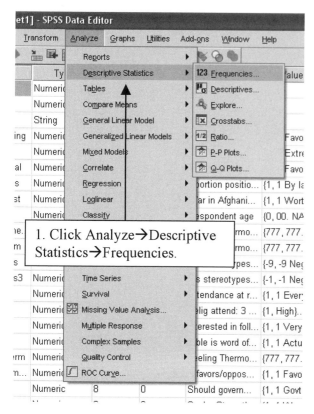
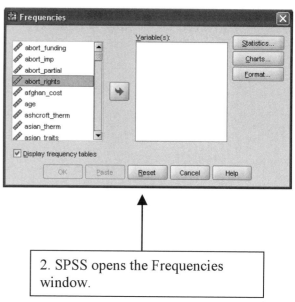

1. Click Analyze→Descriptive Statistics→Frequencies.

2. SPSS opens the Frequencies window.

Figure 1–6 Retrieving Coding Information

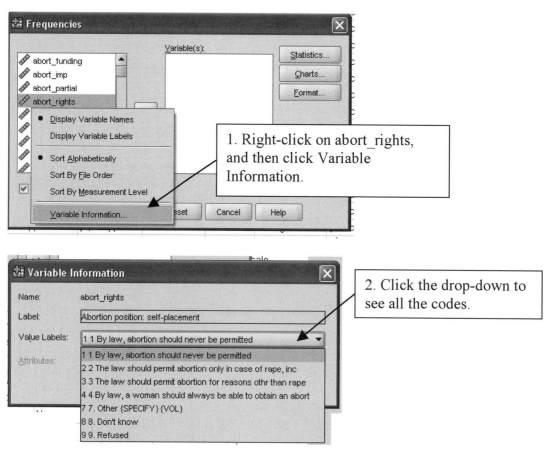

1. Right-click on abort_rights, and then click Variable Information.

2. Click the drop-down to see all the codes.

Figure 1–7 SPSS Viewer: Outline Pane and Contents Pane

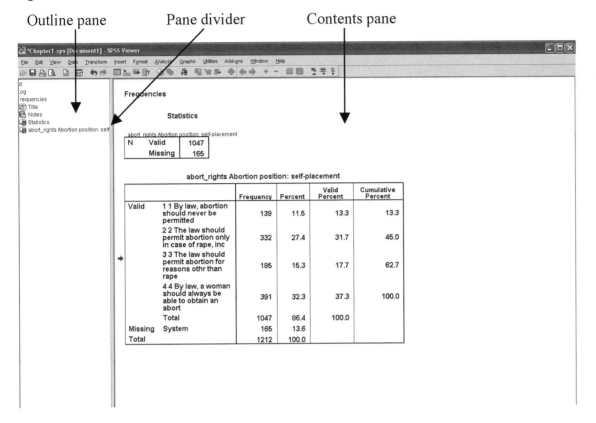

the results in the Viewer (Figure 1-7). The Viewer has two panes. In the Outline pane, SPSS keeps a running log of the analyses you are performing. The Outline pane references each element in the Contents pane, which reports the results of your analyses. In this book we are interested exclusively in the Contents pane. Reduce the size of the Outline pane by first placing the cursor on the Pane divider. Click and hold the left button of the mouse and then move the Pane divider over to the left-hand border of the Viewer. The Viewer should now look like Figure 1-8. The output for abort_rights shows you the frequency distribution for all valid responses, with value codes labeled. In Chapter 2 we discuss frequency analysis in more detail. Our immediate purpose is to become familiar with SPSS output.

Here are some key facts about the Viewer. First, the Viewer is a separate file, created by you during your analysis of the data. It is completely distinct from the data file. Whereas SPSS data files all have the file extension *.sav, Viewer files have the file extension *.spv.[2] The output can be saved, under a name that you choose, and then reopened later. Second, the output from each succeeding analysis does not overwrite the file. Rather, it appends new results to the Viewer file. If you were to run another analysis for a different variable, SPSS would dump the results in the Viewer below the analysis you just performed. Third, there are a number of ways to toggle between the Viewer and the Data Editor. (See Figure 1-8.) Clicking the Go To Data icon on the toolbar takes you to the Data Editor. Clicking Window on the main menu bar and checking "NES2004.sav - SPSS Data Editor" also returns you to the Data Editor. Clicking Window on the main menu bar in the Data Editor and checking "*.spv - SPSS Viewer" sends you to your output file. And, of course, Windows accumulates icons for all open files along the bottom Taskbar. Finally, you may select any part of the output file, print it, or copy and paste it into a word processing program.

Many of the exercises in this workbook will ask you to print the results of your SPSS analyses, so let's cover the print procedure. We'll also address a routine necessity: saving output.

Selecting, Printing, and Saving Output

Printing desired results requires, first, that you select the output or portion of output you want to print. A quick and easy way to select a single table or chart is to place the cursor anywhere on the desired object and click once. Let's say you want to print the abort_rights frequency distribution. Place the cursor on the frequency table and click. A red arrow appears in the left-hand margin next to the table (Figure 1-9). Now

Figure 1–8 SPSS Viewer: Outline Pane Minimized

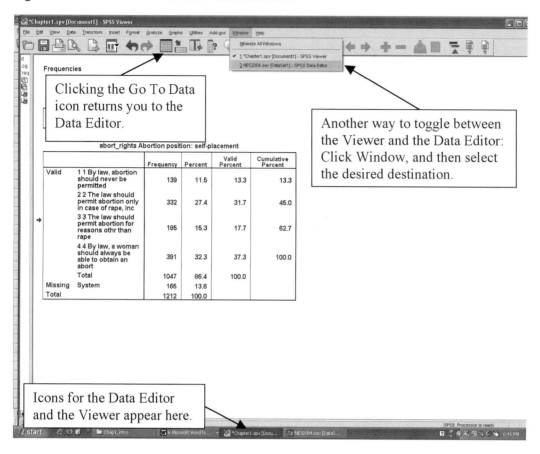

Figure 1–9 Selecting, Printing, and Saving Output

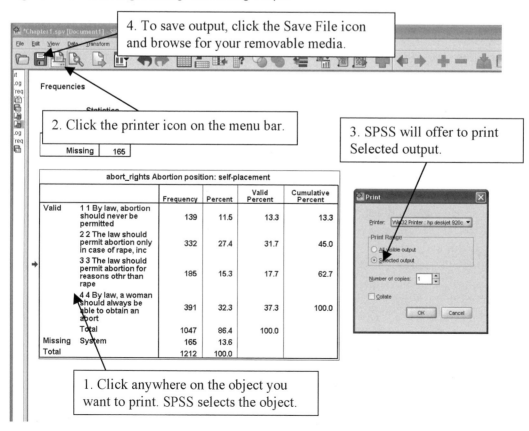

click the printer icon on the Viewer menu bar. The Print window opens. In the window's Print Range panel, the radio button next to "Selected output" should already be clicked. Clicking OK would send the frequency table to the printer. To select more than one table or graph, hold down the Control key (Ctrl) while selecting the desired output with the mouse. Thus, if you wanted to print the frequency table and the statistics table, first click on one of the desired tables. While holding down the Ctrl key, click on the other table. SPSS will select both tables.

To save your output, simply click the familiar floppy-disk-shaped icon on the Viewer menu bar (refer to Figure 1-9). Browse for an appropriate location—most likely the removable media containing your copy of the datasets. Invent a file name (but preserve the .spv extension), such as "chap1.spv," and click Save. SPSS saves all of the information in the Viewer to the file chap1.spv. Saving your output protects your work. Plus, the output file can always be reopened later. Suppose you are in the middle of a series of SPSS analyses and you want to stop and return later. You can save the Viewer file, as described here, and exit SPSS. When you return, you start SPSS and load a data file (like NES2004) into the Data Editor. In the main menu bar of the Data Editor, you click File → Open → Output, find your .spv file, and open it. Then you can pick up where you left off.

EXERCISES

1. (Dataset: NES2004. Variables: abort_funding, gender.) Earlier we spent some time using the Data View and the Variable View to describe NES2004 Respondent #1 (the person coded 1 on V040001). In this exercise you will use your familiarity with the Data Editor to determine Respondent 1's gender and opinions on government funding of abortions.

   A. With NES2004 open, go to the Data View. What numeric code does Respondent 1 have on abort_funding? A code of (fill in the blank) _____. Go to the Variable View. Just as you did earlier in this chapter, find abort_funding and click in the Values cell. On the question of government funding for abortions, Respondent 1 (circle one)

   favors strongly.   favors not strongly.   opposes not strongly.   opposes strongly.

   B. Return to the Data View. What is Respondent 1's code on the variable gender. A code of (fill in the blank) _____. Go to the Variable View. Respondent 1's gender is (circle one)

   male.   female.

2. Suppose that you have just opened World, States, or GSS2006 for the first time. The first thing you do is to click Edit → Options and consider the Variable Lists panel of the General tab. You must make sure that which two choices are selected and applied? (check two)

   ❏ Display labels   ❏ File

   ❏ Display names   ❏ Measurement level

   ❏ Alphabetical

NOTES

1. These are examples of user-defined missing values. The author has set most of the variables in the datasets you will be using in this book to *system missing*. A system missing code appears as a dot (.) in the Data View. SPSS automatically treats these codes as missing.
2. In SPSS 15 or earlier, output files have the *.spo extension. SPSS 16 cannot read output files created by earlier releases. SPSS Viewer files generally are not backward compatible. For example, release 13 output files cannot be opened using release 12.

2

Descriptive Statistics

Procedures Covered

Analyze → Descriptive Statistics → Frequencies
Analyze → Reports → Case Summaries

Analyzing descriptive statistics is the most basic—and sometimes the most informative—form of analysis you will do. Descriptive statistics reveal two attributes of a variable: its typical value (central tendency) and its spread (degree of dispersion or variation). The precision with which you can describe central tendency for any given variable depends on the variable's level of measurement. For nominal-level variables you can identify the *mode*, the most common value of the variable. For ordinal-level variables, those whose categories can be ranked, you can find the mode and the *median*—the value of the variable that divides the cases into two equal-size groups. For interval-level variables you can obtain the mode, median, and arithmetic *mean*, the sum of all values divided by the number of cases.

In this chapter you will use Analyze → Descriptive Statistics → Frequencies to obtain appropriate measures of central tendency, and you will learn to make informed judgments about variation. With the correct prompts, the Frequencies procedure also provides valuable graphic support—bar charts and (for interval variables) histograms. These tools are essential for distilling useful information from datasets having hundreds of anonymous cases, such the National Elections Studies (NES2004) or General Social Survey (GSS2006). For smaller datasets with aggregated units, such as the States and World datasets, SPSS offers an additional procedure: Analyze → Reports → Case Summaries. Case Summaries lets you see firsthand how specific cases are distributed across a variable that you find especially interesting.

INTERPRETING MEASURES OF CENTRAL TENDENCY AND VARIATION

Finding a variable's central tendency is ordinarily a straightforward exercise. Simply read the computer output and report the numbers. Describing a variable's degree of dispersion or variation, however, often requires informed judgment.[1] Here is a general rule that applies to any variable at any level of measurement: A variable has no dispersion if all the cases—states, countries, people, or whatever—fall into the same value of the variable. A variable has maximum dispersion if the cases are spread evenly across all values of the variable. The number of cases in one category equals the number of cases in every other category.

Central tendency and variation work together in providing a complete description of any variable. Some variables have an easily identified typical value and show little dispersion. For example, suppose you were to ask a large number of U.S. citizens what sort of economic system they believe to be the best: capitalism, communism, or socialism. What would be the modal response, the economic system preferred by most people? Capitalism. Would there be a great deal of dispersion, with large numbers of people choosing the alternatives, communism or socialism? Probably not.

In other instances, however, you may find that one value of a variable has a more tenuous grasp on the label *typical.* And the variable may exhibit more dispersion, with the cases spread out more evenly across the variable's other values. For example, suppose a large sample of voting-age adults were asked, in the weeks preceding a presidential election, how interested they are in the campaign: very interested, somewhat interested, or not very interested. Among your own acquaintances you probably know a number of people who fit into each category. So even if one category, such as "somewhat interested," is the median, many people will likely be found at the extremes of "very interested" and "not very interested." In this instance the amount of dispersion in a variable—its degree of spread—is essential to understanding and describing it.

These and other points are best understood by working through some guided examples. For the next several analyses, you will analyze GSS2006. Open the dataset by double-clicking the GSS2006 icon. (If you are using *SPSS Student Version,* open GSS2006A_Student.) In the Data Editor, click Edit → Options and then click on the General tab. Just as you did with NES2004 in Chapter 1, make sure that the radio buttons in the Variable Lists area are set for Display names and Alphabetical. (If these options are already set, click Cancel. If they are not set, select them, click Apply, and then click OK. Now you are ready to go.)

DESCRIBING NOMINAL VARIABLES

First, you will obtain a frequency distribution and bar chart for a nominal-level variable, zodiac, which records respondents' astrological signs. Click Analyze → Descriptive Statistics → Frequencies. Scroll down to the bottom of the left-hand list until you find zodiac. Click zodiac into the Variable(s) panel. To the right of the Variable(s) panel, click the Charts button (Figure 2-1). The Frequencies: Charts window appears. In

Figure 2–1 Obtaining Frequencies and a Bar Chart (nominal variable)

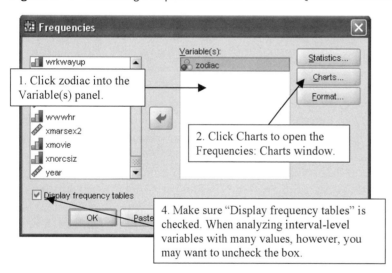

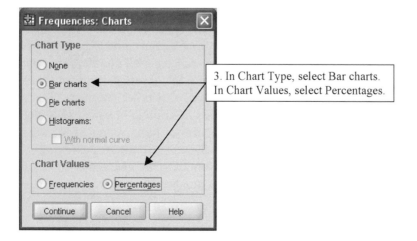

Figure 2–2 Frequencies Output (nominal variable)

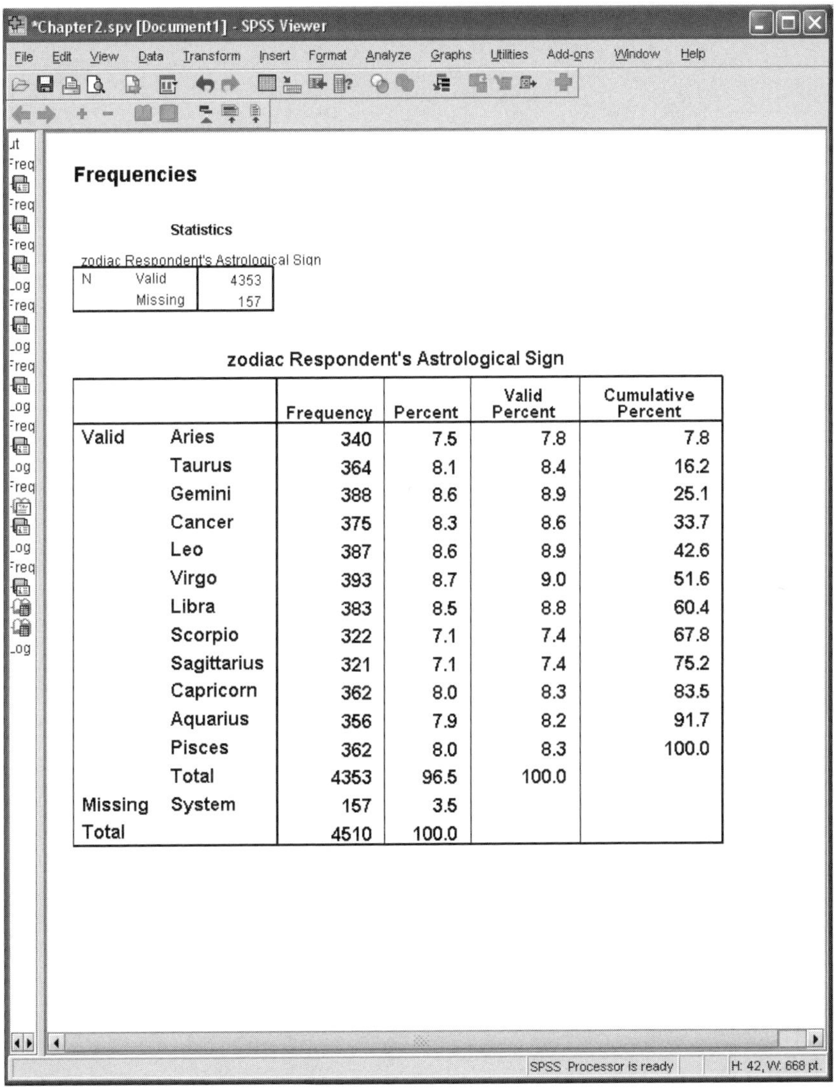

Frequencies

Statistics

zodiac Respondent's Astrological Sign

N	Valid	4353
	Missing	157

zodiac Respondent's Astrological Sign

		Frequency	Percent	Valid Percent	Cumulative Percent
Valid	Aries	340	7.5	7.8	7.8
	Taurus	364	8.1	8.4	16.2
	Gemini	388	8.6	8.9	25.1
	Cancer	375	8.3	8.6	33.7
	Leo	387	8.6	8.9	42.6
	Virgo	393	8.7	9.0	51.6
	Libra	383	8.5	8.8	60.4
	Scorpio	322	7.1	7.4	67.8
	Sagittarius	321	7.1	7.4	75.2
	Capricorn	362	8.0	8.3	83.5
	Aquarius	356	7.9	8.2	91.7
	Pisces	362	8.0	8.3	100.0
	Total	4353	96.5	100.0	
Missing	System	157	3.5		
Total		4510	100.0		

Chart Type select Bar charts. In Chart Values select Percentages. Click Continue, which returns you to the main Frequencies window. Click OK. SPSS runs the analysis.

SPSS has produced two items of interest in the Viewer: a frequency distribution of respondents' astrological signs and a bar chart of the same information. Examine the frequency distribution (Figure 2-2). The value labels for each astrological code appear in the leftmost column, with Aries occupying the top row of numbers and Pisces the bottom row. There are four numeric columns: Frequency, Percent, Valid Percent, and Cumulative Percent. What does each column mean? The Frequency column shows raw frequencies, the actual number of respondents having each zodiac sign. Percent is the percentage of *all* respondents, including missing cases, in each category of the variable. Ordinarily the Percent column can be ignored, because we generally are not interested in including missing cases in our description of a variable. Valid Percent is the column to focus on. Valid Percent tells us the percentage of nonmissing responses in each value of zodiac. Finally, Cumulative Percent reports the percentage of cases that fall in *or below* each value of the variable. For ordinal or interval variables, as you will see, the Cumulative Percent column can provide valuable clues about how a variable is distributed. But for nominal variables, which cannot be ranked, the Cumulative Percent column provides no information of value.

Now consider the Valid Percent column more closely. Scroll between the frequency distribution and the bar chart, which depicts the zodiac variable in graphic form (Figure 2-3). What is the mode, the most common astrological sign? For nominal variables, the answer to this question is (almost) always an easy call: Simply find the value with the highest percentage of responses. Virgo is the mode. Does this variable have

Figure 2–3 Bar Chart (nominal variable)

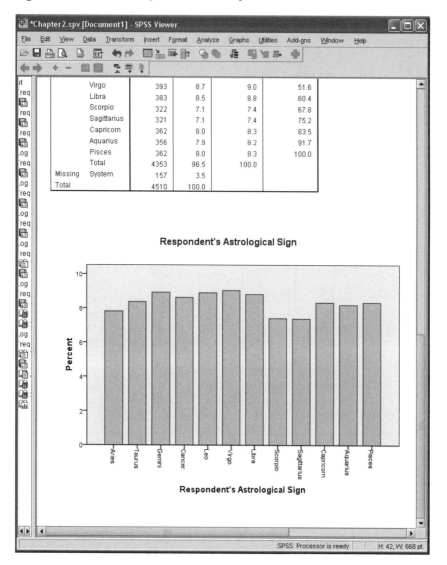

little dispersion or a lot of dispersion? Again study the Valid Percent column and the bar chart. Apply the following rule: A variable has no dispersion if the cases are concentrated in one value of the variable; a variable has maximum dispersion if the cases are spread evenly across all values of the variable. Are most of the cases concentrated in Virgo, or are there many cases in each value of zodiac? Because respondents show great heterogeneity in astrological signs, you would conclude that zodiac has a high level of dispersion.

DESCRIBING ORDINAL VARIABLES

Next, you will analyze and describe two ordinal-level variables, one of which has little variation and the other of which is more spread out. Along the top menu bar of the Viewer, click Analyze → Descriptive Statistics → Frequencies. SPSS "remembers" the preceding analysis, so zodiac is still in the Variable(s) list. Click zodiac back into the left-hand list. Scroll through the list until you find these variables: helppoor and helpsick. Each of these is a 5-point ordinal scale. Helppoor asks respondents to place themselves on a scale between 1 ("The government should take action to help poor people") and 5 ("People should help themselves"). Helpsick, using a similar 5-point scale, asks respondents about government responsibility or individual responsibility for medical care. Click helppoor and helpsick into the Variable(s) list. SPSS retained your earlier settings for Charts, so accompanying bar charts will appear in the Viewer. Click OK.

SPSS runs the analysis for each variable and produces two frequency distributions, one for helppoor and one for helpsick, followed by two bar charts of the same information. First, let's focus on helppoor. To

Figure 2–4   Frequencies Output (ordinal variables)

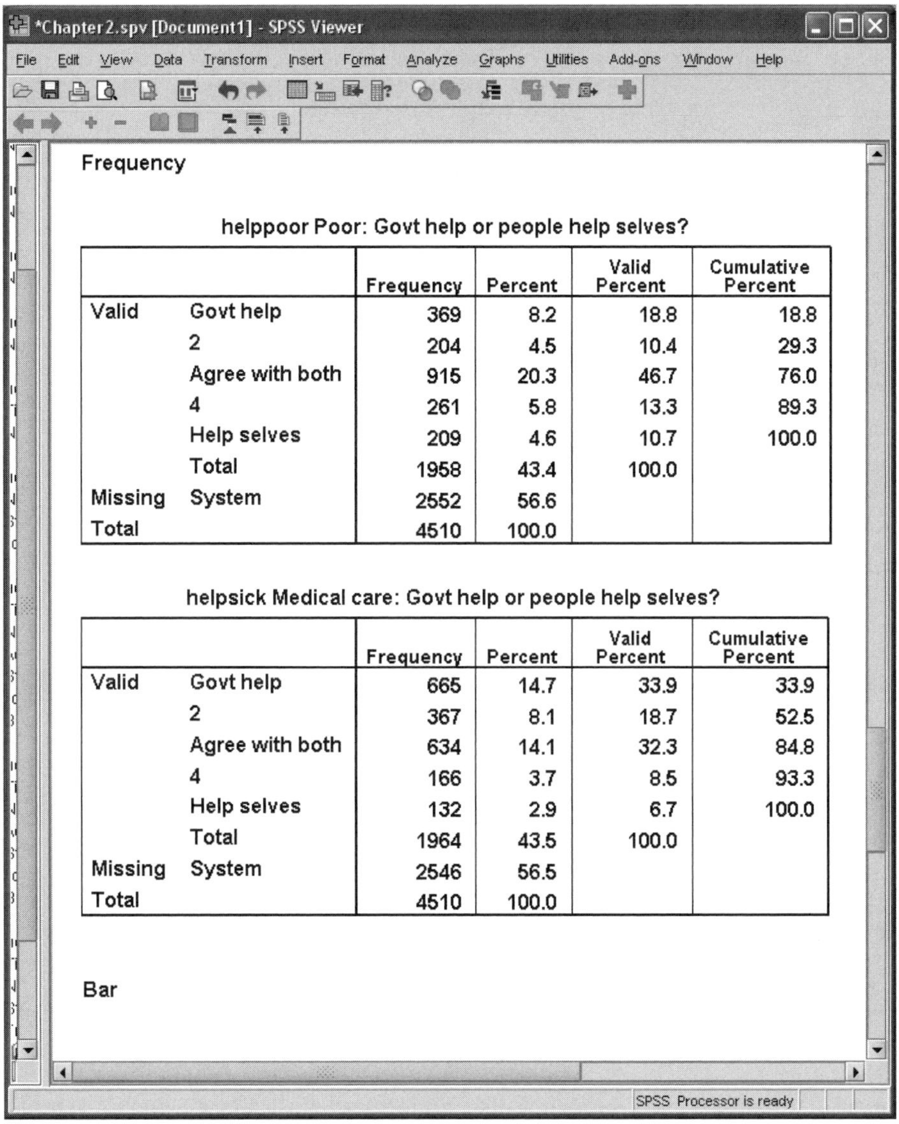

```
*Chapter2.spv [Document1] - SPSS Viewer                          [_][□][✕]
File   Edit   View   Data   Transform   Insert   Format   Analyze   Graphs   Utilities   Add-ons   Window   Help
```

Frequency

helppoor Poor: Govt help or people help selves?

		Frequency	Percent	Valid Percent	Cumulative Percent
Valid	Govt help	369	8.2	18.8	18.8
	2	204	4.5	10.4	29.3
	Agree with both	915	20.3	46.7	76.0
	4	261	5.8	13.3	89.3
	Help selves	209	4.6	10.7	100.0
	Total	1958	43.4	100.0	
Missing	System	2552	56.6		
Total		4510	100.0		

helpsick Medical care: Govt help or people help selves?

		Frequency	Percent	Valid Percent	Cumulative Percent
Valid	Govt help	665	14.7	33.9	33.9
	2	367	8.1	18.7	52.5
	Agree with both	634	14.1	32.3	84.8
	4	166	3.7	8.5	93.3
	Help selves	132	2.9	6.7	100.0
	Total	1964	43.5	100.0	
Missing	System	2546	56.5		
Total		4510	100.0		

Bar

```
                                                    SPSS Processor is ready
```

get a feel for this variable, scroll back and forth between the frequency distribution (Figure 2-4) and the bar chart (Figure 2-5). How would you describe the central tendency and dispersion of this variable? Because helppoor is an ordinal variable, you can report both its mode and its median. Its mode, clearly enough, is the response "Agree with both." What about the median? This is where the Cumulative Percent column of the frequency distribution comes into play. *The median for any ordinal (or interval) variable is the category below which 50 percent of the cases lie.* Is the first category, "Govt help," the median? No, this code contains fewer than half the cases. How about the next higher category? No, again. The Cumulative Percent column still has not reached 50 percent. The median occurs in the "Agree with both" category.

Does helppoor have a high or low degree of dispersion? The dispersion of an ordinal variable can be evaluated in two complementary ways. One way is to take a close look at the bar graph. If helppoor had a high level of variation, the bars would have roughly equal heights, much like the zodiac variable that you analyzed earlier. If helppoor had no dispersion, then all the cases would fall into one category—there would be only one bar showing in the graphic. Another way to evaluate variation is to compare the mode and the median. If the mode and the median fall in the same category, the variable has lower dispersion than if the mode and the median fall in different values of the variable. You sometimes have to exercise judgment in determining variation, but it seems clear that helppoor is a variable with a fairly low degree of dispersion. The fence-straddling response, "Agree with both," is prominent in the bar graph. What is more, both the mode and the median are within this response category.

Figure 2–5 Bar Chart (ordinal variable with low dispersion)

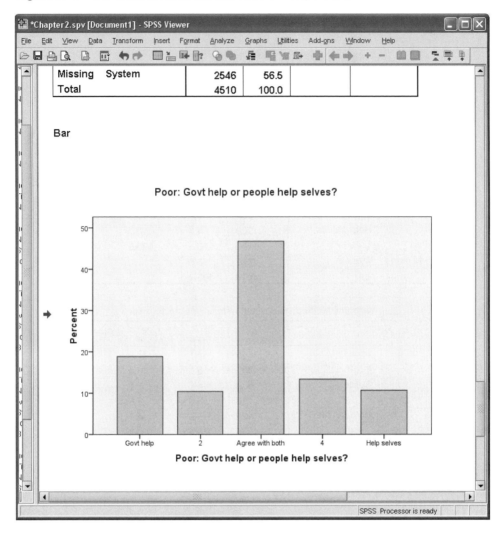

Now turn your attention to the frequency distribution for helpsick (see Figure 2-4) and the accompanying bar chart (Figure 2-6). What is the mode? The most common response is "Govt help." So "Govt help" is the mode. Use the Cumulative Percent column to find the median. Is the first value, "Govt help," the median? No, this value contains fewer than half the cases. Now go up one value to the category labeled "2" on the 5-point ordinal scale. Is this the median? Yes, it is. According to the Cumulative Percent column, more than 50 percent of the cases fall in or below this value. So "Govt help" is the mode, but the median falls within response category 2, which lies a bit more toward the "Help selves" side of the variable.

Which variable, helppoor or helpsick, has higher variation? Notice that, unlike helppoor, respondents' values on helpsick are more spread out, with sizable numbers of cases falling in the first response category ("Govt help") and the middle category ("Agree with both"). Indeed, these two quite different responses are close rivals for the distinction of being the modal opinion on this issue. And, unlike helppoor, helpsick's mode and median are different, providing a useful field mark of higher variation. Thus helpsick has more variation—greater dispersion—than helppoor.

DESCRIBING INTERVAL VARIABLES

Let's now turn to the descriptive analysis of interval-level variables. An interval-level variable represents the most precise level of measurement. Unlike nominal variables, whose values stand for categories, and ordinal variables, whose values can be ranked, the values of an interval variable *tell you the exact quantity of the characteristic being measured*. For example, age qualifies as an interval-level variable because its values impart each respondent's age in years.

Figure 2–6 Bar Chart (ordinal variable with high dispersion)

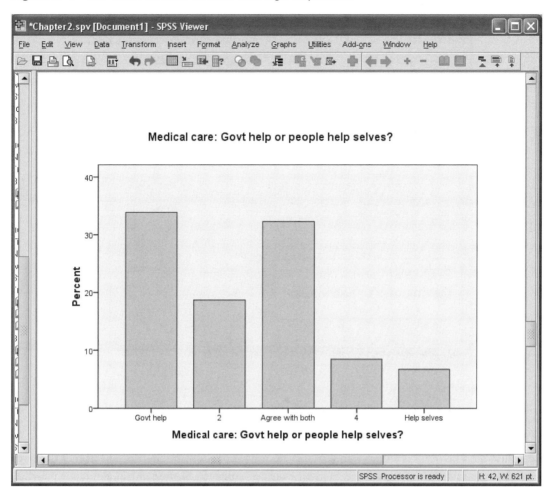

Because interval variables have the most precision, they can be described more completely than can nominal or ordinal variables. For any interval-level variable, you can report its mode, median, and arithmetic average, or *mean*. In addition to these measures of central tendency, you can make more sophisticated judgments about variation. Specifically, you can determine if an interval-level distribution is *skewed*. What is skewness, and how do you know it when you see it?

Skewness refers to the symmetry of a distribution. If a distribution is not skewed, the cases tend to cluster symmetrically around the mean of the distribution, and they taper off evenly for values above and below the mean. If a distribution is skewed, by contrast, one tail of the distribution is longer and skinnier than the other tail. Distributions in which some cases occupy the higher values of an interval variable—distributions with a skinnier right-hand tail—are said to have a *positive skew*. By the same token, if the distribution has some cases at the extreme lower end—the distribution has a skinnier left-hand tail—then the distribution has a *negative skew*. Skewness has a predictable effect on the mean. A positive skew tends to "pull" the mean upward; a negative skew pulls it downward. However, skewness has less effect on the median. Because the median reports the middle-most value of a distribution, it is not tugged upward or downward by extreme values. *For badly skewed distributions, it is a good practice to use the median instead of the mean in describing central tendency.*

A step-by-step analysis of a GSS2006 variable, age, will clarify these points. Click Analyze → Descriptive Statistics → Frequencies. If helppoor and helpsick are still in the Variable(s) list, click them back into the left-hand list. Click age into the Variable(s) list. Click the Charts button. Make sure that Bar charts (under Chart Type) and Percentages (under Chart Values) are selected. Click Continue, which returns you to the main Frequencies window. (You may wish to refer back to Figure 2-1, which shows this window.)

So far, this procedure is the same as in your analysis of zodiac, helppoor, and helpsick. When running a frequencies analysis of an interval-level variable, however, you need to do two additional things. One of

Figure 2–7 Frequencies: Statistics Window (modified)

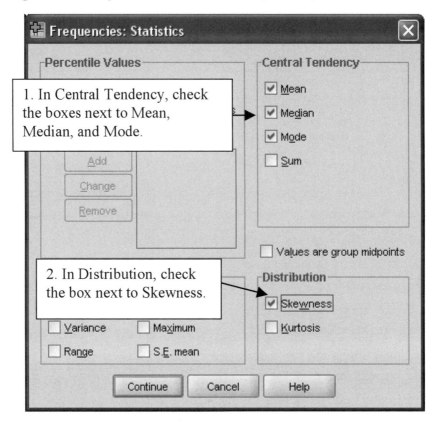

these is a must-do. The other is a may-want-to-do. The must-do: Click the Statistics button at the bottom of the Frequencies window. The Frequencies: Statistics window appears. In the Central Tendency panel, click the boxes next to Mean, Median, and Mode. In the Distribution panel, click Skewness. The Frequencies: Statistics window should look like Figure 2-7. Click Continue, returning to the main Frequencies window. The may-want-to-do: *Un*check the box next to Display frequency tables, appearing at the foot of the left-hand list.[2] Click OK.

SPSS runs the analysis of age and dumps the requested statistics and bar chart into the SPSS Viewer (Figure 2-8). Most of the entries in the Statistics table are familiar to you: valid number of cases (N); number of missing cases; and mean, median, and mode. In addition, SPSS has reported values for skewness and for something called standard error of skewness that are as precise as they are mysterious. What do these numbers mean? When a distribution is perfectly symmetrical—no skew—it has skewness equal to 0. If the distribution has a skinnier right-hand tail—positive skew—then skewness will be a positive number. A skinnier left-hand tail, logically enough, returns a negative number for skewness. Just about all distributions will have some degree of skewness. How much is too much? That's where the standard error of skewness comes into play. Follow this simple rule: Divide skewness by its standard error. If the result has a magnitude (absolute value) of greater than 2, then the distribution is significantly skewed, and you should use the median as the best measure of central tendency. If you divide skewness by its standard error and get a number whose magnitude is 2 or less, the distribution is not significantly skewed. In this case, use the mean as the best measure of central tendency.

The age variable plainly has a positive skew. Compare the mean age with the median age. Recall that a positive skew pulls the mean upward. Remember also that the mean is susceptible to skewness, and the median less so. Is the skewness more than twice its standard error? Yes, and then some. Obviously, a number of cases in the upper reaches of this variable have pulled the mean off the exact 50-50 center of the distribution. The bar chart confirms this suspicion. The skinnier right-hand tail is the telltale sign of a positive skew.[3]

All of the guided examples thus far have used bar charts for graphic support. For nominal and ordinal variables, a bar chart should always be your choice. For interval variables, however, you may want to ask SPSS to produce a histogram instead. What is the difference between a bar chart and a histogram? When is

Figure 2–8 Statistics and Bar Chart (interval variable)

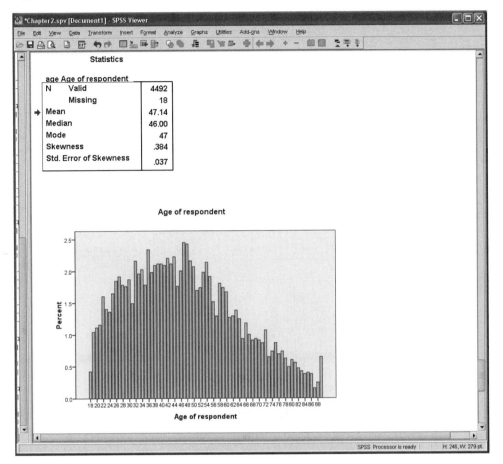

one preferred over the other? A bar chart displays each value of a variable and shows you the percentage (alternatively, the raw number) of cases that fall into each category. A histogram is similar, but instead of displaying each discrete value, it collapses categories into ranges (called bins), resulting in a compact display. Histograms are sometimes more readable and elegant than bar charts. Most of the time a histogram will work just as well as a bar chart in summarizing an interval-level variable. For interval variables with a large number of values, a histogram is the graphic of choice. (Remember: For nominal or ordinal variables, you always want a bar chart.)

So that you can become familiar with histograms, let's run the analysis of age once again—only this time we'll ask SPSS to produce a histogram instead of a bar chart. Click Analyze → Descriptive Statistics → Frequencies. Make sure age is still in the Variable(s) list. Click Statistics, and then uncheck all the boxes: Mean, Median, Mode, and Skewness. Click Continue. Click Charts, and then select the Histograms radio button in Chart Type. Click Continue. For this analysis, we do not need a frequency table. In the Freqencies window, uncheck the Display frequency tables box. (Refer to Figure 2-1.) Click OK.

This is a bare-bones run. SPSS reports its obligatory count of valid and missing cases, plus a histogram for age (Figure 2-9). On the histogram's horizontal axis, notice the hash marks, which are spaced at 20-year intervals. SPSS has compressed the data so that each bar represents about 2 years of age rather than 1 year of age. Now scroll up the Viewer to the bar chart of age, which you produced in the preceding analysis. Notice that the histogram has smoothed out the nuance and choppiness of the bar chart, though it still captures the essential qualities of the age variable.

OBTAINING CASE-LEVEL INFORMATION WITH CASE SUMMARIES

When you analyze a large survey dataset, as you have just done, you generally are not interested in how respondent x or respondent y answered a particular question. Rather, you want to know how the entire sample of respondents distributed themselves across the response categories of a variable. Sometimes,

Figure 2–9 Histogram (interval variable)

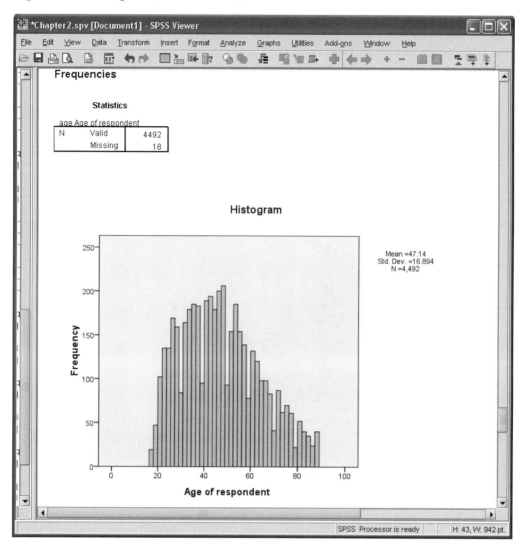

however, you gather data on particular cases because the cases are themselves inherently important. The States dataset (50 cases) and World dataset (191 cases) are good examples. With these datasets, you may want to push the descriptions beyond the relative anonymity of Frequencies analysis and find out where particular cases "are" on an interesting variable. Analyze → Reports → Case Summaries is readymade for such elemental insights. Before beginning this guided example, close GSS2006 and open States.

Suppose that you are interested in studying state laws that regulate abortions. The States dataset contains the variable abortlaw, which records the number of restrictions imposed by each state, from no restrictions (a value of 0 on abortlaw) to 10 restrictions (abortlaw value of 10). If you were to run a Frequencies analysis of abortlaw, you would find that one state has 0 restrictions, two states have 10, and a fair amount of variation exists among states. But exactly which states are the least restrictive? Which are the most restrictive? Where does your state fall on the list? Case Summaries can quickly answer questions like these. SPSS will sort the states on the basis of a "grouping variable" (in this example, the number of abortion restrictions) and then produce a report telling you which states are in each group.

With the States dataset open, click Analyze → Reports → Case Summaries. The Summarize Cases window opens (Figure 2-10). You need to do three things here. First, click the variable containing the cases' identities into the Variables window. In the States dataset, this variable is named state, an alphabetic descriptor of each state's name. (In World, the variable named country contains the name of each case.) Second, click the variable of interest, abortlaw, into the Grouping Variable(s) window. Finally, uncheck the Limit cases box. If this box is left checked, SPSS will limit the analysis to the first 100 cases, which in many instances (such as the States dataset) will work fine, but in other instances (such as the World dataset) will not. Click OK. SPSS gives you a table reporting which states go with which values of abortlaw (Figure 2-11).

Figure 2–10 The Summarize Cases Window (modified)

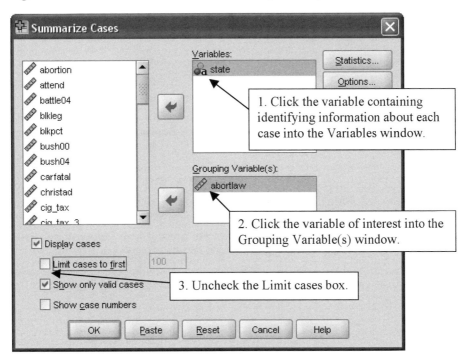

Figure 2–11 Case Summaries Output

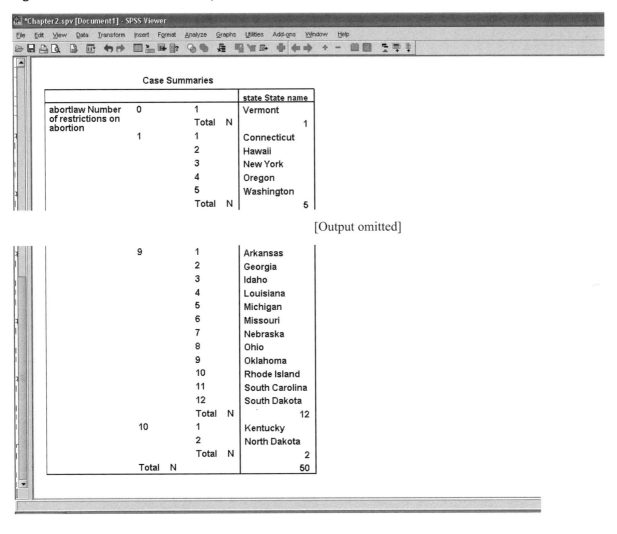

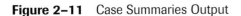

Vermont has no restrictions; a varied group of states (Connecticut, Hawaii, New York, Oregon, and Washington) have one restriction. At the other end of the scale, a large contingent of twelve states have nine restrictions each, and two (Kentucky and North Dakota) have ten.

EXERCISES

For Exercises 1 through 4 you will analyze GSS2006. For Exercise 5 you will analyze World, for Exercise 6 you will work with NES2004, and for Exercises 7 and 8 you will analyze States. After you have done the analyses for these exercises, make sure you save the output file to your hard drive or to removable media. Give the output file a descriptive name, such as chap2exercises.

1. (Dataset: GSS2006. Variables: stoprndm, tapphone, wotrial.) We frequently describe public opinion by referring to how citizens distribute themselves on a political issue. *Consensus* is a situation in which just about everyone, 80–90 percent of the public, holds the same position (or very similar positions) on an issue. *Dissensus* is a situation in which opinion is spread out pretty evenly across all positions on an issue. *Polarization* refers to a configuration of opinion in which people are split between two extreme poles of an issue, with only a few individuals populating the more moderate, middle-of-the-road positions. In this exercise you will decide whether consensus, dissensus, or polarization best describes public opinion on each of three civil liberties issues.

   The 2006 General Social Survey asked respondents to "suppose [that] the government suspected that a terrorist act was about to happen." Respondents were then asked whether "the authorities should have the right to" do each of three things: stop and search people in the street at random (stoprndm), tap people's telephone conversations (tapphone), or detain people for as long as they want without putting them on trial (wotrial). For each variable, responses are coded 0 (authorities definitely should have the right), 1 (authorities probably should have the right), 2 (authorities probably should not have the right), and 3 (authorities definitely should not have the right). So lower codes denote stronger pro-security opinions and higher codes denote stronger pro-civil liberties opinions.

   A. Open GSS2006. Click Analyze → Descriptive Statistics → Frequencies. Click stoprndm, tapphone, and wotrial into the Variables panel. Click Charts. In Chart Types, select Bar charts, and in Chart Values, select Percentages. Refer to the Valid Percent column of the frequency distributions of stoprndm, tapphone, and wotrial. In the table that follows, write the appropriate valid percent next to each question mark (?):

Should authorities have right?	stoprndm Valid Percent	tapphone Valid Percent	wotrial Valid Percent
Definitely yes	?	?	?
Probably yes	?	?	?
Probably no	?	?	?
Definitely no	?	?	?
Total	100.0	100.0	100.0

   B. Examine the percentages in part A. Examine the bar charts that you created. Which of the following statements is supported by your analysis? (check one)

   ❑ The public is polarized on the question of whether the authorities should have the right to stop and search people in the street.

❏ Dissensus exists on the question of whether the authorities should have the right to detain people for as long as they want without putting them on trial.

❏ On all three civil liberties questions, most people respond, "Probably yes," when asked whether authorities should have the right.

C. Suppose that a group of civil libertarians sought to build a strong pro-civil liberties consensus on one of these three issues. On which issue would the group stand the best chance of succeeding? (circle one)

   Stop and search people   Tap telephone conversations   Detain people without trial

   Briefly explain your reasoning. _____

   _____

   _____

D. Print the bar chart for the variable you chose in part C.

2. (Data set: GSS2006. Variable: science_quiz.) The late Carl Sagan once lamented: "We live in a society exquisitely dependent on science and technology, in which hardly anyone knows anything about science and technology." This is a rather pessimistic assessment of the scientific acumen of ordinary Americans. Sagan seemed to be suggesting that the average level of scientific knowledge is quite low and that most people would fail even the simplest test of scientific facts.

   GSS2006 contains science_quiz, which was created from 10 questions testing respondents' knowledge of basic scientific facts. Values on science_quiz range from 0 (the respondent did not answer any of the questions correctly) to 10 (the respondent correctly answered all 10).[4]

A. Consider three possible scenarios for the distribution of science_quiz. All three scenarios assume that science_quiz has a median value of 6 on the 10-item scale. In scenario X, science_quiz has no skew. In scenario Y, science_quiz has a positive skew. In scenario Z, science_quiz has a negative skew. Below are three graphic shells, labeled Scenario X, Scenario Y, and Scenario Z. Sketch a curved line or a set of bar-chart bars within each shell, depicting what the distribution of science_quiz would look like if that scenario were accurate.

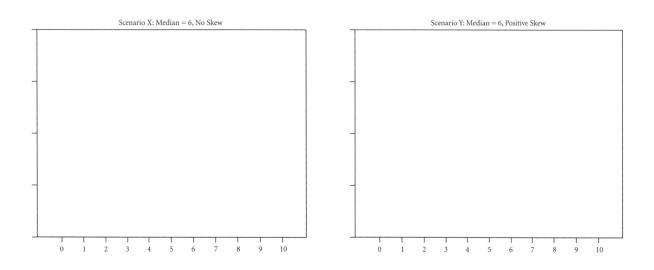

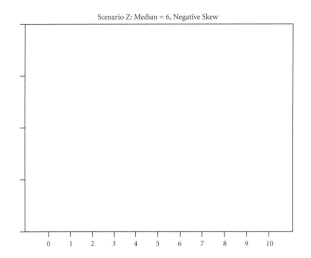

Scenario Z: Median = 6, Negative Skew

```
    0   1   2   3   4   5   6   7   8   9   10
```

B. Run Frequencies to obtain a frequency distribution of science_quiz. In Charts, request a bar chart with percentages. Fill in the table that follows. (In your output, be sure to focus on Valid Percent instead of Percent.)

science_quiz	Frequency	Valid Percent	Cumulative Percent
0	?	?	?
1	?	?	?
2	?	?	?
3	?	?	?
4	?	?	?
5	?	?	?
6	?	?	?
7	?	?	?
8	?	?	?
9	?	?	?
10	?	?	100.0
Total	?	100.0	

C. Examine the frequency distribution and the bar chart. Based on your analysis, which scenario does the distribution of science_quiz most closely approximate? (circle one)

Scenario X   Scenario Y   Scenario Z

Briefly explain your reasoning. _____

_____

_____

_____

D. According to conventional academic standards, any science_quiz score of 5 or lower would be an F, a failing grade. A score of 6 would be a grade of D, a 7 would be a C, an 8 a B, and scores of 9 or 10 would be an A. Based on these standards, about what percentage of people got passing grades on science_quiz? (circle one)

About 30 percent   About 40 percent   About 50 percent   About 60 percent

What percentage got a C or better? (circle one)

About 30 percent   About 40 percent   About 50 percent   About 60 percent

E. Print the chart that you created for this exercise.

3. (Dataset: GSS2006. Variable: fem_role). Two pundits are arguing about how the general public views the role of women in the home and in politics.

Pundit 1: "Our society has a sizable minority of traditionally minded individuals who think that the proper 'place' for women is taking care of the home and caring for children. This small but vocal group of traditionalists aside, the typical adult supports the idea that women belong in work and in politics."

Pundit 2: "Poppycock! It's just the opposite. The extremist 'women's liberation' crowd has distorted the overall picture. The typical view among most citizens is that women should be in the home, not in work and politics."

A. GSS2006 contains fem_role, an interval-level variable that measures respondents' attitudes toward women in society and politics. Scores can range from 0 (women belong in the home) to 12 (women belong in work and politics).

If pundit 1 is correct, fem_role will have (circle one)

a negative skew.   no skew.   a positive skew.

If pundit 2 is correct, fem_role will have (circle one)

a negative skew.   no skew.   a positive skew.

If pundit 1 is correct, fem_role's mean will be (circle one)

lower than its median.   the same as its median.   higher than its median.

If pundit 2 is correct, fem_role's mean will be (circle one)

lower than its median.   the same as its median.   higher than its median.

B. Click Analyze → Descriptive Statistics → Frequencies. Click fem_role into the Variable(s) panel. In Statistics, obtain the mean, median, and mode, as well as the skewness. Click Charts. In Chart Types, select Bar charts, and in Chart Values, select Percentages. Examine the results in the SPSS Viewer. Fill in the table that follows.

Statistics for fem_role	
Mean	?
Median	?
Mode	?
Skewness	?
Std. Error of Skewness	?

C. Which of the following bar charts—X, Y, or Z—most closely resembles the bar chart you obtained? (circle one)

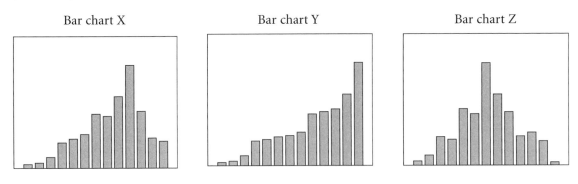

Bar chart X   Bar chart Y   Bar chart Z

D. Based on your analysis, whose assessment is more accurate? (circle one)

Pundit 1's   Pundit 2's

Briefly explain your reasoning.

_____

_____

_____

_____

_____

_____

_____

_____

E. Print the output from this exercise.

4. (Dataset: GSS2006. Variable: attend.) The General Social Survey provides a rich array of variables that permit scholars to study religiosity among the adult population. GSS2006 contains attend, a 9-point ordinal scale that measures how often respondents attend religious services. Values can range from 0 ("Never attend") to 8 ("Attend more than once a week").

A. The shell of a bar chart is given below. The categories of attend appear along the horizontal axis. What would a bar chart of attend look like if this variable had maximum dispersion? Sketch inside the axes a bar chart that would depict maximum dispersion.

How often R attends religious services

Never   Once a year   Once a month   Nearly every week   > Once a week
   < Once/year   Several times/year   2–3 times/month   Every week

B. What would a bar chart of attend look like if this variable had no dispersion? Sketch inside the axes a bar chart that would depict no dispersion.

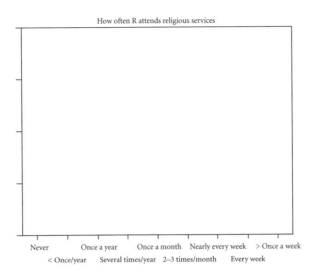

How often R attends religious services

Never Once a year Once a month Nearly every week > Once a week
< Once/year Several times/year 2–3 times/month Every week

C. Obtain frequencies output and a bar chart for attend. In the main Frequencies window, make sure that the Display frequency tables box is checked. In Statistics, see that all the boxes are unchecked. In Charts, request Bar charts with Percentages. Based on your examination of the frequency distribution,

the mode of attend is _____.   the median of attend is _____.

D. Based on your examination of the frequency distribution and bar chart, you would conclude that attend has (circle one)

low dispersion.   high dispersion.

E. Print the output from this exercise.

5. (Dataset: World. Variables: women05, country.) As noted earlier, for Exercise 5 you will analyze World. From the main menu bar in the SPSS Data Editor, click File → Open → Data. (Alternatively, click the open file icon on the menu bar.) Double-click the World icon.

What percentage of members of the U.S. House of Representatives are women? In 2005 the number was 15.0 percent, according to the Inter-Parliamentary Union, an international organization of parliaments.[5] How does the United States compare to other democratic countries? Is 15 percent comparatively low, comparatively high, or about average for a typical national legislature?

A. World contains women05, the percentage of women in the lower house of the legislature in each of 111 democracies. Perform a frequencies analysis on women05. In Statistics, obtain the mean, median, mode, and skewness. In the main Frequencies window, make sure that the Display frequency tables box is checked. (In Charts, Chart Type, make sure None is selected.) Fill in the table that follows.

Statistics for women05	
Mean	?
Median	?
Mode	?
Skewness	?
Std. Error of Skewness	?

B. Based on the statistics you obtained, which is the better measure of central tendency to use, the mean or the median? (circle one)

Mean   Median

Briefly explain your answer. _____

_____

C. Examine the Cumulative Percent column of the frequency distribution. Recall that 15.0 percent of U.S. House members are women. According to the frequency distribution, which of the following statements *most accurately describes* the U.S. House? (check one)

❏ The percentage of women in the U.S. House is below the median of women05.

❏ The U.S. House is in the lowest one-tenth of democracies in the percentage of women.

❏ Using the appropriate measure of central tendency, the percentage of women in the U.S. House can be described as "above average."

D. Suppose a women's advocacy organization vows to support female congressional candidates so that the U.S. House might someday "be ranked among the top one-fourth of democracies in the percentage of female members." According to the frequency distribution, to meet this goal women would need to constitute about what percentage of the House? (circle one)

About 20 percent   About 30 percent   About 40 percent

E. Run Analyze → Reports → Case Summaries. Click country into the Variables box and women05 into the Grouping Variable(s) box. Examine the output.

Which five countries have the lowest percentages of women in their legislatures?

_____

_____

_____

Which five countries have the highest percentages of women in their legislatures?

_____

_____

_____

6. (Dataset: NES2004. Variables: bigbus_therm, buspeople.) Survey respondents can be quite sensitive to words or phrases appearing in questions. To cite a classic example, the public expresses much greater support for government "assistance to the poor" than for "people on welfare."[6] Now consider these two business-related terms: "big business" and "business people." Do you think that respondents generally view business people in a more positive light than big business? In this exercise you will investigate the idea that the term "business people" evokes more favorable opinions than does the term "big business."

A. NES2004 contains the variables bigbus_therm and buspeople. Both variables are 100-point *feeling thermometers,* scales for which respondents are asked to rate groups or political personalities from 0 (cold or negative ratings) to 100 (warm or positive ratings). Run frequencies for bigbus_therm (the feeling thermometer of "big business") and buspeople (the feeling thermometer of "business people"). In the main Frequencies window, make sure that the Display frequency tables box is checked. In Statistics, obtain the mean and median, as well as the skewness. (For this exercise you do not need to find the mode.) In Charts, select None. Write the appropriate statistics for each variable in the table below.

	bigbus_therm	buspeople
Mean	?	?
Median	?	?
Skewness	?	?
Std. Error of Skewness	?	?

B. Two of the following statements are supported by the results in part A. They are (check two)

❏ The business people thermometer has a significant negative skew.

❏ The mean of the big business thermometer is higher than its median.

❏ A comparison of the appropriate measures of central tendency reveals that respondents rate business people about 13 points higher than they rate big business.

❏ A comparison of the appropriate measures of central tendency reveals that respondents rate business people about 10 points higher than they rate big business.

C. Examine and compare the cumulative frequency distributions for bigbus_therm and buspeople.

What percentage of respondents rated big business 50 degrees or lower? (circle one)

Less than 10 percent   About 25 percent   About 50 percent

What percentage of respondents rated business people 50 degrees or lower? (circle one)

Less than 10 percent   About 25 percent   About 50 percent

7. (Dataset: States. Variables: defexpen, state.) Here is the conventional political wisdom: Well-positioned members of Congress from a handful of states are successful in getting the federal government to spend revenue in their states—defense-related expenditures, for example. The typical state, by contrast, receives far fewer defense budget dollars.

A. Suppose you had a variable that measured the amount of defense-related expenditures in each state. The conventional wisdom says that, when you look at how all 50 states are distributed on this variable, a few states would have a high amount of defense spending. Most states, however, would have lower values on this variable.

If the conventional wisdom is correct, the distribution of defense-related expenditures will have (circle one)

a negative skew.   no skew.   a positive skew.

If the conventional wisdom is correct, the mean of defense-related expenditures will be (circle one)

lower than its median.   the same as its median.   higher than its median.

B. States contains the variable defexpen, defense expenditures per capita for each of the 50 states. Perform a frequencies analysis of defexpen. In Statistics, obtain the mean and median, as well as the skewness. (You do not need to obtain the mode for this exercise.) In the main Frequencies window, uncheck the Display frequency tables box. In Charts, select Histograms. Examine the results in the Viewer. Record the mean, median, skewness, and standard error of skewness in the table that follows.

Statistics for defexpen	
Mean	?
Median	?
Skewness	?
Std. Error of Skewness	?

C. Which is the better measure of central tendency? (circle one)

Mean   Median

Briefly explain your answer. _____

_____

_____

D. Based on your analysis, would you say that the conventional wisdom is accurate or inaccurate? (check one)

❑ The conventional wisdom is accurate.

❑ The conventional wisdom is inaccurate.

E. Run Case Summaries, putting state in the Variables box and using defexpen as the grouping variable.

The state with the lowest defense spending per capita is (fill in the blank) _____,

and the state with the highest defense spending per capita is _____.

F. Print the histogram you produced in part B.

8. (Dataset: States. Variables: blkpct, hispanic.) Two demographers are arguing over how best to describe the racial and ethnic composition of the "typical" state.

Demographer 1: "The typical state is 10.3 percent black and 8.8 percent Hispanic."

Demographer 2: "The typical state is 7.2 percent black and 5.8 percent Hispanic."

A. Run frequencies for blkpct (the percentage of each state's population that is African American) and hispanic (the percentage of each state's population that is Hispanic). In Statistics, obtain the mean and median, as well as the skewness. (You do not need to obtain the mode for this exercise.) In Charts,

Chart Type, select None. In the main Frequencies window, uncheck the Display frequency tables box. Record the appropriate statistics for each variable in the table that follows.

	blkpct	hispanic
Mean	?	?
Median	?	?
Skewness	?	?
Std. Error of Skewness	?	?

B. Based on your analysis, which demographer is more accurate? (circle one)

Demographer 1   Demographer 2

Write a few sentences explaining your reasoning. _____

_____

_____

_____

C. Run Case Summaries. Click state into Variables and click hispanic into Grouping Variable(s).

Which five states have the lowest percentages of Hispanics?

_____

_____

_____

Which five states have the highest percentages of Hispanics?

_____

_____

_____

That concludes the exercises for this chapter. Before exiting SPSS, be sure to save your output file.

NOTES

1. In this chapter we use the terms *dispersion, variation,* and *spread* interchangeably.
2. For interval-level variables that have a large number of categories, as does age, a frequency distribution can run to several output pages and is not very informative. Unchecking the Display frequency tables box suppresses the frequency distribution. A general guide: If the interval-level variable you are analyzing has 15 or fewer categories, go ahead and obtain the frequency distribution. If it has more than 15 categories, suppress the frequency distribution.
3. For demographic variables that are skewed, median values rather than means are often used to give a clearer picture of central tendency. One hears or reads reports, for example, of median family income or the median price of homes in an area.
4. Science_quiz was created by summing the number of correct responses to the following questions (all are in true-false format, except for earthsun): The center of the Earth is very hot (General Social Survey variable, hotcore); It is

the father's gene that decides whether the baby is a boy or a girl (boyorgrl); Electrons are smaller than atoms (electron); The universe began with a huge explosion (bigbang); The continents on which we live have been moving their locations for millions of years and will continue to move in the future (condrift); Human beings, as we know them today, developed from earlier species of animals (evolved); Does the Earth go around the Sun, or does the Sun go around the Earth? (earthsun); All radioactivity is man-made (radioact); Lasers work by focusing sound waves (lasers); Antibiotics kill viruses as well as bacteria (viruses).

5. See Inter-Parliamentary Union web site (www.ipu.org/english/home.htm).

6. Tom W. Smith, "That Which We Call Welfare by Any Other Name Would Smell Sweeter: An Analysis of the Impact of Question Wording on Response Patterns," *Public Opinion Quarterly* 51 (Spring 1987): 75–83.

3

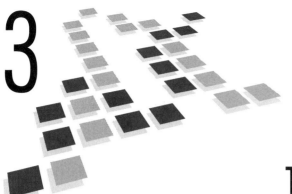

Transforming Variables

Procedures Covered

Transform → Recode into Different Variables

Transform → Visual Binning

Transform → Compute Variable

Transform → Recode into Same Variables

Political researchers sometimes must modify the variables they want to analyze. Generally speaking, such *variable transformations* become necessary or desirable in two common situations. Often a researcher wants to collapse a variable, combining its values or codes into a smaller number of useful categories. The researcher can do so through the Recode transformation feature or the Visual Binning procedure. In other situations a dataset may contain several variables that provide similar measures of the same concept. In these instances the researcher may want to combine the codes of different variables, creating a new and more precise measure. The Compute transformation feature is designed for this task.

In this chapter you will learn how to use the Recode, Visual Binning, and Compute commands. The chapter contains four guided examples, all of which use NES2004. The variables you modify or create in this chapter (and in this chapter's exercises) will become permanent variables in the datasets. After you complete each guided example, be sure to save the dataset.

USING RECODE

With Recode, you can manipulate any variable at any level of measurement—nominal, ordinal, or interval. But you should exercise vigilance and care. Follow three guidelines. First, before using Recode, you must obtain a frequency distribution of the variable you intend to manipulate. Second, after using Recode, it is important to check your work. Third, you should properly label the new variable and its values. Open NES2004, and let's work through the first example.

Recoding a Nominal-level Variable

NES2004 contains marital, a demographic variable that measures marital status in six categories:

Marital Status	Code
Married	1
Widowed	2
Divorced	3
Separated	4
Never married	5
Partnered, not married	6

A Frequencies analysis of marital produced the following result:

marital Marital status

		Frequency	Percent	Valid Percent	Cumulative Percent
Valid	1 Married	625	51.6	51.6	51.6
	2 Widowed	92	7.6	7.6	59.2
	3 Divorced	153	12.6	12.6	71.8
	4 Separated	42	3.5	3.5	75.3
	5 Never married	276	22.8	22.8	98.1
	6 Partnered, not married	23	1.9	1.9	100.0
	Total	1211	99.9	100.0	
Missing	System	1	.1		
Total		1212	100.0		

Now, think about research questions for which you might want to make fine distinctions among people—comparing, for example, the 153 divorced individuals with the 276 individuals who never married. Much of the time, however, you might be after a simpler comparison—the 51.6 percent of the valid cases who are married (code 1) and the remaining 48.4 percent of the sample who are unmarried (codes 2 through 6). How would you collapse the codes of marital into two categories and still preserve the potentially useful values of the original variable?

On the main menu bar, click Transform and consider the array of choices (Figure 3-1). Notice that SPSS presents two recoding options: Recode into Same Variables and Recode into Different Variables. When you recode a variable into the same variable, SPSS replaces the original codes with the new codes. The original information is lost. When you recode a variable into a different variable, SPSS uses the original codes to create a new variable. The original variable is retained. In some situations (discussed later) you will want to pick Recode into Same Variables. Most of the time, however, you should use the second option, Recode into Different Variables.

Figure 3–1 The Transform Drop-down Menu

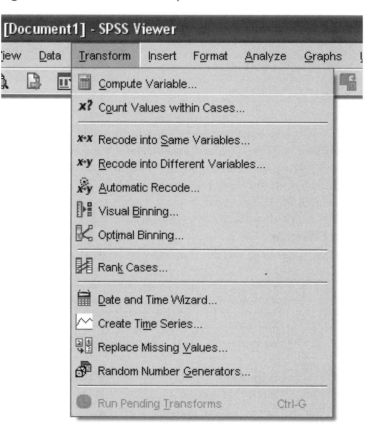

Figure 3–2 Recode into Different Variables Window

Click Recode into Different Variables. The Recode into Different Variables window opens (Figure 3-2). Scroll down the left-hand variable list and find marital. Click marital into the Input Variable → Output Variable box. SPSS puts marital into the box, with this designation: "marital → ?" This is SPSS-speak for "What do you want to name the new variable you are creating from marital?" Click in the Name box and type "married" (without quotation marks). Let's take this opportunity to give the new variable, married, a descriptive label. Click in the Label box and type "Is R married?" Click the Change button. The Recode into Different Variables window should now look like Figure 3-3.

Figure 3–3 Recoding a Nominal-level Variable

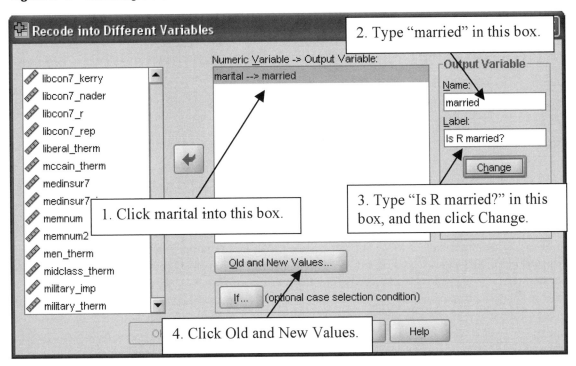

Figure 3–4   Recode into Different Variables: Old and New Values Window (default)

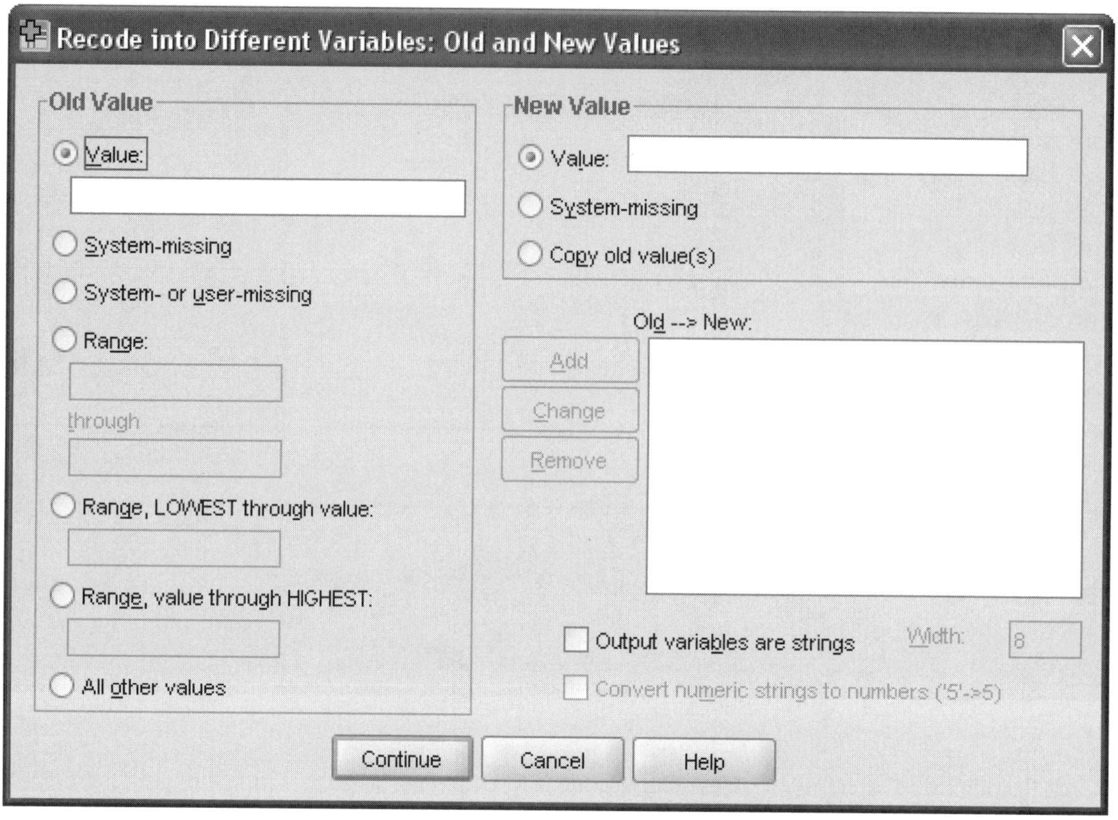

Now let's do the recoding. Click Old and New Values. The Recode into Different Variables: Old and New Values window pops up (Figure 3-4). There are two main panels. In the left-hand, Old Value panel, we will tell SPSS how to combine the original codes for marital. In the right-hand, New Value panel, we will assign codes for the new variable, which we have named married.

How do we want things to end up? Let's say that we want the new variable, married, to have two codes: code 1 for married respondents and code 0 for unmarried respondents. Plus we need to make sure that any respondents who have missing values on marital also have missing values on married. So we need to instruct SPSS to follow this recoding protocol:

Marital status	Old value (marital)	New value (married)
Married	1	1
Widowed	2	0
Divorced	3	0
Separated	4	0
Never married	5	0
Partnered, not married	6	0
	Missing	Missing

Make sure that the top radio button in the Old Value panel is selected (the default), click the cursor in the box next to "Value," and type "1." Move the cursor directly across to the right-hand New Value panel, make sure the top radio button is selected (again, the default setting), and type "1" in the Value box. Click the Add button. In the Old → New box, SPSS records your instruction with "1 → 1," meaning "All respondents coded 1 on marital will be coded 1 on married." Now return to the left-hand Old Value panel and select the uppermost Range button, the one simply labeled "Range." The two boxes beneath "Range" are activated. In the upper Range box, type "2." In the lower Range box, type "6." Move the cursor to the New Value panel and type "0" in the Value box. Click Add. SPSS responds, "2 thru 6 → 0," letting you know that all respon-

Figure 3–5 Recoding a Nominal-level Variable into Categories

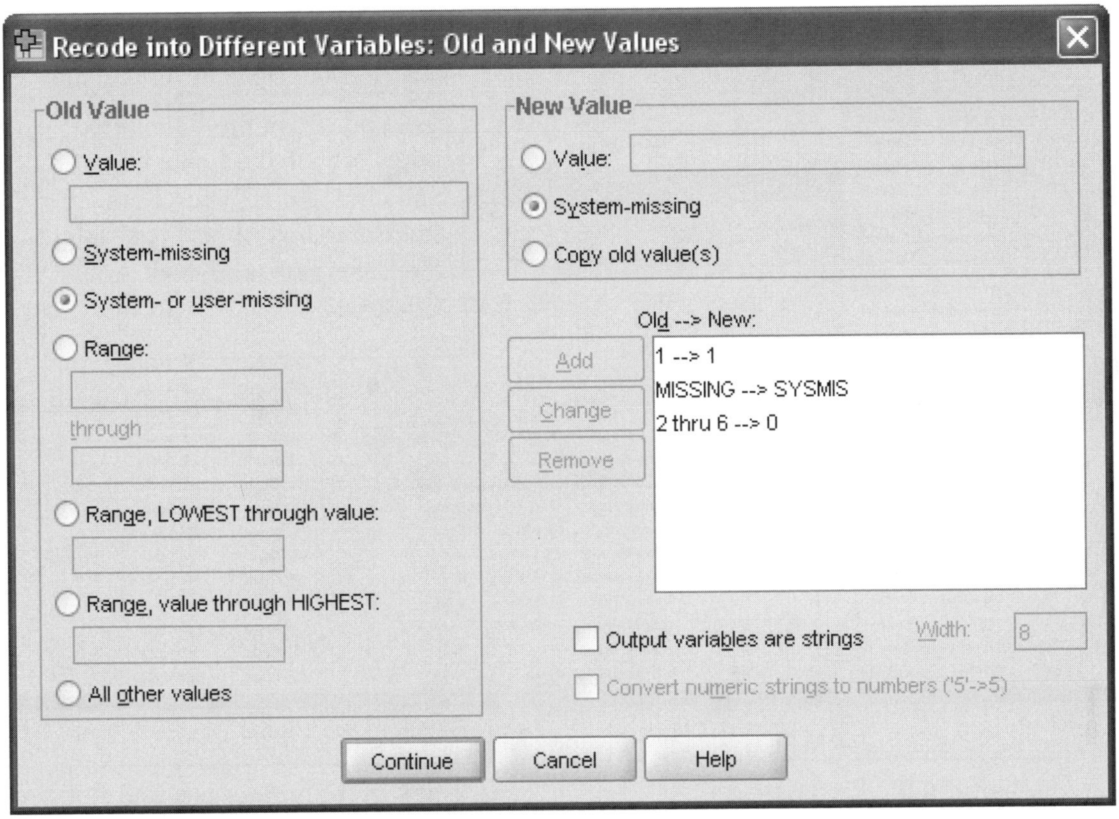

dents coded 2, 3, 4, 5, or 6 on marital will be coded 0 on married. One last loose end: In the Old Value panel, click the radio button next to "System- or user-missing." In the New Value panel, click the radio button next to "System-missing." Click Add. SPSS records your instruction as "MISSING → SYSMIS," meaning that any respondents having missing values on marital will be assigned missing values on married. The Recode into Different Variables: Old and New Values window should now look like Figure 3-5. Click Continue, returning to the main Recode into Different Variables window. Click OK. SPSS runs the recode.

Did the recode work correctly? This is where the check-your-work guideline takes effect. Run Frequencies on married to ensure that you did things right:

married Is R married?

		Frequency	Percent	Valid Percent	Cumulative Percent
Valid	0	586	48.3	48.4	48.4
	1	625	51.6	51.6	100.0
	Total	1211	99.9	100.0	
Missing	System	1	.1		
Total		1212	100.0		

The frequency table displays the label for the newly minted variable. More important, the valid percentages check out: 48.4 percent coded 0 and 51.6 percent coded 1. The recode worked as planned. However, we still need to make sure that the numeric codes are labeled properly: "No" for numeric code 0 and "Yes" for numeric code 1. To complete the recoding process, one more step is required.

In the Data Editor, make sure that the Variable View tab is clicked. Scroll down to the bottom of the Data Editor, where you will find married (Figure 3-6). (SPSS always puts newly created variables on the bottom row of the Variable View.) While we are doing the essential work of assigning value labels, we will also tidy up the formatting of the variable we just created. Click in the Decimals cell, which shows "2," and

Figure 3–6 Assigning Value Labels to a Recoded Variable

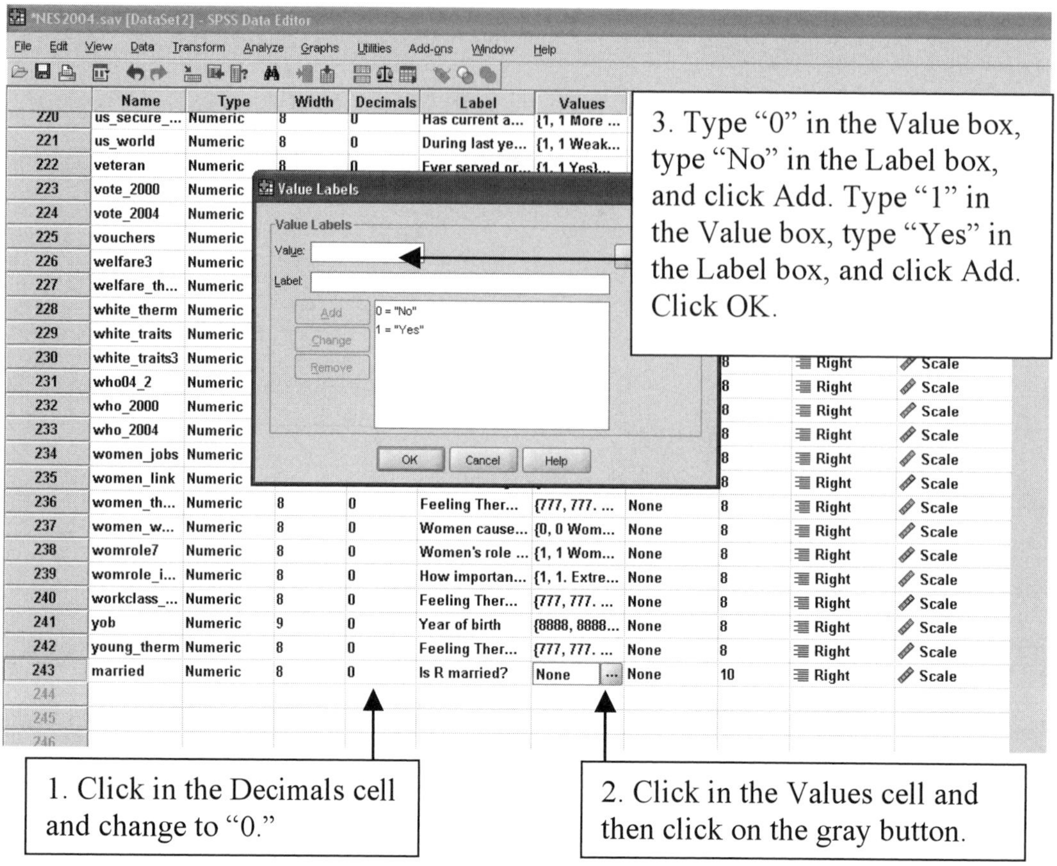

3. Type "0" in the Value box, type "No" in the Label box, and click Add. Type "1" in the Value box, type "Yes" in the Label box, and click Add. Click OK.

1. Click in the Decimals cell and change to "0."

2. Click in the Values cell and then click on the gray button.

change this value to "0." Next, click in the Values cell (which currently says "None"), and then click on the gray button that appears. The Value Labels window presents itself. In the box next to "Value," type "0." In the box next to "Label," type "No." Click Add. Repeat the process for code 1, typing "1" in the Value box and "Yes" in the Label box. Click Add. Click OK. Looks good.

You have just invested your time in recoding an original variable into a new variable and, in the process, made NES2004 better and more usable. Before going on to the next example, make sure you save the dataset.

Recoding an Interval-level Variable

Collapsing the values of a categorical variable, as you have just done, is perhaps the most common use of the Recode transformation feature. The original variable may be nominal level, such as marital. Or it may be ordinal level. For example, it might make sense to collapse four response categories such as "strongly agree," "agree," "disagree," and "strongly disagree" into two, "agree" and "disagree." At other times the original variable is interval level, such as age or income. In such cases you could use Recode to create a new variable having, say, three or four ordinal-level categories. Let's pursue this route.

NES2004 contains the variable yob, which records the year of birth for each respondent. Our goal here is to collapse yob into three theoretically useful categories: respondents born before 1950, those born between 1950 and 1965, and those born after 1965. How do we proceed? First, of course, we need a frequency distribution for yob. Click Analyze → Descriptive statistics → Frequencies, click yob into the Variable(s) list, and run the analysis. (In the main Frequencies window, make sure the Display frequency tables box is checked.) The frequency distribution is a real monster:

yob Year of birth

		Frequency	Percent	Valid Percent	Cumulative Percent
Valid	1914	2	.2	.2	.2
	1916	4	.3	.3	.5
	1917	1	.1	.1	.6
	1918	4	.3	.3	.9
	1919	1	.1	.1	1.0
	1920	4	.3	.3	1.3
	1921	5	.4	.4	1.8
	1922	10	.8	.8	2.6
	1923	5	.4	.4	3.0
	1924	4	.3	.3	3.3
	1925	6	.5	.5	3.8
	1926	8	.7	.7	4.5
	1927	16	1.3	1.3	5.8
	1928	10	.8	.8	6.7
	1929	8	.7	.7	7.3
	1930	9	.7	.8	8.1
	1931	10	.8	.8	8.9
	1932	9	.7	.8	9.7
	1933	9	.7	.8	10.4
	1934	10	.8	.8	11.2
	1935	10	.8	.8	12.1
	1936	11	.9	.9	13.0
	1937	14	1.2	1.2	14.2
	1938	16	1.3	1.3	15.5
	1939	19	1.6	1.6	17.1
	1940	21	1.7	1.8	18.8
	1941	27	2.2	2.2	21.1
	1942	21	1.7	1.8	22.8
	1943	20	1.7	1.7	24.5
	1944	13	1.1	1.1	25.6
	1945	14	1.2	1.2	26.8
	1946	23	1.9	1.9	28.7
	1947	26	2.1	2.2	30.8
	1948	27	2.2	2.2	33.1
	1949	23	1.9	1.9	35.0
	1950	19	1.6	1.6	36.6
	1951	22	1.8	1.8	38.4
	1952	25	2.1	2.1	40.5
	1953	25	2.1	2.1	42.6
	1954	25	2.1	2.1	44.7
	1955	22	1.8	1.8	46.5
	1956	24	2.0	2.0	48.5
	1957	23	1.9	1.9	50.4
	1958	27	2.2	2.2	52.7
	1959	24	2.0	2.0	54.7
	1960	26	2.1	2.2	56.8
	1961	20	1.7	1.7	58.5
	1962	19	1.6	1.6	60.1
	1963	26	2.1	2.2	62.2
	1964	18	1.5	1.5	63.8
	1965	25	2.1	2.1	65.8
	1966	23	1.9	1.9	67.8
	1967	16	1.3	1.3	69.1
	1968	17	1.4	1.4	70.5
	1969	23	1.9	1.9	72.4
	1970	19	1.6	1.6	74.0
	1971	21	1.7	1.8	75.8
	1972	14	1.2	1.2	76.9
	1973	23	1.9	1.9	78.8
	1974	15	1.2	1.2	80.1
	1975	23	1.9	1.9	82.0
	1976	28	2.3	2.3	84.3
	1977	20	1.7	1.7	86.0
	1978	18	1.5	1.5	87.5
	1979	23	1.9	1.9	89.4
	1980	22	1.8	1.8	91.2
	1981	25	2.1	2.1	93.3
	1982	19	1.6	1.6	94.9
	1983	15	1.2	1.2	96.2
	1984	17	1.4	1.4	97.6
	1985	21	1.7	1.8	99.3
	1986	8	.7	.7	100.0
	Total	1200	99.0	100.0	
Missing	System	12	1.0		
Total		1212	100.0		

Figure 3–7 Recoding an Interval-level Variable

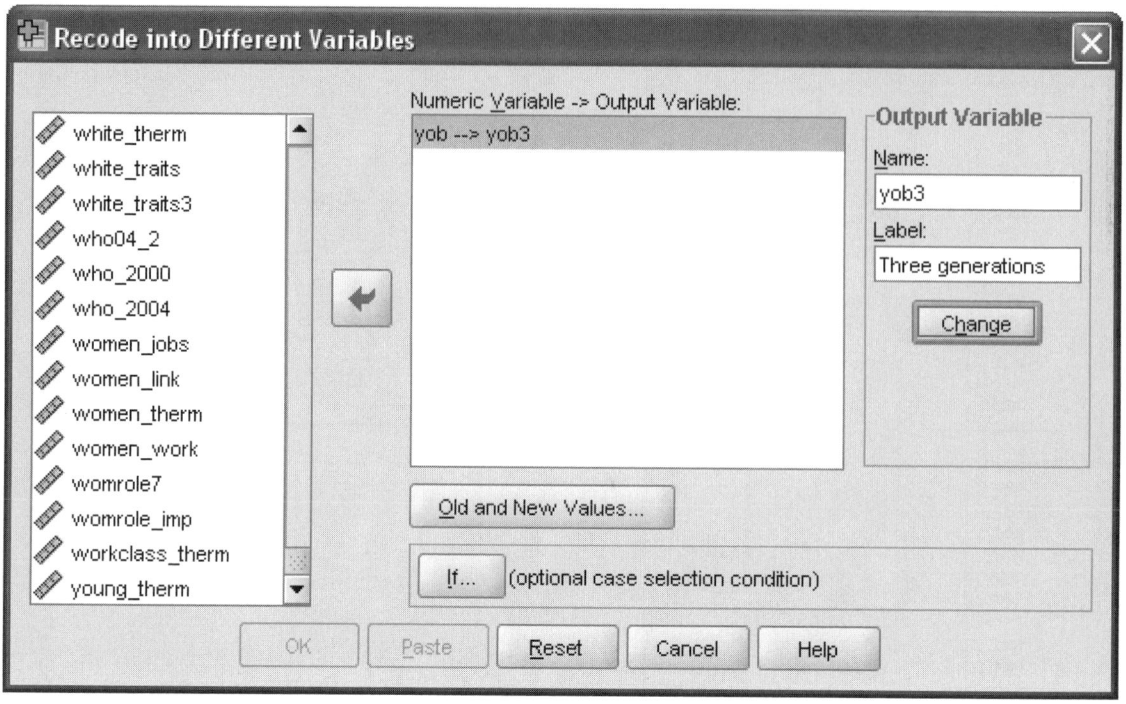

Let's use this distribution to get an idea of what the recoded variable should look like. To do this, focus on the Cumulative Percent column. What percentage of the sample falls into the oldest category—born before 1950? That's easy: 35.0 percent. What percentage of the sample falls *in or below* the middle category, people born between 1950 and 1965? Well, 65.8 percent of the sample was born in 1965 or earlier, so 65.8 percent of the sample should fall into the first two categories of the recoded variable. These two numbers, 35.0 percent and 65.8 percent, will help us to verify that our recode was performed properly.

Now do the recode. Click Transform → Recode into Different Variables. Click Reset to clear the panels. Click yob into the Input Variable → Output Variable box. Type "yob3" in the Name box. Type "Three generations" in the Label box and click Change (Figure 3-7). Click Old and New Values. Let's first create the oldest category for yob3. In the Old Value panel, select the radio button next to "Range, LOWEST through value"; doing so activates the box. Type "1949" in the box. In the New Value panel, type "1" in the Value box and click Add. SPSS translates the instruction as "Lowest thru 1949 → 1," lumping all respondents between the lowest value of yob (1914) and a value of 1949 on yob into code 1 of the new variable, yob3. In the Old Value panel, select the "Range" button and type "1950" in the upper box and "1965" in the lower box. Type "2" in the Value box in the New Value panel and click Add. That's the middle "baby boomer" age group, now coded 2 on yob3. In the Old Value panel, select the radio button next to "Range, value through HIGHEST" and type "1966" in the box. Type "3" in the Value box in the New Value panel and click Add. That puts the youngest generation into code 3 on yob3. Complete the recode by clicking the System- or user-missing button in the Old Value panel and the System-missing button in the New Value panel. Click Add. The Recode into Different Variables: Old and New Values window should now look like Figure 3-8. Click Continue. Click OK. Check your work by running Frequencies on yob3 and examining the output:

yob3 Three generations

		Frequency	Percent	Valid Percent	Cumulative Percent
Valid	1	420	34.7	35.0	35.0
	2	370	30.5	30.8	65.8
	3	410	33.8	34.2	100.0
	Total	1200	99.0	100.0	
Missing	System	12	1.0		
Total		1212	100.0		

Figure 3–8   Collapsing an Interval-level Variable into Categories

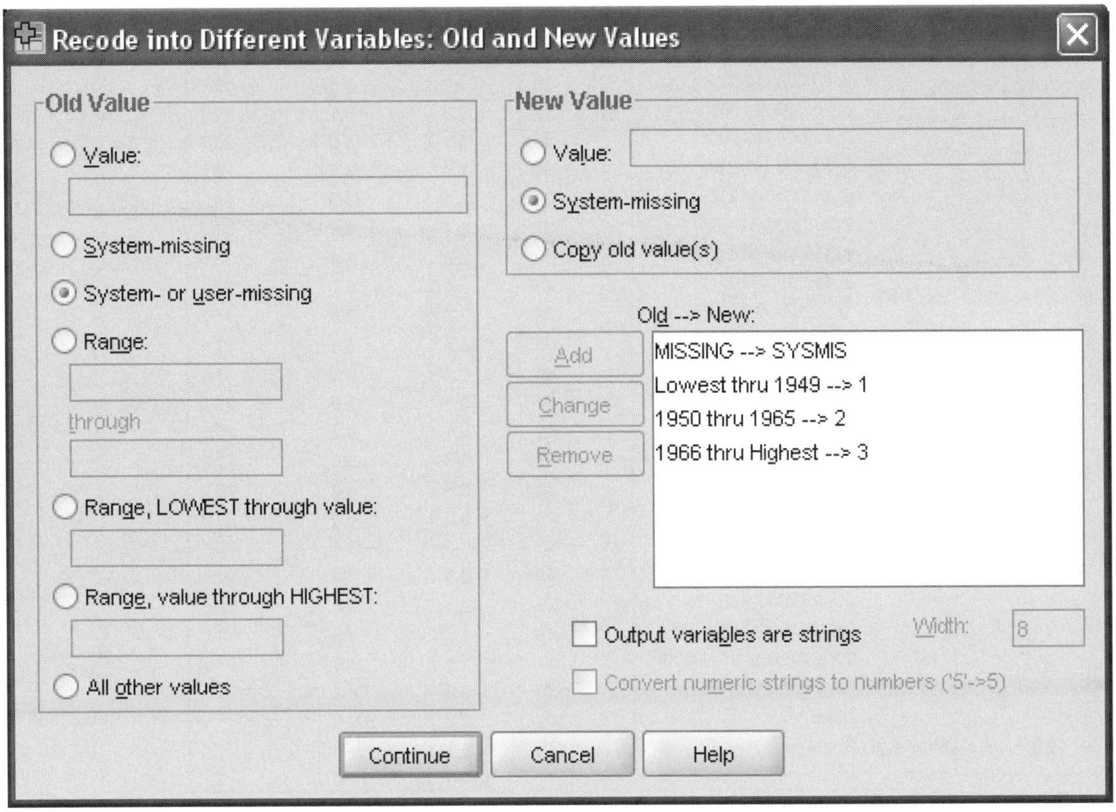

The cumulative percent markers, 35.0 percent and 65.8 percent, are just where they are supposed to be. Yob3 checks out. Before proceeding, scroll to the bottom of the Variable View in the Data Editor and make two changes to yob3. First, change Decimals to 0. Second, click in the Values cell and label yob3's values as follows:

Value	Value label
1	Before 1950
2	1950-1965
3	After 1965

USING VISUAL BINNING[1]

For nominal and ordinal variables, Recode is fast and easy to use. However, for interval variables, as we have just seen, it can be a bit more cumbersome. For collapsing interval variables, SPSS's more obscure (and underappreciated) Visual Binning procedure provides an attractive alternative to Recode. Visual Binning is as good as Recode for creating variables, such as yob3, for which the researcher has selected theoretically meaningful cutpoints for defining the categories of the new variable. And it is superior to Recode for situations in which you want to quickly collapse an interval-level variable into a handful of values, each containing roughly equal numbers of cases. Let's work through a guided example using NES2004's income_r, an interval measure of each respondent's income. You will use Visual Binning to create a three-category ordinal, income_r3. Income_r3 will break income_r into thirds: the lowest-income third ("Low"), the middle-income third ("Middle"), and the highest-income third ("High").

Collapsing an Interval-level Variable with Visual Binning

An obligatory Frequencies run on income_r produces the following output:

income_r Respondent income

		Frequency	Percent	Valid Percent	Cumulative Percent
Valid	1. LT $2,999	118	9.7	10.7	10.7
	2. $3,000 -$4,999	36	3.0	3.3	14.0
	3. $5,000 -$6,999	34	2.8	3.1	17.1
	4. $7,000 -$8,999	40	3.3	3.6	20.7
	5. $9,000 -$10,999	45	3.7	4.1	24.8
	6. $11,000-$12,999	38	3.1	3.4	28.2
	7. $13,000-$14,999	37	3.1	3.4	31.6
	8. $15,000-$16,999	32	2.6	2.9	34.5
	9. $17,000-$19,999	41	3.4	3.7	38.2
	10. $20,000-$21,999	39	3.2	3.5	41.7
	11. $22,000-$24,999	65	5.4	5.9	47.6
	12. $25,000-$29,999	65	5.4	5.9	53.5
	13. $30,000-$34,999	84	6.9	7.6	61.2
	14. $35,000-$39,999	62	5.1	5.6	66.8
	15. $40,000-$44,999	64	5.3	5.8	72.6
	16. $45,000-$49,999	44	3.6	4.0	76.6
	17. $50,000-$59,999	70	5.8	6.4	82.9
	18. $60,000-$69,999	43	3.5	3.9	86.8
	19. $70,000-$79,999	40	3.3	3.6	90.5
	20. $80,000-$89,999	26	2.1	2.4	92.8
	21. $90,000-$104,999	16	1.3	1.5	94.3
	22. $105,000-$119,000	17	1.4	1.5	95.8
	23. $120,000 and over	46	3.8	4.2	100.0
	Total	1102	90.9	100.0	
Missing	System	110	9.1		
Total		1212	100.0		

If you were using Recode to collapse this variable into three equally sized groups, you would create the "Low" group by collapsing codes 1 through 8 (cumulative percent, 34.5), the "Middle" group by combining codes 9 through 14 (cumulative percent, 66.8), and the "High" group by collapsing codes 15 through 23. That's a fair amount of Recode drudgery. Visual Binning will accomplish the same task with fewer clicks and less typing.

Click Transform → Visual Binning, opening the Visual Binning window. Scroll the variable list, find income_r, and click it into the Variables to Bin panel, as shown in Figure 3-9. Click Continue. To light up the panels, click on income_r in the Scanned Variable List (Figure 3-10). There are three panels to the right of the Scanned Variable List: the Name panel (top), a graphic display of the selected variable (middle), and the Grid panel (bottom). Click in the Binned Variable box in the Name panel. This is where you provide a name for the variable you are about to create. Type "income_r3." Click in the box beneath Label, where SPSS has supplied a default name, "Respondent income (Binned)". Modify the label to read, "Respondent income: 3 categories." Later you will attend to the Grid panel, but first you need to create income_r3's categories. Click Make Cutpoints and consider the Make Cutpoints window (Figure 3-11). Because you want equal-sized groups, select the radio button next to Equal Percentiles Based on Scanned Cases. And because you want three groups, click in the Number of Cutpoints box and type "2." Why "2"? Here is the rule: If you wish to create a variable having k categories, then you must request k–1 cutpoints. (Reassuringly, after you type "2," SPSS automatically puts "33.3" in the Width(%) box.) Click Apply, returning to the continuation window (Figure 3-12). Now, notice the values "8," "14," and "HIGH" that SPSS has entered in the Grid panel. Earlier, when we inspected the frequency distribution of income_r, we knew that these cutpoints would divide respondents into nearly equal groups.[2]

Figure 3–9 Visual Binning Opening Window

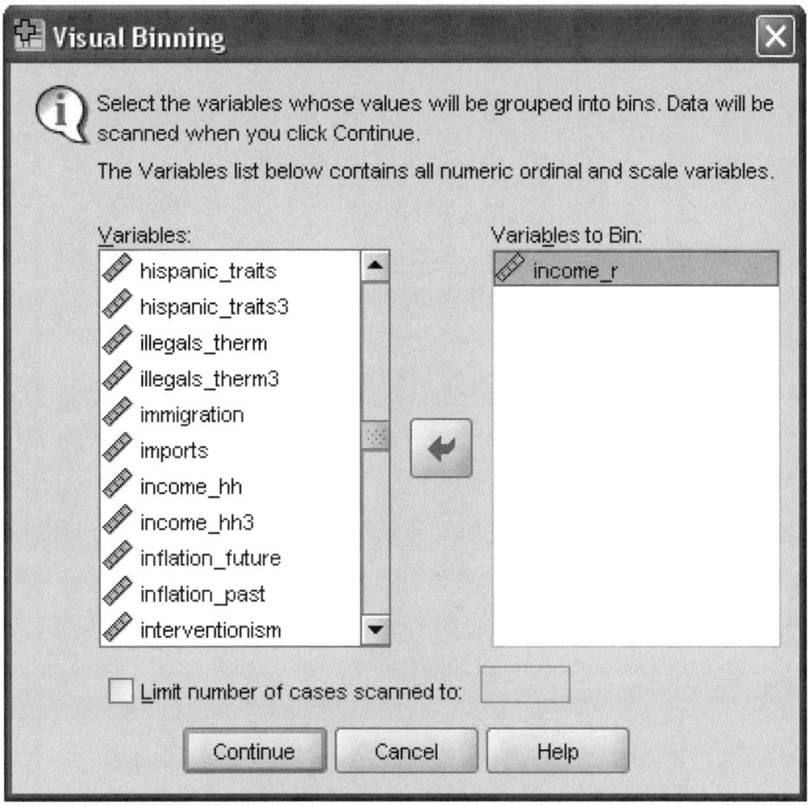

Figure 3–10 Visual Binning Continuation Window

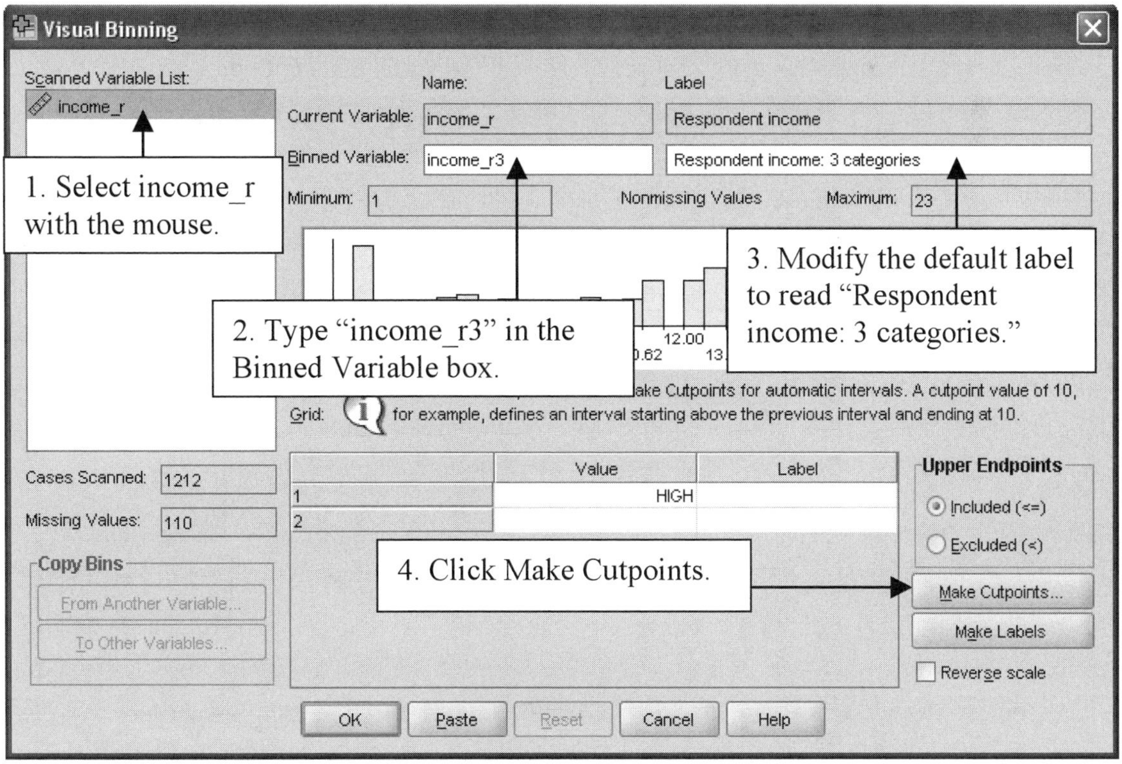

Figure 3–11 Visual Binning: Make Cutpoints Window

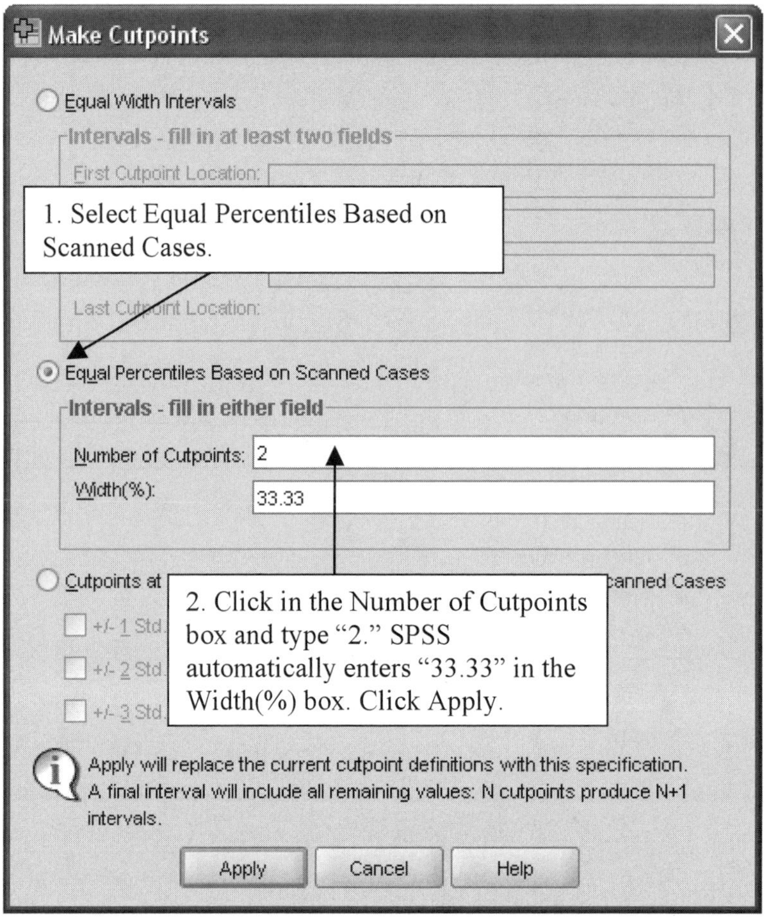

Figure 3–12 Labeling Values of Collapsed Variable in the Visual Binning Continuation Window

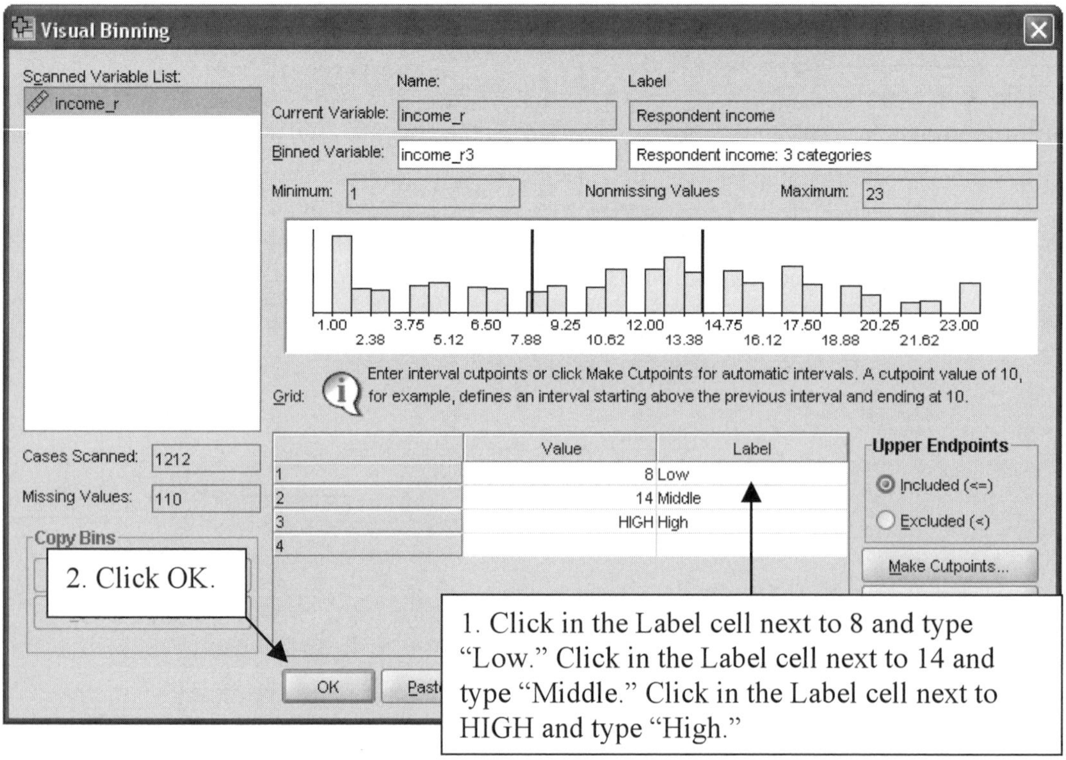

Let's finish the job by typing labels in the Values cells next to each cutpoint number, as shown in Figure 3-12: "Low" next to "8," "Middle" next to "14," and "High" next to "HIGH."[3] All set. Click OK. (Click OK again when SPSS issues an odd warning, "Binning specifications will create 1 variables.") Check your work by running Frequencies on income_r3:

income_r3 Respondent income: 3 categories

		Frequency	Percent	Valid Percent	Cumulative Percent
Valid	Low	380	31.4	34.5	34.5
	Middle	356	29.4	32.3	66.8
	High	366	30.2	33.2	100.0
	Total	1102	90.9	100.0	
Missing	System	110	9.1		
Total		1212	100.0		

This is a nice-looking three-category ordinal. Save the dataset, and let's move to the next topic.

USING COMPUTE

As you have seen, SPSS Recode requires a certain amount of discernment and care. SPSS Compute, however, allows the creative juices to flow. Although SPSS permits the creation of new variables through a dizzying variety of complex transformations, the typical use of Compute is pretty straightforward. By and large, Compute is typically used to create a simple *additive index* from similarly coded variables. Consider a simple illustration. Suppose you have three variables, each of which measures whether or not a respondent engaged in each of the following activities during an election campaign: tried to convince somebody how to vote, put a campaign bumper sticker on his or her car, or gave money to one of the candidates or parties. Each variable is coded identically: 0 if the respondent did not engage in the activity and 1 if he or she did. Now, each of these variables is interesting in its own right, but you might want to add them together, creating an overall measure of campaigning: People who did not engage in any of these activities would end up with a value of 0 on the new variable; those who engaged in one activity, a code of 1; two activities, a code of 2; and all three activities, a code of 3.

Here are some suggested guidelines to follow in using Compute to create a simple additive index. First, before running Compute, make sure that each of the variables is coded identically. In the preceding illustration, if the "bumper sticker" variable were coded 1 for no and 2 for yes, and the other variables were coded 0 and 1, the resulting additive index would be incorrect. Second, make sure that the variables are all coded in the same *direction*. If the "contribute money" variable were coded 0 for yes and 1 for no, and the other variables were coded 0 for no and 1 for yes, the additive index would again be incorrect.[4] Third, after running Compute, obtain a frequency distribution of the newly created variable. Upon examining the frequency distribution, you may decide to use Recode to collapse the new variable into more useful categories. Suppose, for example, that you add the three campaign acts together and get the following frequency distribution for the new variable:

Additive index: Number of campaign acts

Value label	Value	Percentage of sample
Engaged in none	0	60
Engaged in one	1	25
Engaged in two	2	13
Engaged in three	3	2
Total		100

It looks like a Recode run may be in order—collapsing respondents coded 2 or 3 into the same category.

These points are best understood firsthand. Dataset NES2004 contains the following four variables, each of which measures respondents' attitudes toward gay rights: whether the respondent favors gay adoptions (gay_adopt), anti-discrimination laws (gay_discrim), gay marriage (gay_marriage), and gays in the military (gay_military). For each of these variables, respondents who oppose gay rights are coded 0, and those who favor gay rights are coded 1. We are going to add these variables together, using the expression "gay_adopt+ gay_discrim+gay_marriage+gay_military." Think about this expression for a moment. If a respondent were coded as opposing gay rights on all four measures, what would be his or her score on an additive index? It would be 0 + 0 + 0 + 0 = 0. What if the respondent were coded as favoring gay rights on all four? In that case, 1 + 1 + 1 + 1 = 4. Thus, we know from the get-go that the values of the new variable will range from 0 to 4.

Let's get SPSS to compute a new variable, which we will name gay_rights, by summing the codes of gay_adopt, gay_discrim, gay_marriage, and gay_military. Click Transform → Compute, invoking the Compute Variable window (Figure 3-13). A box labeled "Target Variable" is in the window's upper left-hand corner. This is where we name the new variable. Click in the Target Variable box and type "gay_rights," as shown in Figure 3-14. The large box on the right side of the window, labeled "Numeric Expression," is where we tell SPSS which variables to use and how to combine them. Scroll down the left-hand variable list until you find gay_adopt. Click gay_adopt into the Numeric Expression box. Using the keyboard (or the calculator pad beneath the Numeric Expression box), type or click a plus sign (+) to the right of gay_adopt. Returning to the variable list, click gay_discrim into the Numeric Expression box. Repeat this process for the remaining variables, until the Numeric Expression box reads, "gay_adopt+gay_discrim+gay_marriage+ gay_military," as shown in Figure 3-14. Before we create gay_rights, let's give it a descriptive label. Click the

Figure 3–13 Compute Variable Window

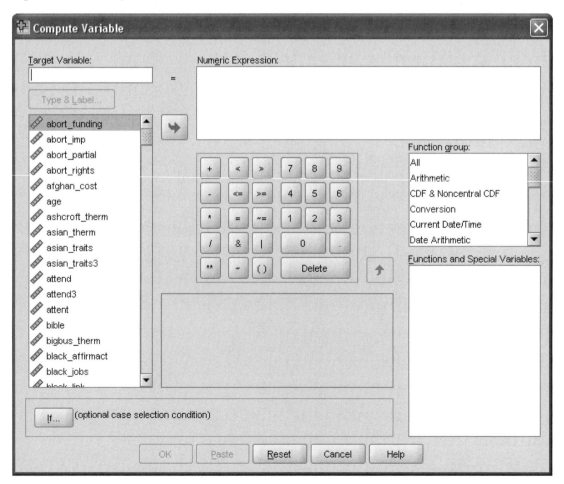

Figure 3–14 Computing a New Variable

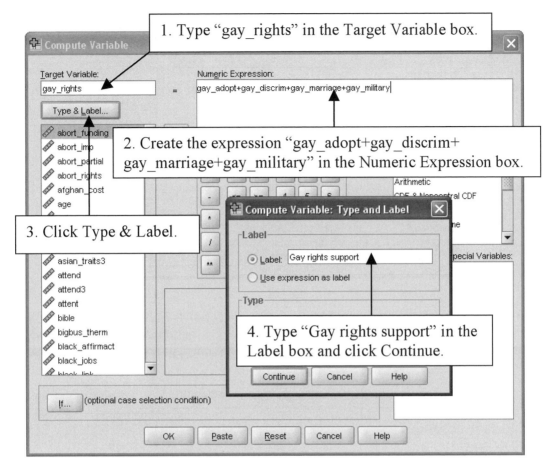

Type & Label button, which opens the Compute Variable: Type and Label window (see Figure 3-14). Type "Gay rights support" in the Label box and click Continue. You are ready to run the compute. Click OK. SPSS does its work. What does the new variable, gay_rights, look like? To find out, run Frequencies on gay_rights:

gay_rights Gay rights support

		Frequency	Percent	Valid Percent	Cumulative Percent
Valid	0	88	7.3	10.0	10.0
	1	157	13.0	17.8	27.8
	2	203	16.7	23.0	50.8
	3	163	13.4	18.5	69.3
	4	271	22.4	30.7	100.0
	Total	882	72.8	100.0	
Missing	System	330	27.2		
Total		1212	100.0		

   This is an interesting variable. Notice the fairly high variation in respondents' opinions about gay rights. The median value of gay_rights falls in code 2 (cumulative percentage, 50.8), the code received by respondents who held pro-gay rights opinions on two of the questions. However, the mode of gay_rights falls in code 4 (30.7 percent of the cases), the strongest pro-gay rights category. Suppose (for the sake of learning a new skill) that you wanted to collapse gay_rights into three categories: "Low" support, "Moderate" support, and "High" support. You might collapse codes 0 and 1 (the "Low" category), codes 2 and 3 ("Moderate"), and code 4 as a standalone ("High").

Figure 3–15   Recoding a New Variable

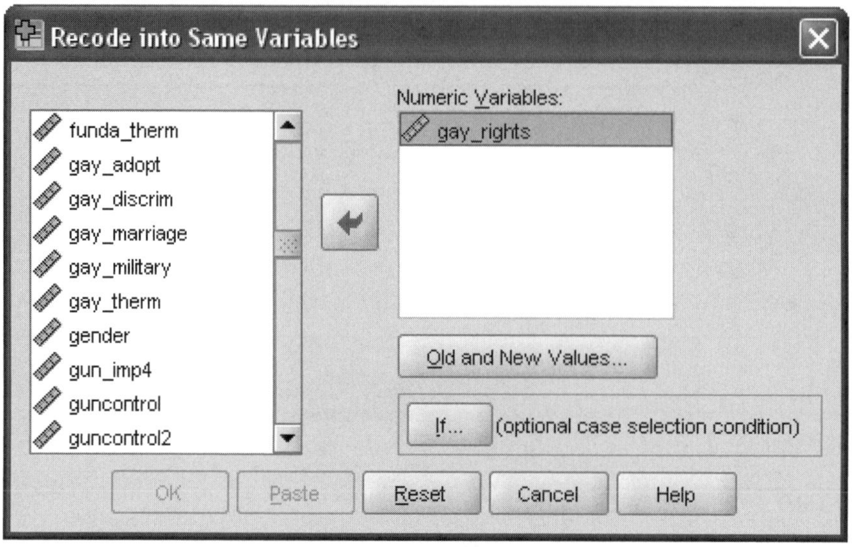

This is a situation in which Recode into Same Variables is appropriate.[5] Click Transform → Recode into Same Variables. In the Recode into Same Variables window, click gay_rights into the Numeric Variables box, as shown in Figure 3-15. Click Old and New Values. Follow this recoding protocol:

Old value	New value
Range: 0 through 1	1
Range: 2 through 3	2
4	3
System- or user-missing	System-missing

The Recode into Same Variables: Old and New Values window should look like Figure 3-16. Click Continue. Click OK. Again run Frequencies on gay_rights to check the recode:

gay_rights Gay rights support

		Frequency	Percent	Valid Percent	Cumulative Percent
Valid	1	245	20.2	27.8	27.8
	2	366	30.2	41.5	69.3
	3	271	22.4	30.7	100.0
	Total	882	72.8	100.0	
Missing	System	330	27.2		
Total		1212	100.0		

A flawless recode is a thing of beauty. Scroll to the bottom of the Variable View of the Data Editor and perform the usual housekeeping tasks with gay_rights. First, change Decimals to 0. Second, click in the Values cell and assign these value labels:

Value	Value label
1	Low
2	Moderate
3	High

Before proceeding with the exercises, be sure to save the dataset.

Figure 3–16 Collapsing a New Variable into Categories

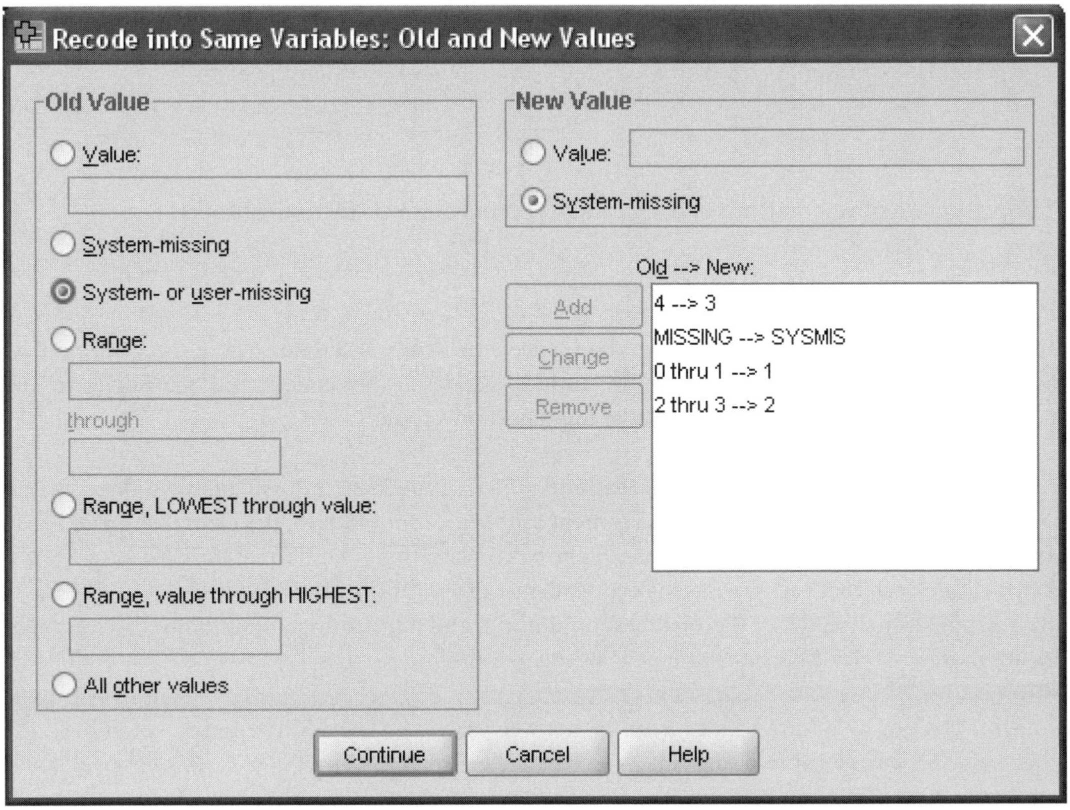

EXERCISES

1. (Dataset: GSS2006. Variable: polviews.) GSS2006 contains polviews, which measures political ideology—the extent to which individuals "think of themselves as liberal or conservative." Here is how polviews is coded:

Value	Value label
1	Extremely liberal
2	Liberal
3	Slightly liberal
4	Moderate
5	Slightly conservative
6	Conservative
7	Extremely conservative

A. Run Frequencies on polviews. The percentage of respondents who are either "extremely liberal," "liberal," or "slightly liberal" is (fill in the blank)

_____ percent.

The percentage of respondents who are either "slightly conservative," "conservative," or "extremely conservative" is (fill in the blank)

_____ percent.

B. Use polviews and Recode into Different Variables to create a new variable named polview3. Give polview3 this label: "Ideology: 3 categories." Collapse the three liberal codes into one category (coded 1 on polview3), put the moderates into their own category (coded 2 on polview3), and collapse the three conservative codes into one category (coded 3 on polview3). (Don't forget to recode missing values on polviews into missing values on polview3.) Run Frequencies on polview3.

The percentage of respondents who are coded 1 on polview3 is (fill in the blank)

_____ percent.

The percentage of respondents who are coded 3 on polview3 is (fill in the blank)

_____ percent.

C. In the Variable View of the Data Editor, change Decimals to 0, and then click in the Values cell and supply the appropriate labels: "Liberal" for code 1, "Moderate" for code 2, and "Conservative" for code 3. Run Frequencies on polview3. Print the Frequencies output.

2. (Dataset: GSS2006. Variables: stoprndm, tapphone, wotrial.) In Chapter 2 you analyzed three civil liberties issues: Whether people think that government authorities should have the right to stop and search people on the street (stoprndm), to record telephone conversations (tapphone), or hold suspects without trial (wotrial). Recall that each variable is coded identically: 0 (authorities "definitely should" have the right), 1 ("probably should"), 2 ("probably should not"), and 3 ("definitely should not"). Thus, higher codes are more pro–civil liberties and lower codes are more pro-security.

A. Imagine creating an additive index from these three variables. The additive index would have scores that range between what two values?

Between a score of _____ and a score of _____.

B. Suppose a respondent takes the strongest pro-security position on two of the issues but takes the "probably should" position on the third issue. What score would this respondent have?

A score of _____

C. Use Compute to create an additive index from stoprndm, tapphone, and wotrial. Name the new variable civlibs. Give civlibs this label: "Pro-civil liberties scale." Run Frequencies on civlibs. Referring to your output, fill in the table that follows.

civlibs Pro-civil liberties scale

Score on civlibs	Frequency	Valid Percent
?	?	?
?	?	?
?	?	?
?	?	?
?	?	?
?	?	?
?	?	?
?	?	?
?	?	?
?	?	?
Total	?	100.0

3. (Dataset: GSS2006. Variables: tol_communist, tol_racist.) GSS2006 contains two variables, each of which measures respondents' levels of tolerance toward a specific group: communists (tol_communist) and racists (tol_racist). Each of these variables is coded identically in the same direction: 0 (least tolerant), 1 (middle), and 2 (most tolerant). When you create an additive index from tol_communist and tol_racist, the index will range from a score of 0 for the least tolerant respondents to a score of 4 for the most tolerant respondents.

A. Compute an additive index from tol_communist and tol_racist. Name the new variable tolerance, and give tolerance this label: "Communists/Racists." Run Frequencies on tolerance. Refer to your output, and fill in the table that follows.

tolerance Communists/Racists

Score on tolerance	Frequency	Valid Percent
0	?	?
1	?	?
2	?	?
3	?	?
4	?	?
Total	?	100.0

B. Let's say that you want to collapse tolerance into three ordinal categories. The first category will combine respondents scoring 0 or 1, the second will combine respondents scoring 2 or 3, and the third will include those scoring 4. What percentage of respondents will be in each of the collapsed categories?

Combined percentage of respondents scoring 0 or 1: _____.

Combined percentage of respondents scoring 2 or 3: _____.

Percentage of respondents scoring 4: _____.

C. Use Transform → Recode into Same Variables to collapse tolerance into three categories. Code the new categories 1, 2, and 3, as shown in this recoding protocol:

Old value	New value
Range: 0 through 1	1
Range: 2 through 3	2
4	3
System- or user-missing	System-missing

Run Frequencies on tolerance to ensure that the recoding procedure worked correctly. Fill in the valid percentages in the table that follows.

tolerance Communists/Racists

Score on tolerance	Valid Percent
1	?
2	?
3	?
Total	100.0

D. Ensure that the percentages you recorded in part C match those you calculated in part B. Scroll to the bottom of the Variable View of the Data Editor and make two changes to tolerance. First, change Decimals to 0. Second, click in the Values cell and assign these new values: 1 "Low," 2 "Middle," and 3 "High."

4. (Dataset: GSS2006. Variable: income06.) In this chapter you learned to use Visual Binning by collapsing an NES2004 measure of income into three roughly equal ordinal categories. In this exercise, you will use Visual Binning to collapse a very similar variable from GSS2006, income06. Just as you did with income_r, you will collapse income06 into income06_3, a three-category ordinal measure of income.

A. Refer back to this chapter's visual binning guided example and retrace the steps. Here is new information you will need:

Variable to bin	income06
Binned variable name	income06_3
Binned variable label	Total income: 3 categories
Number of cutpoints	2
Labels for Value cells	Low, Middle, High

B. Run Frequencies on income06_3. Refer to your output. Fill in the table that follows.

income06_3 Total family income: 3 categories

income06_3	Frequency	Valid Percent
1 Low	?	?
2 Middle	?	?
3 High	?	?
Total	?	100.0

By performing the exercises in this chapter, you have added four variables to GSS2006: polview3, civlibs, tolerance, and income06_3. Before exiting SPSS, make sure to save the dataset.

NOTES

1. In earlier releases of SPSS, this feature is called Visual Bander.
2. Suppose you wanted to collapse yob into yob3 using Visual Binning instead of Recode. You would follow these steps: 1. Click Transform → Visual Binning and scan yob. 2. In the Scanned Variable List of the continuation window, select yob. 3. In the Name panel, supply the name "yob3" and the label "Three generations." 4. In the Value cells of the Grid panel, type "1949" in the topmost cell and "1965" in the next lower cell. (SPSS automatically supplies the word "HIGH" in the lowest of the three cells.) 5. In the Label cells of the Grid panel, supply value labels for each value ("Before 1950" goes with "1949," "1950-1965" goes with "1965," and "After 1965" goes with "HIGH"). 6. Click OK. Yob3 will be created and correctly labeled.
3. The SPSS-supplied numbers in the Value cells of the Grid panel are not the numeric codes that SPSS assigns to the categories. The numeric codes are numbered sequentially, beginning with 1. So income_r3 has numeric codes 1 ("Low"), 2 ("Middle"), and 3 ("High").
4. Survey datasets are notorious for reverse-coding. Survey designers do this so that respondents don't fall into the trap of response-set bias, or automatically giving the same response to a series of questions.
5. Recode → Into Same Variables is an appropriate choice because the original variables are not being replaced or destroyed in the process. If the recode goes badly and gay_rights gets fouled up, you can always use the original variables to compute gay_rights again.

4

Making Comparisons

Procedures Covered

Analyze → Descriptive Statistics → Crosstabs
Analyze → Compare Means → Means
Graphs → Legacy Dialogs → Line
Graphs → Legacy Dialogs → Bar

All hypothesis testing in political research follows a common logic of comparison. The researcher separates subjects into categories of the independent variable and then compares these groups on the dependent variable. For example, suppose you think that gender (independent variable) affects opinions about gun control (dependent variable) and that women are more likely than men to favor gun control. You would divide subjects into two groups on the basis of gender, women and men, and then compare the percentage of women who favor gun control with the percentage of men who favor gun control. Similarly, if you hypothesize that Republicans have higher incomes than do Democrats, you would divide subjects into partisanship groups (independent variable), Republicans and Democrats, and compare the average income (dependent variable) of Republicans with that of Democrats.

Although the logic of comparison is always the same, the appropriate method depends on the level of measurement of the independent and dependent variables. In this chapter you will learn how to use SPSS to address two common hypothesis-testing situations: those in which both the independent and the dependent variables are categorical (nominal or ordinal) and those in which the independent variable is categorical and the dependent variable is interval level. You will also learn to add visual support to your hypothesis testing by creating and editing bar charts and line charts.

USING CROSSTABS

Cross-tabulations are the workhorse vehicles for testing hypotheses for categorical variables. When setting up a cross-tabulation, you must observe the following three rules. First, put the independent variable on the columns and the dependent variable on the rows. Second, always obtain column percentages, not row percentages. Third, test the hypothesis by comparing the percentages of subjects who fall into the same category of the dependent variable.

Consider this hypothesis: In a comparison of individuals, those who are younger will pay less attention to political campaigns than will those who are older. NES2004 contains the variable attent, which measures respondents' levels of interest in the presidential campaign: very much, somewhat, or not much interested. This will serve as the dependent variable. A variable you created in Chapter 3, yob3, is the independent variable. Recall that yob3 categorizes respondents into three generational cohorts by year of birth: before 1950, between 1950 and 1965, and after 1965.

Open NES2004, and let's test the hypothesis. In the Data Editor, click Analyze → Descriptive Statistics → Crosstabs. The Crosstabs window appears, sporting four panels. For now, focus on the two upper right-hand

Figure 4-1 Crosstabs Window (modified)

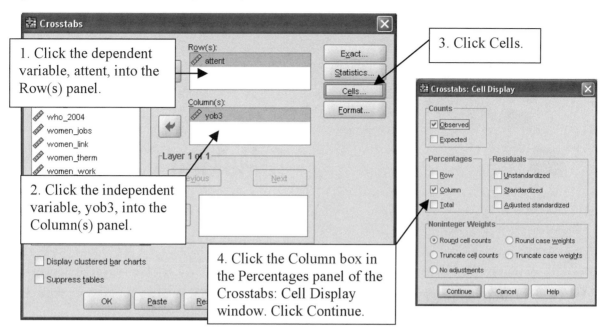

panels: Row(s) and Column(s). (The oddly labeled Layer 1 of 1 panel comes into play in Chapter 5.) This is where we apply the first rule for a properly constructed cross-tabulation: The independent variable defines the columns, and the dependent variable defines the rows. Because attent is the dependent variable, click it into the Row(s) panel, as shown in Figure 4-1. Find yob3 in the left-hand variable list and click it into the Column(s) panel.

Now for the second rule of cross-tab construction: Always obtain column percentages. On the right-hand side of the Crosstabs window, click the Cells button (refer to Figure 4-1). SPSS displays the available options for Counts, Percentages, and Residuals. Left to its own defaults, SPSS will produce a cross-tabulation showing only the number of cases ("observed" counts) in each cell of the table. That's fine. But to follow the second rule, we also want column percentages—the percentage of each category of the independent variable falling into each category of the dependent variable. Click the Column box in the Percentages panel. Click Continue, which returns you to the Crosstabs window. That's all there is to it. Click OK.

SPSS runs the analysis and displays the results in the Viewer—a case processing summary followed by the requested cross-tabulation:

Interested in following campaigns? * Three generations Crosstabulation

			Before 1950	1950-1965	After 1965	Total
			Three generations			
Interested in following campaigns?	1 Very much	Count	213	161	120	494
		% within Three generations	50.7%	43.5%	29.3%	41.2%
	3 Somewhat	Count	159	157	207	523
		% within Three generations	37.9%	42.4%	50.5%	43.6%
	5 Not much	Count	48	52	83	183
		% within Three generations	11.4%	14.1%	20.2%	15.2%
Total		Count	420	370	410	1200
		% within Three generations	100.0%	100.0%	100.0%	100.0%

By convention, SPSS identifies its Crosstabs output with the label of the dependent variable, followed by an asterisk (*), and then the label of the independent variable. In fact, when SPSS runs Crosstabs, it produces a set of side-by-side frequency distributions of the dependent variable—one for each category of the independent variable—plus an overall frequency distribution for all analyzed cases. Accordingly, the table has four columns of numbers: one for respondents born before 1950, one for those born between 1950 and

1965, one for those born after 1965, and a total column showing the distribution of all cases across the dependent variable. And, as requested, each cell shows the number (count) and column percentage.

What do you think? Does the cross-tabulation fit the hypothesis? The third rule of cross-tabulation analysis is easily applied. Focusing on the "very much" value of the dependent variable ("Interested in following campaigns?"), we see a clear pattern in the hypothesized direction. A comparison of respondents in the "Before 1950" column with those in the "1950–1965" column reveals a decrease in the "very much" percentage, from 50.7 percent to 43.5 percent. A comparison of the "1950–1965" and "After 1965" columns reveals yet another drop in campaign interest, from 43.5 percent to 29.3 percent. Yes, the analysis supports the hypothesis.

USING COMPARE MEANS

We now turn to another common hypothesis-testing situation: when the independent variable is categorical and the dependent variable is interval level. The logic of comparison still applies—divide cases on the independent variable and compare values of the dependent variable—but the method is different. Instead of comparing percentages, we now compare means.

To illustrate, let's say that you are interested in explaining this dependent variable: attitudes toward Hillary Clinton. Why do some people have positive feelings toward her whereas others harbor negative feelings? Here is a plausible idea: Partisanship (independent variable) will have a strong effect on attitudes toward Hillary Clinton (dependent variable). The hypothesis: In a comparison of individuals, those who are Democrats will have more favorable attitudes toward Hillary Clinton than will those who are Republicans.

NES2004 contains hillary_therm, a 100-point feeling thermometer. Each respondent was asked to rate Clinton on this scale, from 0 (cold or negative) to 100 (warm or positive). This is the dependent variable. NES2004 also has partyid7, which measures partisanship in seven categories, from Strong Democrat (coded 0) to Strong Republican (coded 6). (The intervening codes capture gradations between these poles: Weak Democrat, Independent-Democrat, Independent, Independent-Republican, and Weak Republican.) This is the independent variable. If the hypothesis is correct, we should find that Strong Democrats have the highest mean scores on hillary_therm and that mean scores decline systematically across categories of partyid7, hitting bottom among respondents who are strong Republicans. Is this what happens?

Click Analyze → Compare Means → Means. The Means window pops into view. Scroll down the left-hand variable list until you find hillary_therm, and then click it into the Dependent List panel, as shown in Figure 4-2. Now scroll to partyid7 and click it into the Independent List panel. In the Means window, click Options. The Means: Options window (also shown in Figure 4-2) permits you to select desired statistics

Figure 4–2 Means Window (modified)

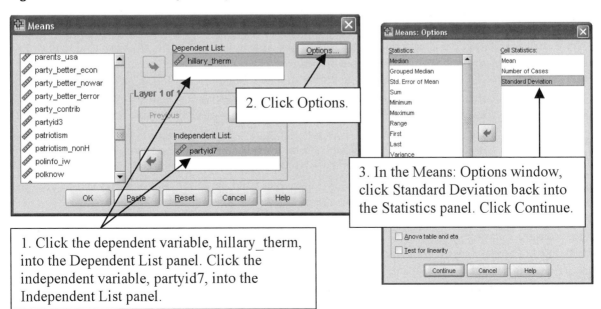

from the left-hand Statistics panel and click them into the right-hand Cell Statistics panel. Alternatively, you can remove statistics from Cell Statistics by clicking them back into the left-hand panel. Unless instructed otherwise, SPSS will always report the mean, number of cases, and standard deviation of the dependent variable for each category of the independent variable. Because at present we are not interested in obtaining the standard deviation, select it with the mouse and click it back into the left-hand Statistics panel. Our mean comparison table will report only the mean value of hillary_therm and the number of cases for each category of partyid7. Click Continue, returning to the Means window. Click OK.

Compared with cross-tabulations, mean comparison tables are models of minimalism:

Feeling Thermometer: Hillary Clinton

Summary: R party ID	Mean	N
Str Dem	81.85	201
Wk Dem	72.63	179
Ind Dem	67.26	208
Indep	57.07	112
Ind Rep	42.08	137
Wk Rep	37.53	154
Str Rep	23.27	191
Total	55.69	1182

The label of the dependent variable, "Feeling Thermometer: Hillary Clinton," appears along the top of the table. The label for the independent variable, "Summary: R party ID," defines the left-most column, which shows all seven categories, from Strong Democrat at the top to Strong Republican at the bottom. Beside each category, SPSS has calculated the mean of hillary_therm and reported the number of respondents falling into each value of partisanship. (The bottom row, "Total," gives the mean for the whole sample.)

Among Strong Democrats the mean for hillary_therm is pretty high—about 82 degrees. Does the mean decline as attachment to the Democratic Party weakens and identification with the Republican Party strengthens? Well, the mean drops sharply among Weak Democrats (who average about 73 degrees), shows a not-so-precipitous drop among Independent-Democratic leaners (about 67), and then continues to decline predictably, with another mild drop between Independent-Republican leaners and Weak Republicans. Strong Republicans, who average close to 23 degrees on the thermometer, have the chilliest response to Clinton. Indeed, on average, the dependent variable declines by nearly 60 degrees between Strong Democrats at one pole to Strong Republicans at the other extreme. On the whole, then, the data support the hypothesis.

GRAPHING RELATIONSHIPS

We have already seen that bar charts and histograms can be a great help in describing the central tendency and dispersion of a *single* variable. SPSS graphic procedures are also handy for illustrating relationships *between* variables. It will come as no surprise that SPSS supports a large array of graphic styles. To get a flavor of this variety, click Graphs → Legacy Dialogs and consider the choices (Figure 4-3). The legacy charts, as the name implies, use interfaces developed in earlier releases of SPSS. Even so, the Legacy Dialogs are still the best way to create graphics in SPSS.[1] In this chapter you will learn to use Bar and Line. (In Chapter 8 you will work with Scatter/Dot.) A bar chart is useful for summarizing the relationship between two categorical variables. A line chart adds clarity to the relationship between a categorical independent variable and an interval-level dependent variable. Line charts are elegant and parsimonious, and they can be used to display the relationship between two categorical variables as well.

To get an idea of how SPSS produces a line chart, let's begin by creating one of our own, using the results from the hillary_therm-partyid7 example. Turn your attention to Figure 4-4, an empty graphic "shell." The horizontal axis, called the *category axis,* displays values of the independent variable, party ID. Each partisanship category is represented by a hash mark, from Strong Democrat on the left to Strong Republican on the right. The vertical axis, called the *summary axis,* represents mean values of the dependent variable, Hillary Clinton thermometer ratings. Now, with a pen or pencil, make a dot directly above each

Figure 4–3 Graphs Drop-down Menu

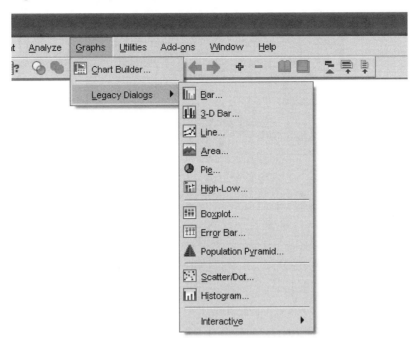

Figure 4–4 Line Chart Shell: Mean Values of Hillary Clinton Thermometer, by Party Identification

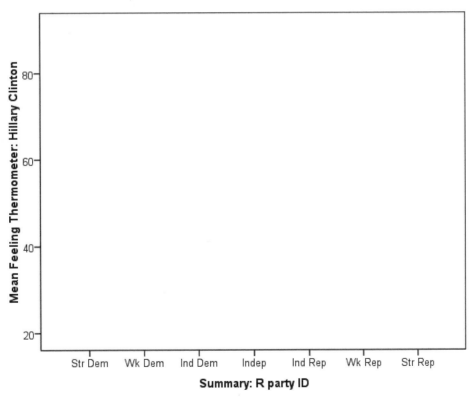

category of the independent variable, recording the mean of hillary_therm for each partisan category. Above the Strong Democrat hash mark, for example, place a dot at 82 on the summary axis. Go to the right along the category axis until you reach the hash mark for Weak Democrat and make a dot directly above the hash mark, at about 73. Do the same for the remaining partisan groups, placing a dot vertically above each hash mark at the mean value of hillary_therm. (Don't worry about being precise. Just get the dots close to the

Figure 4–5 Line Charts Window (default)

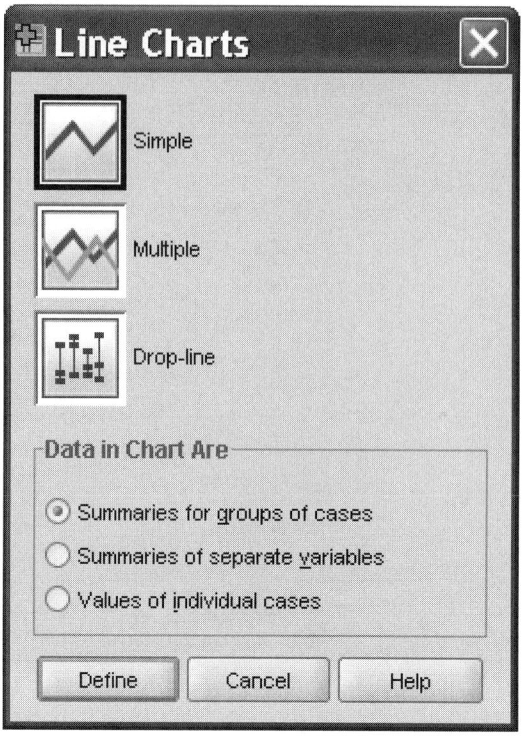

mean values.) Using a straight edge, connect the dots. Voilà! You've created a line chart for the relationship, a visual summary that is easy to interpret and present.

Using Line Chart

Now let's get SPSS to do the work for us. Click Graphs → Legacy Dialogs → Line. The Line Charts window opens (Figure 4-5). Make sure that the icon next to "Simple" is clicked and that the radio button next to "Summaries for groups of cases" is selected.[2] Click Define. The Define Simple Line window appears (Figure 4-6). The two top-most boxes—the (currently grayed out) Variable box in the Line Represents panel and the Category Axis box—are where we tailor the line chart to our specifications. (There are two additional boxes in the Panel by area, one labeled "Rows" and one labeled "Columns," as shown in Figure 4-6. For our purposes in this book, these boxes may be safely ignored.)

We know that we want partyid7 to define the category (horizontal) axis. Scroll down to partyid7 and click it into the Category Axis box. We also want to graph the mean values of hillary_therm for each category of partyid7. To do this, we've got to give SPSS instructions.[3] In the Line Represents panel, select the Other statistic radio button, as shown in Figure 4-7. The box beneath "Variable" is activated. Now scroll the left-hand variable list until you find hillary_therm, and then click hillary_therm into the Variable box. SPSS moves hillary_therm into the Variable box and gives it the designation "MEAN(hillary_therm)." In Line Chart, whenever you request Other statistic and click a variable into the Variable box (as we have just done), SPSS assumes that you want to graph the mean values of the requested variable (as, in this case, we do).[4] So this default serves our current needs. All set. Click OK.

A line chart of the hillary-partyid7 relationship appears in the Viewer (Figure 4-8). Line charts are at once simple and informative. You can immediately see the negative linear relationship between the independent and dependent variables. Note too the milder declines between the independent leaners and weak partisans of each party.

Using Bar Chart

In most respects, Bar Chart is similar to Line Chart. However, we do need to dwell on a key difference between the two. Also, we will take an excursion into the Chart Editor. In this guided example, you will obtain a bar chart of the relationship you analyzed earlier between campaign interest (the dependent variable attent) and generational cohort (the independent variable yob3).

Figure 4–6   Define Simple Line Window (default)

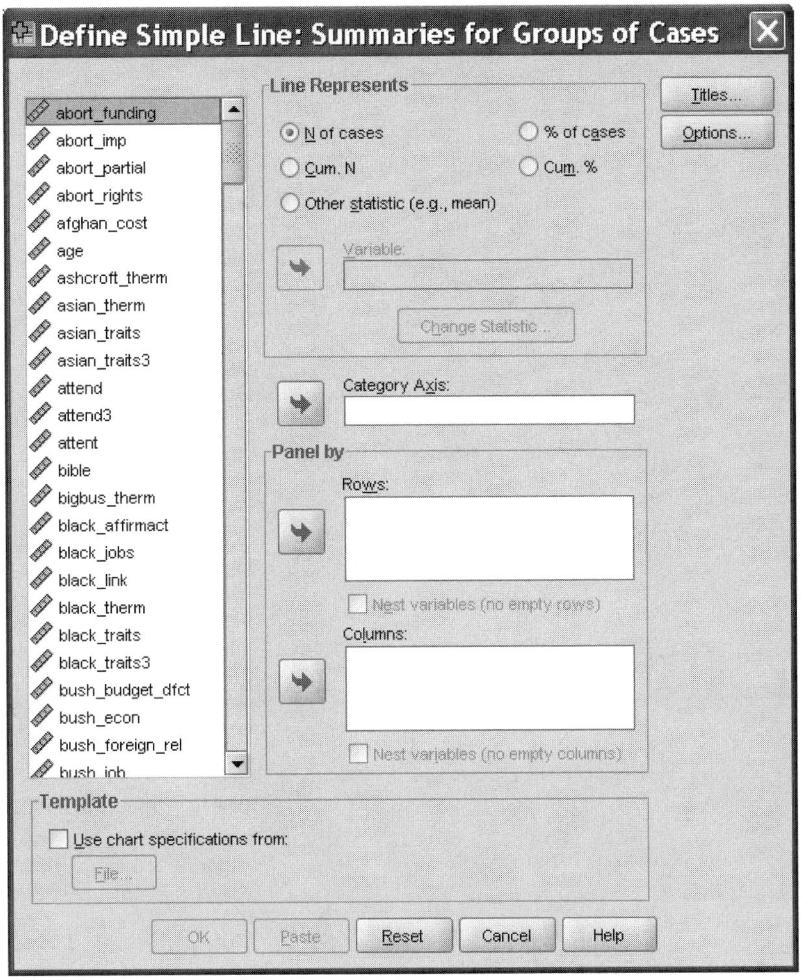

Figure 4–7   Define Simple Line Window (modified)

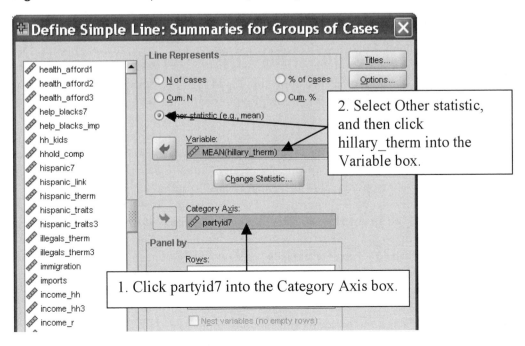

Figure 4–8 Line Chart Output: Mean Values of Hillary Clinton Thermometer, by Party Identification

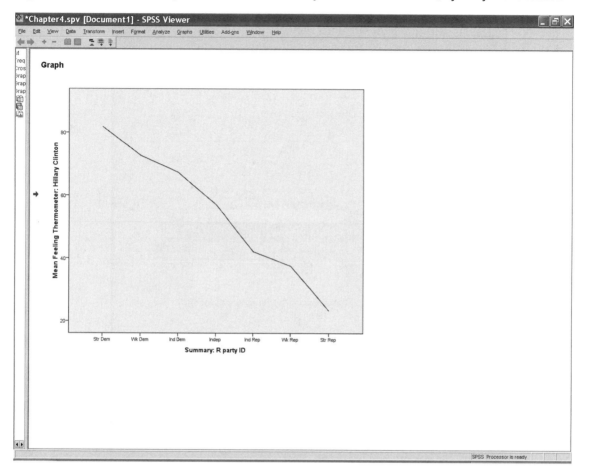

Click Graphs → Legacy Dialogs → Bar. The Bar Charts window gives you the same set of choices as the Line Charts window. Ensure that the same choices are selected: Simple *and* Summaries for groups of cases. Click Define. The Define Simple Bar window opens, and it too is identical to the Define Simple Line window in every detail. As a substantive matter, we want to depict the percentage of respondents in each category of yob3 who said they were "very much" interested in the 2004 campaigns. Because yob3 is the independent variable, it goes in the Category Axis box. (Scroll to yob3 and click it over.) So far this is the same as before. At this point, however, the peculiarities of Bar Chart require that we refamiliarize ourselves with specific coding information about the dependent variable, attent.

Find attent in the left-hand variable list, place the cursor pointer on it, and then *right*-click. Click on Variable Information and review the numeric codes. Respondents saying that they were "very much" interested are coded 1, those "somewhat" interested are coded 3, and the "not much" interested are coded 5 (Figure 4-9). Commit this fact to short-term memory: Respondents who are "very much" interested are coded 1 on the dependent variable, attent.

Now return to the Bars Represent panel. Select the Other statistic radio button, and click attent into the Variable box. The designation "MEAN(attent)" appears in the Variable box, as shown in Figure 4-10. Just as it did in Line Chart, SPSS assumes that we are after the mean of attent. This default is fine for mean comparisons, but in this case it won't do. Click the Change Statistic button.[5]

The Statistic window presents itself (Figure 4-10). The radio button for the default, Mean of values, is currently selected. However, we are interested in obtaining the percentage of cases in code 1 ("very much" interested) on attent. How do we get SPSS to cooperate with this request? Click the radio button at the bottom on the left, the one labeled "Percentage inside," as shown in Figure 4-11. The two boxes, one labeled "Low" and the other labeled "High," go active. Our request is specific and restrictive: We want the percentage of respondents in code 1 only. Expressed in terms that SPSS can understand, we want the percentage of cases "inside" a coded value of 1 on the low side and a coded value of 1 on the high side. Click the cursor in the Low box and type a "1." Click the cursor in the High box and type a "1."[6] The Statistic window should

Figure 4–9 Reviewing Numeric Codes

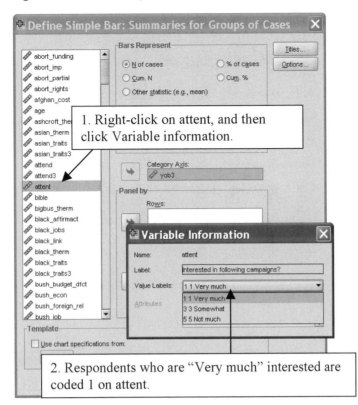

1. Right-click on attent, and then click Variable information.

2. Respondents who are "Very much" interested are coded 1 on attent.

Figure 4–10 Define Simple Bar Window and Statistic Window (default)

1. Select Other statistic, and then click attent into the Variable box.

2. Click Change Statistic.

By default, SPSS will display mean values of attent.

Figure 4–11 Statistic Window (modified)

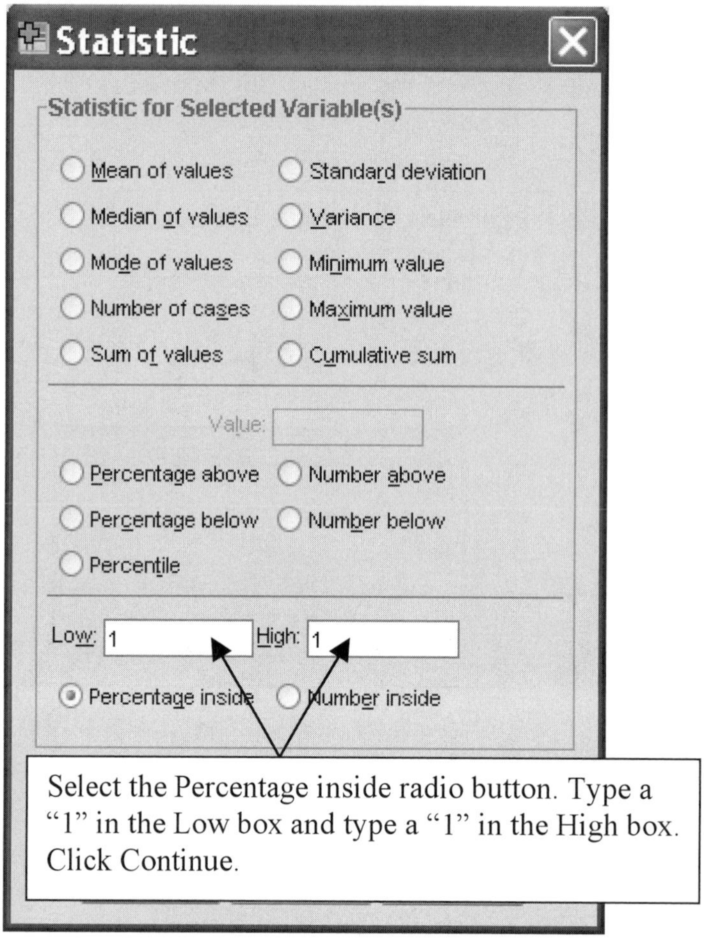

now look like Figure 4-11. Click Continue, returning to the Define Simple Bar window. The Define Simple Bar window should now look like Figure 4-12. Click OK.

All of your point-and-click drudgery has paid off. SPSS displays a bar chart of the relationship between generation and campaign interest (Figure 4-13). The category axis is nicely labeled, and the heights of the bars clearly depict this pattern: As the generations change from older to younger, the percentage of people "very much" interested in the campaign declines. At least *we* know what the bars represent, because we did the analysis. An interested observer (such as your instructor), however, might do a double-take at the title on the vertical axis, "%in(1,1) Interested in following campaigns?" SPSS is relentlessly literal. We asked it to graph the percentages of people between code 1 and code 1 on attent, so that is how SPSS has titled the axis. This chart is not ready for prime time. We need to give the vertical axis a more descriptive title, and perhaps make other appearance-enhancing changes.

USING THE CHART EDITOR

SPSS permits the user to modify the content and appearance of any tabular or graphic object it produces in the Viewer. The user invokes the SPSS Editor, makes any desired changes, and then returns to the Viewer. The changes made in the Chart Editor are recorded automatically in the Viewer. In this section we describe how to retitle the vertical axis of the bar chart you just created. We'll also change the color of the bars. (The SPSS default color is rather uninspired, and it doesn't print well.)

In the Viewer, place the cursor anywhere on the bar chart and double-click. SPSS opens the Chart Editor (Figure 4-14). As with any editing software, the Chart Editor recognizes separate elements within an object. It recognizes some elements as text. These elements include the axis titles and the value labels for the categories of yob3. It recognizes other elements as graphic, such as the bars in the bar chart. First we will edit a text element, the title on the vertical axis. Then we will modify a graphic element, the color of the bars.

Figure 4–12 Define Simple Bar Window (modified)

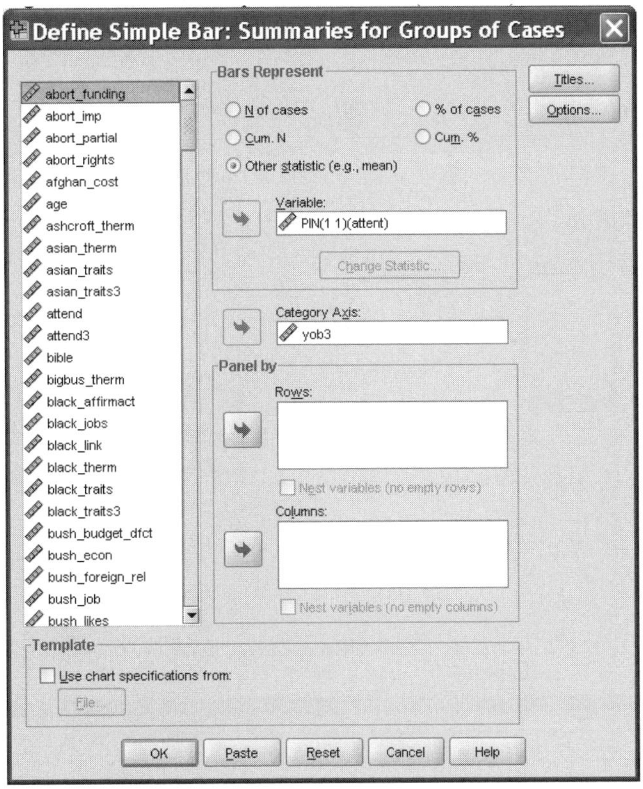

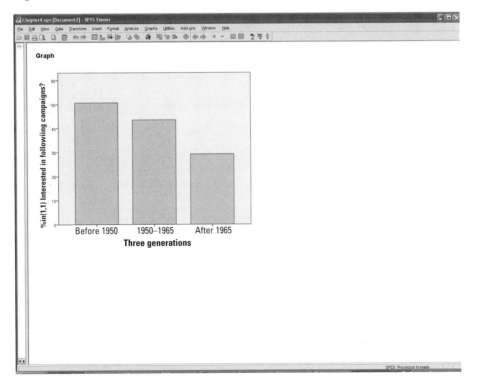

Figure 4–13 Bar Chart Output

Place the cursor anywhere on the title "%in(1,1) Interested in following campaigns?" and single-click. SPSS selects the axis title. With the cursor still placed on the title, single-click again. SPSS moves the text into editing mode inside the chart (Figure 4-15). Delete the current text. In its place type the title "Percent 'Very much' interested in campaign." Now click on one of the bars. (As soon as you click off the axis title, it

Figure 4–14 Chart Editor

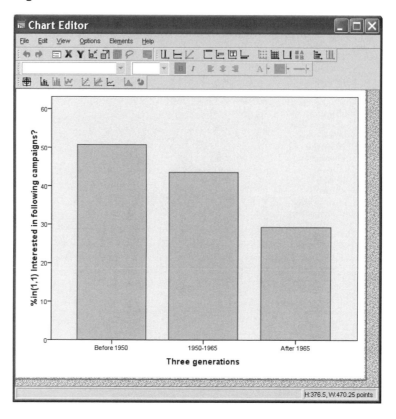

Figure 4–15 Chart Axis Title Ready for Editing

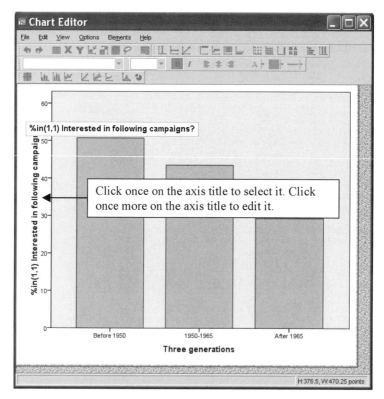

Click once on the axis title to select it. Click once more on the axis title to edit it.

returns to its rightful position in the chart.) SPSS selects all of the bars (see Figure 4-16). Click on the Properties icon located near the upper-left corner of the Chart Editor window. This opens the Properties window, the most powerful editing tool in the Chart Editor's arsenal. (Special note: If you plan to do a lot of editing, it is a good idea to open the Properties window soon after you enter the Chart Editor. Each time

Figure 4–16 Using the Properties Window to Change the Bar Code

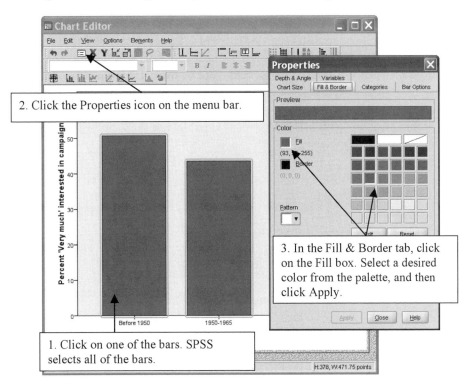

Figure 4–17 Edited Bar Chart in the SPSS Viewer

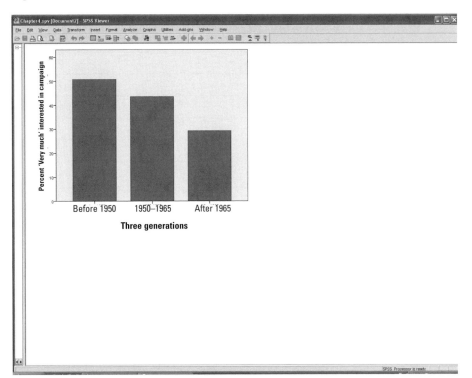

you select a different text or graphic element with the mouse, the Properties window changes, displaying the editable properties of the selected element.) Click on the Fill & Border tab, as shown in Figure 4-16. In the Color panel, click in the Fill box. In the color palette, click on a desirable hue, and then click Apply. SPSS makes the change. Okay. Close the Properties window and exit the Chart Editor. The finished product appears in the Viewer (Figure 4-17).

EXERCISES

1. (Dataset: NES2004. Variables: medinsur7, partyid7.) Here is a widely remarked difference between Democrats and Republicans: Democrats favor government-funded medical insurance and Republicans prefer private insurance plans. Is this difference borne out by the data? Dataset NES2004 contains the variable medinsur7, a 7-point scale that measures respondents' opinions on this issue. Respondents indicate their opinions by choosing any position on this scale, from 1 (government plan) at one end to 7 (private plan) at the other end. This is the dependent variable. Use the 7-point party identification scale (partyid7) as the independent variable.

A. If Democrats are more likely than Republicans to favor government-funded medical insurance, then Democrats will have (check one)

❏ a higher mean than do Republicans on medinsur7.

❏ about the same mean as do Republicans on medinsur7.

❏ a lower mean than do Republicans on medinsur7.

B. Using Analyze → Compare Means → Means, obtain a mean comparison that shows the mean score on the dependent variable, medinsur7, and the number of cases for each category of the independent variable, partyid7. Write the results in the table that follows.

Summary of Govt/private medical insurance scale: self-placement

Party ID	Mean	N
Strong Democrat	?	?
Weak Democrat	?	?
Independent-Dem	?	?
Independent	?	?
Independent-Rep	?	?
Weak Republican	?	?
Strong Republican	?	?
Total	?	?

C. Does your analysis support the idea that Democrats are more likely than Republicans to support government-funded medical insurance? (circle one)

Yes   No

Briefly explain your answer. _____

_____

_____

D. Obtain a line chart of the relationship. Remember to put the independent variable, partyid7, on the Category Axis and the dependent variable, medinsur7, in the Variable box of the Line Represents panel. Print the line chart.

2. (Dataset: NES2004. Variables: gay_marriage, attend3, gender, libcon3_r.) Should gay couples be legally permitted to marry? This controversial issue gained center stage during the 2004 elections, when amendments regarding the legal definition of marriage appeared on the ballot in 11 states. One can imagine several characteristics that divide people on this issue. People who are religiously observant might be more likely to oppose gay marriage than are the less observant. Men may be more opposed than women are. Or conservatives might be less open to the idea than are liberals.

Dataset nes2004.dta contains gay_marriage, which is coded 0 (the respondent thinks gay marriage should not be allowed) or 1 (the respondent thinks gay marriage should be allowed). This is the dependent variable that you will use to test each of the following hypotheses:

Hypothesis 1. In a comparison of individuals, people who have high levels of attendance at religious services will be more likely to oppose gay marriage than will people who have lower levels of religious attendance. (The independent variable is attend3.)

Hypothesis 2. In a comparison of individuals, men are more likely than women to oppose gay marriage. (The independent variable is gender.)

Hypothesis 3. In a comparison of individuals, conservatives are more likely than liberals to oppose gay marriage. (The independent variable is libcon3_r.)

A. When using SPSS to obtain a series of cross-tabulations having the same dependent variable but different independent variables, only one Crosstabs run is required. In the Crosstabs window, click gay_marriage into the Row(s) panel. Click attend3, gender, and libcon3_r into the Column(s) panel. (Make sure to request column percentages in the Crosstabs Cell Display.) Run the analysis. In the tables that follow, record the percentages who say "Not allow" when asked about gay marriage:

	Relig attend: 3 cats		
	High	Middle	Low
Percentage "Not allow" gay marriage	83,8% 332?	59% 192?	46,3% 176?
allow	16%	40,4%	53,7%

	Gender	
	1. Male	2. Female
Percentage "Not allow" gay marriage	65%?	62,7%
allow	35%	37,3%

	Ideology: 3 cats		
	Liberal	Moderate	Conservative
Percentage "Not allow" gay marriage	30,6%	56,4%	85? 3
	69,4	43,6	14,7

B. These findings (circle one)

support Hypothesis 1.   do not support Hypothesis 1.

Briefly explain your reasoning. _____

_____

_____

_____

_____

C. These findings (circle one)

                        support Hypothesis 2.   do not support Hypothesis 2.

Briefly explain your reasoning. _____

_____

_____

_____

_____

D. These findings (circle one)

                        support Hypothesis 3.   do not support Hypothesis 3.

Briefly explain your reasoning. _____

_____

_____

_____

_____

E. Obtain a bar chart of the relationship between gay_marriage and libcon3_r. You will want the vertical axis to depict the percentage of respondents in the "Not allow" category of gay_marriage. Remember that those who oppose gay marriage are coded 0 on the dependent variable. Using the Chart Editor, give the vertical axis a more descriptive title. Change the default bar color to a color of your choosing. Print the bar chart you created.

3. (Dataset: NES2004. Variables: who04_2, us_econ_past3.) What factors determine how people vote in presidential elections? Political scientists have investigated and debated this question for many years. A particularly powerful and elegant perspective emphasizes voters' *retrospective* evaluations. According to this view, for example, voters who think that the economy has improved during the year preceding the election are likely to reward the candidate of the incumbent party. Voters who believe that economic conditions have worsened, by contrast, are likely to punish the incumbent party by voting for the candidate of the party not currently in power. As political scientist V. O. Key once famously put it, the electorate plays the role of "rational god of vengeance and reward."[7] Does Key's idea help explain how people voted in the 2004 election?

A. Test this hypothesis: In a comparison of individuals, those who thought the U.S. economy improved during the year preceding the 2004 election were more likely to vote for George W. Bush than were individuals who thought the economy worsened. Use these two variables from NES2004: who04_2 (dependent variable) and us_econ_past3 (independent variable).[8] Obtain a cross-tabulation of the relationship. Record your findings in the table that follows.

Kerry or Bush?*US econ better/worse in last yr? Crosstabulation

US econ better/worse in last yr?

Kerry or Bush?		Better	Same	Worse	Total
1 Kerry	N:	26?	100?	26?	39?3
	Percentage:	12.3%	40.8%	77.6%	49.1%
2 Bush	N:	185?	145?	77	40?1
	Percentage:	87.7%	59.2%	22.4%	50.9%
Total	N:	211	245	344	800
	Percentage:	100.0	100.0	100.0	100.0

B. What do you think? Are the data consistent with the hypothesis?

Write a paragraph explaining your reasoning. _____

_____
_____
_____
_____

4. (Dataset: NES2004. Variables: partyid3, timing2.) Political independents define an increasingly important battleground for presidential campaigns. Here is one thing pollsters and candidates know—or think they know—about people who claim no party affiliation: Independents make up their minds much later in the campaign than do party loyalists.

A. Imagine two variables. One variable measures partisanship in three categories: Democrat, Independent, and Republican. The other variable measures the timing of the vote decision in two categories: early in the campaign or late in the campaign. Mark a check next to the phrase that completes the following hypothesis:

In a comparison of individuals, people who are Independents are (check one)

❏ more likely than Democrats or Republicans to make their vote decision late in the campaign.

❏ less likely than Democrats or Republicans to make their vote decision late in the campaign.

❏ equally as likely as Democrats or Republicans to make their vote decision late in the campaign.

B. Now imagine a bar chart. The three values of the independent variable define the category axis. Democrats are represented by the bar on the left, Independents by the middle bar, and Republicans by the bar on the right. The percentage of each category deciding late in the campaign defines the vertical axis. If the hypothesis were correct, which of the following bar charts—X, Y, or Z—would most closely approximate the relationship? (circle one)

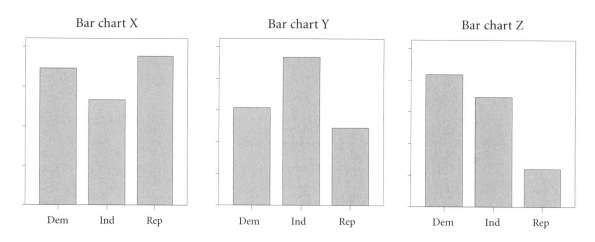

C. Dataset NES2004 contains partyid3, which measures respondents' partisanship by three categories: Democrat, Independent, and Republican. Another variable, timing2, measures how soon before the election respondents made up their minds. Obtain a bar chart of the timing2-partyid3 relationship. Put the independent variable, partyid3, on the category axis. You want the bars to depict the percentages of each partisan group deciding late in the campaign.

Based on this bar chart, would you say that the hypothesis is correct or incorrect? (check one)

❏ The hypothesis appears to be correct.

❏ The hypothesis appears to be incorrect.

D. Using the Chart Editor, give the vertical axis a more descriptive title. Change the bar color. Take some time to explore the Chart Editor. Try to enhance the readability of the graphic. For example, SPSS uses a microscopic 8-point font to label the ticks along the horizontal axis. The labels "Democrat," "Independent," and "Republican" would look better in 10-point font. Figure out how to accomplish this. (Hint: In the Chart Editor, click on one of the labels, and then open the Properties window. Click the Text Style tab.) Print the bar chart you created.

5. (Dataset: GSS2006. Variables: polviews, fem_role). Why do some people hold more traditional views about the role of women in society and politics, whereas others take a less traditional stance? General ideological orientations, liberalism versus conservatism, may play an important role in shaping individuals' opinions on this cultural question. Thus it seems plausible to suggest that ideology (independent variable) will affect opinions about appropriate female roles (dependent variable). The hypothesis: In a comparison of individuals, liberals will be more likely than conservatives to approve of nontraditional female roles.

GSS2006 contains fem_role, a scale that measures opinions about the appropriate role of women. You analyzed this variable in Chapter 2. Recall that fem_role ranges from 0 (women "domestic") to 12 (women in "work, politics"). That is, higher scores denote less traditional beliefs. This is the dependent variable. GSS2006 also has polviews, a 7-point ordinal scale measuring ideology. Scores on polviews can range from 1 ("Extremely liberal") to 7 ("Extremely conservative"). This is the independent variable.

A. According to the hypothesis, as the values of polviews increase, from 1 through 7, mean values of fem_role should (circle one)

decrease.   neither decrease nor increase.   increase.

B. Test the hypothesis using Compare Means → Means. Click fem_role into the Dependent List panel. Click polviews into the Independent List panel. In Options, remove Standard Deviation by selecting it with the mouse and clicking it back into the left-hand Statistics list. Run the analysis and write the results in the table that follows.

Female role: Children, home, politics

Ideological self ID	Mean	N
Extremely liberal	?	?
Liberal	?	?
Slightly liberal	?	?
Moderate	?	?
Slightly conservative	?	?
Conservative	?	?
Extremely conservative	?	?
Total	?	?

C. Do the results support the hypothesis?

Write a few sentences explaining your reasoning. ───────────────────────

──

──

──

──

D. Obtain a line chart of this relationship. Print the line chart you created.

6. (Dataset: GSS2006. Variables: tvhours, news.) Here are two common media fixtures in our lives: newspapers and television. Are people who consume the printed word also avid consumers of broadcast media? Or is it an either-or proposition, with newspaper readers being less likely than newspaper nonreaders to watch television? There isn't much theory to rely on here, so let's test an exploratory hypothesis: In a comparison of individuals, people who read newspapers more often will spend more hours watching television than will those who read newspapers less often.

A. In this hypothesis, the dependent variable is (fill in the blank) _____, and the

independent variable is (fill in the blank) _____.

B. According to this hypothesis, if one compares people who frequently read a newspaper with people who infrequently read a newspaper, (check one)

❏ the mean number of hours spent watching television will be higher among newspaper readers than among the nonreaders.

❏ the mean number of hours spent watching television will be lower among newspaper readers than among the nonreaders.

C. Using GSS2006, test this hypothesis with Compare Means ➜ Means. GSS2006 contains these variables: tvhours and news. The variable tvhours measures the number of hours the respondent watches television each day. And the news variable measures how frequently the respondent reads a newspaper during the week. Lower values on news denote less frequent newspaper reading, and higher values on news denote more frequent newspaper reading. Run the appropriate mean comparison analysis and record your results in the table that follows. Write the value labels of the independent variable in the left-most column. Record the mean values of the dependent variable in the column labeled "Mean" and the number of cases in the column labeled "N."

Values of independent variable	Mean	N
?	?	?
?	?	?
?	?	?
?	?	?
?	?	?
Total	?	?

D. Obtain a line chart of this relationship using Graphs → Line. Print the line chart you created.

E. Examine the mean comparison table and the line chart. Suppose you wanted to collapse the values of the independent variable into only two categories, "Low" and "High." Your goal is to combine values of the independent variable in such a way that respondents scoring "Low" on the collapsed variable share similar values on the dependent variable, and respondents scoring "High" on the independent variable share similar values on the dependent variable. Based on the analysis you just performed, which values of the independent variable would fall into the "High" category? (circle all that apply)

Never   Less than once/week   Once/week   Few times/week   Every day

7. (Dataset: GSS2006. Variables: intrace_2, affrmact2, natrace.) Untruthful answers by survey respondents can create big headaches for public opinion researchers. Why might a respondent not tell the truth to an interviewer? Certain types of questions, combined with particular characteristics of the interviewer, can trigger a phenomenon called preference falsification: "the act of misrepresenting one's genuine wants under perceived social pressures."[9] For example, consider the difficulty in gauging opinions on affirmative action, hiring policies aimed at giving preference to black applicants. One might reasonably expect people questioned by an African American interviewer to express greater support for such programs than those questioned by a white pollster. An affirmative action opponent, not wanting to appear racially insensitive to a black questioner, might instead offer a false pro–affirmative action opinion.[10]

GSS2006 contains intrace_2, coded 0 for respondents questioned by a white interviewer and coded 1 for those questioned by a black interviewer. This is the independent variable that will allow you to test two preference falsification hypotheses:

Hypothesis 1. In a comparison of individuals, those questioned by a black interviewer will be more likely to express support for affirmative action than will those questioned by a white interviewer. (The dependent variable is affrmact2, coded 1 for "support" and 2 for "oppose.")

Hypothesis 2. In a comparison of individuals, those questioned by a black interviewer will be more likely to say that we are spending too little to improve the condition of blacks than will those questioned by a white interviewer. (The dependent variable is natrace, which is coded 1 for respondents saying "too little," 2 for those saying "about the right amount," and 3 for "too much.")

A. When you want to obtain two or more cross-tabulations having the same independent variable but different dependent variables, only one Crosstabs run is required. Click the independent variable, intrace_2, into the Column(s) panel. Click affrmact2 and natrace into the Row(s) panel. Request column percentages in the Crosstabs Cell Display. In the table that follows, record the percentages that support affirmative action and the percentages that say we are spending too little to improve the condition of blacks.

	Interviewer's race	
	White	Black
Percent "support" affirmative action	?	?
Percent spending "too little" to improve condition of blacks	?	?

B. These findings (circle one)

        support Hypothesis 1.   do not support Hypothesis 1.

Briefly explain your reasoning. _____

_____

_____

_____

_____

C. These findings (circle one)

        support Hypothesis 2.   do not support Hypothesis 2.

Briefly explain your reasoning. _____

_____

_____

_____

_____

D. Produce a bar chart depicting the percentage of respondents supporting affirmative action for each value of intrace_2. Remember: Those supporting affirmative action are coded 1 on affrmact2. Give the vertical axis a more descriptive title and edit the bar color. Make the chart more readable. Print the chart.

8. (Dataset: GSS2006. Variables: science_gw3, partyid_3, polview3.) The following exchange between George W. Bush and Al Gore during the 2000 campaign illustrates a key feature of the global warming debate: Republicans and conservatives do not perceive scientific agreement on the causes of global warming, while Democrats and liberals do. Bush said, "I don't think we know the solution to global warming yet and I don't think we've got all the facts before we make decisions." Gore disagreed "that we don't know the cause of global warming. I think that we do. It's pollution, carbon dioxide and other chemicals

that are even more potent." Bush persisted: "Some of the scientists, I believe, haven't they been changing their opinion a little bit on global warming? There's a lot of differing opinions. . . ."[11]

At the elite level, at least, the global warming battle line is defined by how much one believes that scientists themselves agree on the issue. Is this line of disagreement evident, as well, among the mass public? Are Democrats more likely to perceive scientific consensus than are Republicans? Are liberals more likely than conservatives to think that scientists agree about the causes of global warming? GSS2006 contains science_gw3, which records respondents' assessments of the extent to which "environmental scientists agree among themselves about the existence and causes of global warming." Responses are coded 1 ("Scientists agree"), 2 ("Middle"), or 3 ("Scientists disagree"). This is the dependent variable. The two independent variables are partyid_3 (coded 1 for Democrats, 2 for Independents, and 3 for Republicans) and polview3 (coded 1 for self-described liberals, 2 for moderates, and 3 for conservatives). (You created polview3 in one of Chapter 3's exercises.)

A. Run the appropriate Crosstabs analyses and examine the output. Which of the following inferences are supported by your analysis? (check all that apply)

❑ Overall, most respondents think that scientists disagree among themselves about the existence and causes of global warming.

❑ Democrats are more likely than Republicans to think that scientists agree among themselves.

❑ Republicans are more likely than Independents to think that scientists agree among themselves.

❑ Liberals are more likely than conservatives to think that scientists agree among themselves.

❑ There is about a 10-point difference between the percentage of liberals who think that scientists agree and the percentage of moderates who think that scientists agree.

❑ There is about a 10-point difference between the percentage of moderates who think that scientists agree and the percentage of conservatives who think that scientists agree.

B. Create a bar chart or a line chart depicting the relationship between science_gw3 and partyid_3. The vertical axis should display the percentage of respondents who think that scientists agree. Give the vertical axis a more descriptive title. Make the chart more readable, and make any other desired appearance-enhancing changes. Print the chart.

9. (Dataset: States. Variables: union, region.) Where are unions stronger? Where are they weaker? Consider the following two claims:

Claim 1: States in the northeastern United States are more likely to have unionized workforces than are states in the South.

Claim 2: States in the Midwest and West have levels of unionization more similar to those in the South than to those in the Northeast.

A. The States dataset contains union, the percentage of each state's workforce who are union members. Another variable, region, is a four-category census classification of the states. Run a mean comparison analysis. Record your results in the table that follows.

Percentage of workers who are union members

Census region	Mean	N
Northeast	?	?
Midwest	?	?
South	?	?
West	?	?
Total	?	50

B. Based on your analysis, would you say that Claim 1 is correct or incorrect? How do you know? Check the appropriate box and complete the sentence that follows.

❏ Claim 1 is correct, because _____

_____.

❏ Claim 1 is incorrect, because _____

_____.

C. Based on your analysis, would you say that Claim 2 is correct or incorrect? How do you know? Check the appropriate box and complete the sentence that follows.

❏ Claim 2 is correct, because _____

_____.

❏ Claim 2 is incorrect, because _____

_____.

D. Produce a bar chart of the relationship you just analyzed. Edit the chart for readability and appearance. Print the chart.

10. (Dataset: States. Variables: cigarettes, cig_tax_3.) Two policy researchers are arguing about whether higher taxes on cigarettes reduce cigarette consumption.

Policy researcher 1: "The demand for cigarettes is highly inelastic—smokers need to consume cigarettes, and they will buy them without regard to the cost. Raising taxes on a pack of cigarettes will have no effect on the level of cigarette consumption."

Policy researcher 2: "Look, any behavior that's taxed is discouraged. If state governments want to discourage smoking, then raising cigarette taxes will certainly have the desired effect. Higher taxes mean lower consumption."

Imagine a line chart of the relationship between cigarette taxes and cigarette consumption. The category axis measures state cigarette taxes in three categories, from lower taxes on the left to higher taxes on the right. The vertical axis records per-capita cigarette consumption. Now consider line charts X, Y, and Z, which follow.

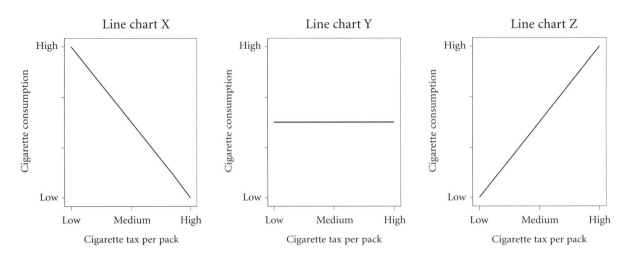

A. If policy researcher 1 were correct, which line chart would most accurately depict the relationship between states' cigarette taxes and cigarette consumption? (circle one)

Line chart X   Line chart Y   Line chart Z

B. If policy researcher 2 were correct, which line chart would most accurately depict the relationship between states' cigarette taxes and cigarette consumption? (circle one)

Line chart X   Line chart Y   Line chart Z

C. The States dataset contains the variables cigarettes and cig_tax_3. Run a mean comparison analysis, using cigarettes as the dependent variable and cig_tax_3 as the independent variable. Record your results in the table that follows.

cigarettes Packs bimonthly per adult pop

cig_tax_3 Cigarette tax per pack: 3 categories	Mean	N
1 $.07–$.64	?	?
2 $.695–$1.36	?	?
3 $1.41–$2.58	?	?
Total	?	50

D. Create and print a line chart of the relationship.

E. Examine the mean comparison table and the line chart. Which policy researcher is more correct? (check one)

❏ Policy researcher 1 is more correct.

❏ Policy researcher 2 is more correct.

11. (Dataset: World. Variables: enpp3_democ, district_size3, frac_eth3.) Two scholars of comparative politics are discussing possible reasons why some democracies have many political parties and other democracies have only a few.

Scholar 1: "It all has to do with the rules of the election game. Some countries, such as the United Kingdom, have single-member electoral districts. Voters in each district elect only one representative. This militates in favor of fewer and larger parties, since small parties have less chance of winning enough votes to gain the seat. Other countries, like Switzerland, have multi-member districts. Because voters choose more than one representative per district, a larger number of smaller parties have a chance to win representation. It doesn't surprise me in the least, then, that the U.K. has fewer political parties than Switzerland."

Scholar 2: "I notice that your explanation fails to mention the single most important determinant of the number of political parties: social structural heterogeneity. Homogeneous societies, those with few linguistic or religious differences, have fewer conflicts and thus fewer parties. Heterogeneous polities, by the same logic, are more contentious and will produce more parties. By the way, the examples you picked to support your case also support mine: the U.K. is relatively homogeneous and Switzerland relatively heterogeneous. It doesn't surprise me in the least, then, that the U.K. has fewer political parties than Switzerland."

A. Scholar 1's hypothesis: In a comparison of democracies, those having single-member districts will have (circle one)

fewer political parties   more political parties

than democracies electing multiple members from each district.

B. State Scholar 2's hypothesis: _____

_____

_____

_____

_____

C. The World dataset variable enpp3_democ measures, for each democracy, the number of effective parliamentary parties: 1–3 parties (coded 1), 4–5 parties (coded 2), or 6–11 parties (coded 3). Use enpp3_democ as the dependent variable to test each hypothesis. For independent variables, test Scholar 1's hypothesis using district_size3, which measures the number of seats per district: countries with single-member districts are coded 1, countries that average more than one but fewer than six members are coded 2, and countries with six or more members per district are coded 3. Test Scholar 2's hypothesis using frac_eth3, which classifies each country's level of ethnic/linguistic fractionalization as low (coded 1), medium (coded 2), or high (coded 3). Countries with higher codes on frac_eth3 have a higher level of ethnic conflict. Run Crosstabs to test the hypotheses. In the table that follows, record the percentages of cases falling into the lowest code of the dependent variable, 1–3 parties.

	Average # of members per district		
	Single-member	> 1 to 5 members	6 or more members
Percentage having 1–3 parties	?	?	?
	Level of ethnic fractionalization		
	Low	Medium	High
Percentage having 1–3 parties	?	?	?

D. Which of the following statements best summarizes your findings? (check one)

❏ Scholar 1's hypothesis is supported by the analysis, but Scholar 2's hypothesis is not supported by the analysis.

❏ Scholar 2's hypothesis is supported by the analysis, but Scholar 1's hypothesis is not supported by the analysis.

❏ Both hypotheses are supported by the analysis.

❏ Neither hypothesis is supported by the analysis.

E. Making specific reference to your findings, write a paragraph explaining your choice in part D.

_____

_____

_____

_____

_____

_____

_____

_____

12. (Dataset: World. Variables: regime_type3, durable.) The two comparative politics scholars are still arguing, only now they're trying to figure out what sort of institutional arrangement produces the longest-lasting, most stable political system. Also, a third scholar joins the interchange of ideas.

Scholar 1: "Presidential democracies, like the United States, are going to be more stable than are any other type of system. In presidential democracies, the executive and the legislature have separate electoral constituencies and separate but overlapping domains of responsibility. The people's political interests are represented both by the president's national constituency and by legislators' or parliament members' more localized constituencies. If one branch does something that's unpopular, it can be blocked by the other branch. The result: political stability."

Scholar 2: "Parliamentary democracies are by far more stable than presidential democracies. In presidential systems, the executive and legislature can be controlled by different political parties, a situation that produces deadlock. Since the leaders of the legislature can't remove the president and install a more compliant or agreeable executive, they are liable to resort to a coup, toppling the whole system. Parliamentary democracies avoid these pitfalls. In parliamentary democracies, all legitimacy and accountability resides with the legislature. The parliament organizes the government and chooses the executive (the prime minister) from among its own leaders. The prime minister and members of parliament have strong incentives to cooperate and keep things running smoothly and efficiently. The result: political stability."

Scholar 3: "You two have made such compelling—if incorrect—arguments that I almost hesitate to point this out: Democracies of any species, presidential or parliamentary, are inherently unstable. Any system that permits the clamor of competing parties or dissident viewpoints is surely bound to fail. If it's stability that you value above all else, then dictatorships will deliver. Strong executives, feckless or nonexistent legislatures, powerful armies, social control. The result: political stability."

The World dataset contains the variable durable, which measures the number of years since the last regime transition. The more years that have passed since the system last failed (higher values on durable), the more stable a country's political system. The variable regime_type3 captures system type: dictatorship, parliamentary democracy, or presidential democracy.

A. Run Compare Means and examine the output. Which of the following inferences are supported by your analysis? (check all that apply)

❏ Parliamentary democracy is the most stable form of government.

❏ On average, parliamentary democracies last about 20 years longer than dictatorships.

❏ Presidential democracies are more stable than dictatorships.

❏ Of the 141 countries included in the analysis, dictatorships are the most prevalent.

❏ Among the 141 countries included in the analysis, the typical regime lasts about 24 years.

B. Create and print a bar chart depicting the relationship between regime durability and regime type.

13. (Dataset: World. Variables: fhrate04_rev, dem_other5.) Why do some countries develop democratic systems whereas others do not? Certainly the transition toward democracy—or away from it—is a complex process. Some scholars emphasize factors internal to countries, such as educational attainment or economic development. Others look to external factors, such as patterns of governance that are prevalent in a country's region of the world. Perhaps governmental systems are like infectious diseases: Similar systems diffuse among countries in close geographic proximity. According to the democratic diffusion hypothesis, countries in regions having fewer democracies are themselves less likely to be democratic than are countries in regions having more democracies.[12]

A. Suppose you had two variables for a large number of countries: a dependent variable that measured democracy along an interval-level scale, with higher scores denoting higher levels of democracy, and an independent variable measuring the number of democracies in each country's region, from fewer democracies to more democracies. According to the democratic diffusion hypothesis, if you were to compare mean values of the dependent variable for countries having different values on the independent variable, you should find (check one)

❏ a lower mean of the dependent variable for countries in regions having fewer democracies than for countries in regions having more democracies.

❏ a higher mean of the dependent variable for countries in regions having fewer democracies than for countries in regions having more democracies.

❏ no difference between the means of the dependent variable for countries in regions having fewer democracies and countries in regions having more democracies.

B. The World dataset contains fhrate04_rev, an interval-level measure of democracy based on the Freedom House rating system. Scores range from 1 (least democratic) to 7 (most democratic).[13] This is the dependent variable. World also has dem_other5, a 5-category ordinal measure that divides countries into five groups, based on the percentage of democracies in each country's geographic region: 10 percent, approximately 40 percent, approximately 60 percent, approximately 90 percent, or 100 percent. This is the independent variable. Run the appropriate mean comparison analysis. Label and record the results in the table that follows.

fhrate04_rev * Freedom House rating of democracy (reversed)

dem_other5 * Percentage of other democracies in region: 5 cats	Mean	N
1 10%	?	?
2 Approx 40%	?	?
3 Approx 60%	?	?
4 Approx 90%	?	?
5 100%	?	?
Total	?	191

C. Create and print a line chart of the relationship you just analyzed.

D. Examine the mean comparison table and the line chart. Which of the following statements are supported by your analysis? (check all that apply)

❏ Countries in regions having fewer democracies are more likely to be democratic than are countries in regions having more democracies.

❏ The relationship between the independent and dependent variables is positive.

❏ The democratic diffusion hypothesis is incorrect.

❏ Countries in regions having fewer democracies are less likely to be democratic than are countries in regions having more democracies.

❏ The relationship between the independent and dependent variables is negative.

14. (Dataset: World. Variables: decentralization4, effectiveness, confidence.) Are decentralized governments more effective than centralized governments? In decentralized systems, local officials have politically autonomous authority to raise public money and administer government programs. This leads to more effective governance and, the argument goes, inspires confidence among citizens. Centralized systems, remote from local problems and burdened by red tape, may be both less effective and less likely to be viewed positively by citizens. Given these putative benefits, it is little wonder that "policy makers and politicians have frequently pushed for decentralisation as a panacea for the ills of poor governance."[14]

The World dataset variable decentralization4 measures, for each country, the level of decentralization by four ordinal categories, from less decentralization (coded 1) to more decentralization (coded 4). The assessments of a panel of expert observers were used to create the variable effectiveness, which measures government effectiveness on a 100-point scale (higher scores denote greater effectiveness). Confidence, also on a 100-point scale, gauges the degree to which a country's citizens have "a great deal" or "quite a lot" of confidence in state institutions (higher scores denote higher confidence).

A. The World variables will permit you to test two hypotheses about the effects of decentralization on effectiveness and confidence. State the two hypotheses.

Effectiveness hypothesis: _____

_____

_____

Confidence hypothesis: _____

_____

_____

B. Test the hypotheses with Compare Means. Write the results in the table that follows.

Level of decentralization		Government effectiveness scale	Confidence in institutions scale
Low	Mean	?	?
	N	?	?
2	Mean	?	?
	N	?	?
3	Mean	?	?
	N	?	?
High	Mean	?	?
	N	?	?
Total	Mean	?	?
	N	?	?

C. The effectiveness hypothesis (circle one)

is supported   is not supported

by the analysis.

Explain your reasoning. _____

_____

_____

D. The confidence hypothesis (circle one)

is supported   is not supported

by the analysis.

Explain your reasoning. _____

_____

_____

That concludes the exercises for this chapter. Before exiting SPSS, be sure to save your output file.

NOTES

1. Chart Builder sits atop of the Graphs drop-down hierarchy, but it is not covered in this book. Introduced in SPSS 14, Chart Builder was meant to enhance ease of use by integrating all chart types into a single drag-and-drop interface. Yet Chart Builder does not permit the flexibility of the Legacy Dialogs. The main problem with Chart Builder is that its performance is tied too closely to a variable's level of measurement, as defined by the Measure attribute in the Data Editor. For example, suppose we want to build a chart using NES2004's partyid7 as the independent variable, with the labels "Str Dem," "Wk Dem," and so on, defining the values of the category axis. Unless the Measure attribute is set to Nominal or Ordinal, Chart Builder will not use partyid7's value labels to define the ticks along the axis. Instead, it will use partyid7's numeric codes: "0," "1," and so on. Setting the Measure attribute to Ordinal or Nominal will produce the hoped-for result: value labels and not numeric codes. But suppose that we want to build another chart using partyid7, this time as the dependent variable, with mean values of partyid7 displayed along the vertical axis. Unless the Measure attribute is set to Scale, Chart Builder will not permit the display of mean values or any other scale-dependent values. So to get value labels on the horizontal axis, you need Nominal or Ordinal, but to get desired statistics on the vertical axis you need Scale. Of course, you could accommodate this discouraging quirk by going into the Data Editor and changing a variable's Measure attribute to fit the project at hand. Or you could use the Legacy Dialogs, which remain happily agnostic about levels of measurement.

2. Because we are graphing one relationship, we want a single line. And because we are comparing groups of partisans, we want SPSS to display a summary measure, the mean, for each group.

3. Unless we modify the Line Represents panel to suit our analysis, SPSS will produce a line chart for the number of cases (N of cases) in each category of partyid7.

4. Of course, you will encounter situations in which you do not want mean values. Later in this chapter we review the procedure for Change Statistic.

5. The Change Statistic button will not be available unless the variable in the Variable box is highlighted. A variable is highlighted automatically when you click it into the Variable box. If you are experimenting and lose the highlighting, simply click directly on the variable in the Variable box. This restores the highlighting.

6. The same result can be achieved by clicking the Percentage below radio button and typing a "3" in the Value box. Because code 1 is the only value of attent lower than 3, SPSS will return the percentage of respondents having code 1 on attent.

7. V. O. Key, *Politics, Parties, and Pressure Groups,* 5th ed. (New York: Crowell, 1964): 568.

8. The variable who04_2 is confined to voters who cast ballots for either Kerry or Bush. Another variable in NES2004, who_2004, records voters' choices among Kerry, Bush, Nader, and Other.

9. Timur Kuran, *Private Truths, Public Lies: The Social Consequences of Preference Falsification* (Cambridge: Harvard University Press, 1995): 3.

10. It may have occurred to you that this effect might be greater for white respondents than for black respondents, with white subjects more likely to hide their true preferences in the presence of a black interviewer. An exercise in Chapter 5 will give you a chance to investigate this possibility.

11. From presidential debate at Wake Forest University, October 11, 2000. In a speech to the Sierra Club on September 9, 2005, Gore again made reference to scientific agreement: "Two thousand scientists, in a hundred countries, engaged in the most elaborate, well organized scientific collaboration in the history of humankind, have produced long-since a consensus that we will face a string of terrible catastrophes unless we act to prepare ourselves and deal with the underlying causes of global warming."

12. See Jeffrey S. Kopstein and David A. Reilly, "Geographic Diffusion and the Transformation of the Postcommunist World," *World Politics,* 53 (2000): 1–37. Noting that "[a]ll of the big winners of postcommunism share the trait of being geographically close to the former border of the noncommunist world" (p. 1), Kopstein and Reilly use geographic proximity to the West as the independent variable in testing the democratic diffusion hypothesis. By and large, the authors find "that the farther away a country is from the West, the less likely it is to be democratic" (p. 10).

13. In raw form, the Freedom House scale ranges from 1 (most democratic) to 7 (least democratic). The "rev" part of fhrate04_rev communicates that the Freedom House scale has been reverse-coded so that higher scores denote higher levels of democracy.

14. Conor O'Dwyer and Daniel Ziblatt, "Does Decentralisation Make Government More Efficient and Effective?" *Commonwealth & Comparative Politics,* 44 (November 2006): 1–18. This quote is from page 2. O'Dwyer and Ziblatt use sophisticated multivariate techniques to test the hypothesis that decentralized systems are more effective. Interestingly, they find that decentralization produces a higher quality of governance in richer countries but a lower quality of governance in poorer countries.

5

Making Controlled Comparisons

Procedures Covered

Analyze → Descriptive Statistics → Crosstabs (with Layers)
Analyze → Compare Means → Means (with Layers)
Graphs → Legacy Dialogs → Clustered Bar
Graphs → Legacy Dialogs → Multiple Line

Political analysis often begins by making simple comparisons using cross-tabulation analysis or mean comparison analysis. Simple comparisons allow the researcher to examine the relationship between an independent variable, X, and a dependent variable, Y. However, there is always the possibility that alternative causes—rival explanations—are at work, affecting the observed relationship between X and Y. An alternative cause is symbolized by the letter Z. If the researcher does not control for Z, then he or she may misinterpret the relationship between X and Y.

What can happen to the relationship between an independent variable and a dependent variable, controlling for an alternative cause? One possibility is that the relationship between the independent variable and the dependent variable is spurious. In a spurious relationship, once the researcher controls for a rival causal factor, the original relationship becomes very weak, perhaps disappearing altogether. The control variable does all of the explanatory work. In another possibility, the researcher observes an additive relationship between the independent variable, the dependent variable, and the control variable. In an additive relationship, two sets of meaningful relationships exist. The independent variable maintains a relationship with the dependent variable, and the control variable helps to explain the dependent variable. A third possibility, interaction, is somewhat more complex. If interaction is occurring, then the effect of the independent variable on the dependent variable depends on the value of the control variable. The strength or tendency of the relationship is different for one value of the control variable than for another value of the control variable.

These situations—a spurious relationship, an additive relationship, and interaction—are logical possibilities. Of course, SPSS cannot interpret a set of controlled comparisons for you. But it can produce tabular analysis and graphics that will give you the raw material you need to evaluate controlled comparisons.

In this chapter you will learn to use Crosstabs to analyze relationships when all three variables—the independent variable, the dependent variable, and the control variable—are nominal or ordinal. Because graphic displays are especially valuable tools for evaluating complex relationships, we will demonstrate how to use SPSS Graphs to obtain bar charts and line charts. These graphics will help you to interpret cross-tabulations with control variables. In this chapter you will also learn to use Compare Means to analyze relationships in which the dependent variable is interval level and the independent and control variables are nominal or ordinal level. These skills are natural extensions of the procedures you learned in Chapter 4.

USING CROSSTABS WITH LAYERS

To demonstrate how to use SPSS Crosstabs to obtain control tables, we will work through an example with GSS2006. This guided example uses one of the variables you created in Chapter 3, polview3.

Consider this hypothesis: In a comparison of individuals, liberals will be more likely to favor the legalization of marijuana than will conservatives. In this hypothesis, polview3, which categorizes respondents as liberal, moderate, or conservative, is the independent variable. GSS2006 contains the variable grass, which records respondents' opinions on the legalization of marijuana. (Code 1 is "Legal," and code 2 is "Not legal.") To stay acquainted with cross-tabulation analysis, let's start by looking at the uncontrolled relationship between polview3 and grass. Then we will add a control variable.

Open GSS2006. Go through the following steps, as covered in Chapter 4: Click Analyze → Descriptive Statistics → Crosstabs. Find the dependent variable, grass, in the left-hand variable list and click it into the Row(s) panel. Find the independent variable, polview3, and click it into the Column(s) panel. Click the Cells button and select the box next to "Column" in the Percentages panel. Click Continue, and then click OK.

grass Should marijuana be made legal? * polview3 Ideology: 3 categories Crosstabulation						
			polview3 Ideology: 3 categories			
			1 Liberal	2 Moderate	3 Conservative	Total
grass Should marijuana be made legal?	1 Legal	Count	245	247	166	658
		% within polview3 Ideology: 3 categories	50.7%	37.5%	26.4%	37.2%
	2 Not legal	Count	238	412	463	1113
		% within polview3 Ideology: 3 categories	49.3%	62.5%	73.6%	62.8%
Total		Count	483	659	629	1771
		% within polview3 Ideology: 3 categories	100.0%	100.0%	100.0%	100.0%

Clearly the hypothesis has merit. Of the liberals, 50.7 percent favor legalization, compared with 37.5 percent of moderates and 26.4 percent of conservatives.

What other factors, besides ideology, might account for differing opinions on marijuana legalization? A plausible answer: whether the respondent has children. Regardless of ideology, people with children may be less inclined to endorse the legalization of an illegal drug than are people who do not have children. And here is an interesting (if complicating) fact: Conservatives are substantially more likely to have children than are liberals.[1] Thus, when we compare the marijuana opinions of liberals and conservatives, as we have just done, we are also comparing people who are less likely to have children (liberals) with people who are more likely to have children (conservatives). It could be that liberals are more inclined to favor legalization, not because they are liberal per se, but because they are less likely to have children. By the same token, conservatives might oppose legalization for reasons unrelated to their ideology: They're more likely to have children. The only way to isolate the effect of ideology on marijuana opinions is to compare liberals who do not have children with conservatives who do not have children, and to compare liberals who have children with conservatives who have children. In other words, we need to control for the effect of having children by holding it constant. SPSS Crosstabs with Layers will perform the controlled comparison we are after.

GSS2006 contains the variable kids, which classifies respondents into one of two categories: those with children (coded 1 and labeled "Yes" on kids) or those without (coded 0 and labeled "No" on kids). Let's run the analysis again, this time adding kids as a control variable.

Again click Analyze → Descriptive Statistics → Crosstabs, returning to the Crosstabs window. You will find the dependent variable, grass, and the independent variable, polview3, just where you left them. To obtain a controlled comparison—the relationship between grass and polview3, controlling for kids—scroll down the variable list until you find kids and click it into the box labeled "Layer 1 of 1," as shown in Figure 5-1. SPSS will run a separate cross-tabulation analysis for each value of the variable that appears in the Layer box. And that is precisely what we want: a cross-tabulation of grass and polview3 for respondents without children and a separate analysis for those with children. Click OK. SPSS returns its version of a control table.

grass Should marijuana be made legal? * polview3 Ideology: 3 categories * kids Does R have children? Crosstabulation

kids Does R have children?					polview3 Ideology: 3 categories			Total
					1 Liberal	2 Moderate	3 Conservative	
0 No	grass Should marijuana be made legal?	1 Legal	Count		104	80	44	228
			% within polview3 Ideology: 3 categories		59.4%	45.2%	30.6%	46.0%
		2 Not legal	Count		71	97	100	268
			% within polview3 Ideology: 3 categories		40.6%	54.8%	69.4%	54.0%
	Total		Count		175	177	144	496
			% within polview3 Ideology: 3 categories		100.0%	100.0%	100.0%	100.0%
1 Yes	grass Should marijuana be made legal?	1 Legal	Count		139	167	122	428
			% within polview3 Ideology: 3 categories		45.4%	34.7%	25.2%	33.6%
		2 Not legal	Count		167	314	363	844
			% within polview3 Ideology: 3 categories		54.6%	65.3%	74.8%	66.4%
	Total		Count		306	481	485	1272
			% within polview3 Ideology: 3 categories		100.0%	100.0%	100.0%	100.0%

Figure 5–1 Crosstabs with Layers

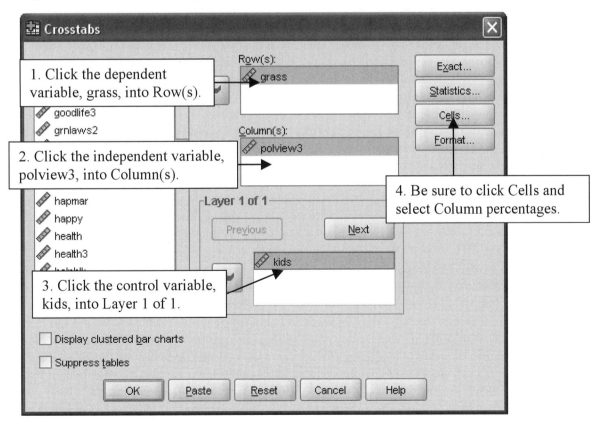

SPSS output using Crosstabs with layers can be a bit confusing at first, so let's consider closely what SPSS has produced. There are two cross-tabulations, appearing as one table. To the left-hand side of the table you will see the label of the control variable, kids: "Does R have children?" The first value of kids, "No," appears beneath that label. So the top cross-tabulation shows the grass-polview3 relationship for people who do not have children. The bottom cross-tabulation shows the relationship for respondents with children, respondents with the value "Yes" on the control variable.

What is the relationship between ideology and support for marijuana legalization among respondents who do not have children? What about for those with children? Is polview3 still related to grass? You can see that ideology is related to marijuana opinions for both values of kids. Among people without children, 59.4 percent of the liberals favor legalization, compared with 45.2 percent of the moderates and 30.6 percent of the conservatives. The more conservative people are, the lower the likelihood that they will favor legalization. The same general pattern holds for people with children: 45.4 percent of the liberals favor legalization, compared with 34.7 percent of moderates and 25.2 percent of conservatives. So, controlling for kids, polview3 is related to grass in the hypothesized way.

One bonus of control tables is that they permit you to evaluate the relationship between the control variable and the dependent variable, controlling for the independent variable. What is the relationship between the control variable, kids, and marijuana attitudes, controlling for ideology? Using the control tables produced by SPSS, you can address this question by jumping between the top cross-tabulation and the bottom cross-tabulation, comparing marijuana opinions of people who share the same ideology but who differ on the control variable, kids. Consider liberals. Are liberals without kids more likely to favor legalization than are liberals with kids? Yes. Among liberals without children, 59.4 percent favor legalization versus only 45.4 percent for liberals with children. How about moderates? Yes, again. There is a noticeable difference between the percentage of moderates without children who favor legalization (45.2 percent) and that of moderates with children who favor legalization (34.7 percent). For conservatives, too, having children somewhat decreases the likelihood of a pro-legalization response, although here the "kid effect" is much weaker than for moderates or liberals: 30.6 percent versus 25.2 percent.

How would you characterize this set of relationships? Does a spurious relationship exist between grass and polview3? Or are these additive relationships, with polview3 helping to explain legalization opinions and kids adding to the explanation? Or is interaction going on? Is the grass-polview3 relationship different for people without children than for people with children? If the grass-polview3 relationship were spurious, then the relationship would weaken or disappear after controlling for kids. Among respondents without children, liberals, moderates, and conservatives would all hold the same opinion about marijuana legalization. Ditto for people with children: Ideology would not play a role in explaining the dependent variable. Because the relationship persists after controlling for kids, we can rule out spuriousness. Now, it is sometimes difficult to distinguish between additive relationships and interaction relationships, so let's dwell on this question. In additive relationships, the effect of the independent variable on the dependent variable is the same or quite similar for each value of the control variable. In interaction relationships, by contrast, the effect of the independent variable on the dependent variable varies in size or tendency for different values of the control variable.

Return to the cross-tabulation output. The grass-polview3 relationship has the same tendency for people with and without children: For both values of the control, liberals are more pro-legalization than are conservatives. But notice that this effect is much larger for people without children. As you compare percentages across the "Legal" row, the percentage drops nearly 30 percentage points, from 59.4 percent among liberals to 30.6 percent among conservatives. Now examine the bottom portion of the control table, the grass-polview3 relationship for people with children. Here we find a weaker effect of ideology on legalization opinions: from 45.4 percent among liberals to 25.2 percent among conservatives, a drop of about 20 percentage points. So the "ideology effect" is about 30 points for people without children and about 20 points for people with children. Another way of describing this set of interaction relationships is to look at the effect of the control variable, kids, at each value of polview3. As mentioned above, the kid effect is sizable for liberals: The pro-legalization percentage is about 15 points lower for liberals with children than for liberals without children. But look at conservatives: having children has a very modest effect (only a 5-point difference) on their opinions. Because polview3 has a much larger effect on grass for one value of the control (respondents without children) than for the other value of the control (respondents with children), this is a set of interaction relationships.

Interaction relationships are quite common. For example, the size of the gender gap—the behavioral and attitudinal differences between women and men—is generally larger for unmarried people than for married people.[2] (Later in this chapter we will look at a different example of interaction effects in the gender gap.) So that you can become comfortable recognizing interaction in cross-tabulations, we will run through another example related to the line of analysis we have been pursuing. We just found that having children weakens the effect of ideology on attitudes toward the legalization of marijuana use—a behavior that, theoretically at least, touches upon the parent-child relationship. Let's see whether the same interaction pattern holds for opinions about another behavior that affects parenting: premarital sex. GSS2006 contains premarsx2, coded 1 for respondents who think it is wrong for a man and woman to "have sex relations before marriage," and coded 2 for respondents who think it is not wrong.

Return to the previous Crosstabs window. Click grass back into the Variable list. Find premarsx2 and click it into the Row(s)panel. Click OK. An SPSS cross-tabulation control table is perhaps more familiar this time around.

premarsx2 Is premarital sex wrong? * polview3 Ideology: 3 categories * kids Does R have children? Crosstabulation

kids Does R have children?				1 Liberal	2 Moderate	3 Conservative	Total
						polview3 Ideology: 3 categories	
0 No	premarsx2 Is premarital sex wrong?	1 Wrong	Count	65	77	95	237
			% within polview3 Ideology: 3 categories	33.3%	47.2%	67.9%	47.6%
		2 Not wrong	Count	130	86	45	261
			% within polview3 Ideology: 3 categories	66.7%	52.8%	32.1%	52.4%
	Total		Count	195	163	140	498
			% within polview3 Ideology: 3 categories	100.0%	100.0%	100.0%	100.0%
1 Yes	premarsx2 Is premarital sex wrong?	1 Wrong	Count	143	290	345	778
			% within polview3 Ideology: 3 categories	43.7%	52.8%	68.5%	56.4%
		2 Not wrong	Count	184	259	159	602
			% within polview3 Ideology: 3 categories	56.3%	47.2%	31.5%	43.6%
	Total		Count	327	549	504	1380
			% within polview3 Ideology: 3 categories	100.0%	100.0%	100.0%	100.0%

Focus on the percentage of respondents saying "not wrong." Clearly enough, ideology has a big effect on the dependent variable for people without children, as well as for people with children. As the independent variable changes from liberal to moderate to conservative, the "not wrong" percentages decline. So, just as with grass, the premarsx2-polview3 relationship has the same tendency for both values of the control, kids. Now take a closer look at the size of this effect. For respondents without kids, the drop is on the order of 35 points, from 66.7 percent to 32.1 percent. For respondents with kids, the decline, though still noteworthy, is nonetheless weaker—about 25 points (from 56.3 percent to 31.5 percent). Keep practicing reading the control table and identifying the interaction pattern. Is the kid effect the same for liberals and conservatives? No, it is not. There is about a 10-point difference on the dependent variable between liberals without kids (66.7 percent) and liberals with kids (56.3 percent). For conservatives, by contrast, having children has no discernible effect on the dependent variable: 32.1 percent of those without kids say "not wrong" compared with 31.5 percent of those with kids. Just as we found with marijuana opinions, the kid effect is larger for liberals than for conservatives.

OBTAINING AND EDITING CLUSTERED BAR CHARTS

In Chapter 4 you learned how to obtain a bar chart or line chart depicting the relationship between an independent variable and a dependent variable. SPSS also produces graphics for controlled comparisons. Here we demonstrate procedures for creating clustered bar charts. In the following guided example, we produce a bar chart that illustrates the relationship between premarsx2 and polview3, controlling for kids.

Click Graphs → Legacy Dialogs → Bar. When the Bar Charts window opens, click Clustered and make sure that the Summaries for Groups of Cases radio button is selected. Click Define. The Define Clustered

Figure 5–2   Opening the Define Clustered Bar Window

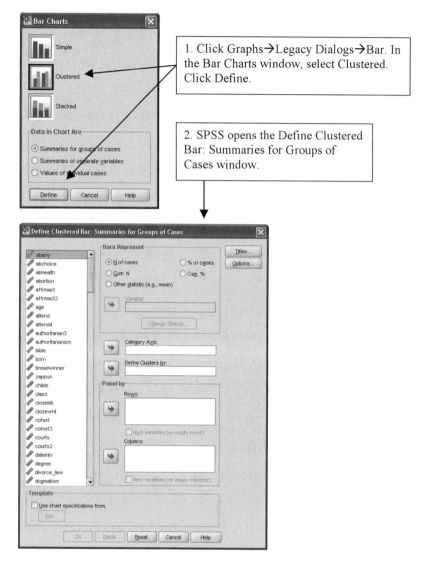

1. Click Graphs→Legacy Dialogs→Bar. In the Bar Charts window, select Clustered. Click Define.

2. SPSS opens the Define Clustered Bar: Summaries for Groups of Cases window.

Bar: Summaries for Groups of Cases window appears (Figure 5-2). What do we want the bar chart to depict? We want to see the percentage of respondents who think premarital sex is "not wrong" (coded 2 on premarsx2) for each value of the independent variable (polview3). But we want to see the premarsx2-polview3 relationship separately for each value of the control variable, kids. In an SPSS clustered bar chart, the values of the independent variable appear along the category axis. Because polview3 is the independent variable, click polview3 into the Category Axis box. For each value of polview3, we want to see the relationship separately for different values of the control variable, kids. In a clustered bar chart, the values of the control variable define the "clusters." The variable kids is the control variable, so click kids into the Define Clusters by box, as shown in Figure 5-3.

Now let's make sure that the bars will represent the percentages of respondents saying "not wrong." In the Bars Represent panel, select the Other statistic radio button, which activates the Variable box. Find premarsx2 in the variable list and then click it into the Variable box. By default, of course, SPSS will display the mean value of premarsx2, "MEAN(premarsx2)," which does not suit our purpose (see Figure 5-3). Click Change Statistic. In the Statistic window, click the Percentage inside radio button. Type "2" in the Low box and "2" in the High box. As in Figure 5-3, these instructions tell SPSS to display the percentage of respondents in one category of the dependent variable, the percentage coded 2 on premarsx2. Click Continue, returning to the Define Clustered Bar window. The Variable box should now read "PIN(2 2)(premarsx2)," meaning "The bars will display the percentages of respondents inside the value of 2 on premarsx2 at the low end and the value of 2 on premarsx2 at the high end." Click OK.

Figure 5–3   Obtaining a Clustered Bar Chart

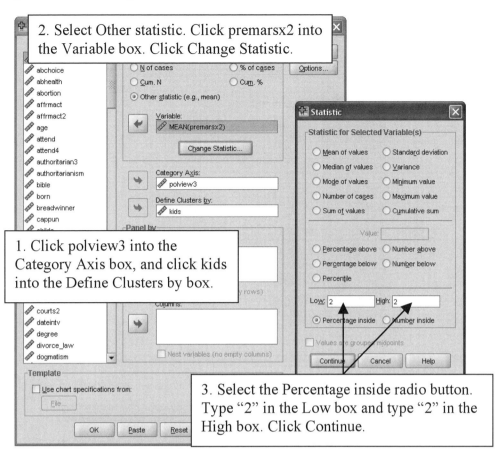

The clustered bar chart, constructed to our specifications, appears in the Viewer (Figure 5-4). This graphic greatly facilitates interpretation of the relationship. The left-hand bars in the clusters (which appear dark gray in Figure 5-4) show the relationship between premarsx2 and polview3 for people without children, and the right-hand bars (light gray) depict the relationship for people with children. Notice that, for both sets of bars, the "not wrong" percentages decline as you move across the values of polview3. Note, too, that within each cluster the without-kids bar is taller than the with-kids bar, revealing the effect of the control variable, kids, on the dependent variable. Finally, as you move from liberal to moderate to conservative, observe the precipitous drop in the heights of the bars for respondents without children and the milder decline among respondents with children. Finally, notice that for liberals the kids-yes bar is about 10 percentage points below the kids-no bar, but for the conservatives the bars are virtually the same height. This is a beautiful bar chart. But let's spruce it up using the Chart Editor.

We will make three changes to the chart: First, we will change the scale axis title. Second, we will label each bar, communicating the percentage of cases each bar represents. Finally, we will change the fill color or fill pattern on one set of bars, clearly distinguishing the category of the control variable, kids. (If you print graphics in black and white, as we do in this book, it is sometimes difficult to tell the difference between bars in the same cluster.)

Place the cursor anywhere on the chart and double-click. This invokes the Chart Editor. To change the scale axis title (Figure 5-5), first select it with a single-click. Single-click again to edit it. Replace the current title with this new title: "Percent saying premarital sex is not wrong." (Clicking anywhere else on the chart returns the axis title to its proper position.) Next let's add data labels to the bars so that each bar displays the percentage of respondents who think premarital sex is not wrong. Click the Data Labels icon on the toolbar above the chart (Figure 5-6). Doing so has the desired consequence of labeling the bars and opening the Data Value Labels tab of the Properties window. Here you can accept SPSS's label positioning choice or you can select a different position, as shown in Figure 5-6. We still have some editing chores left to do, so leave the Properties window open.

Figure 5–4 Clustered Bar Chart with Control Variable

Graph

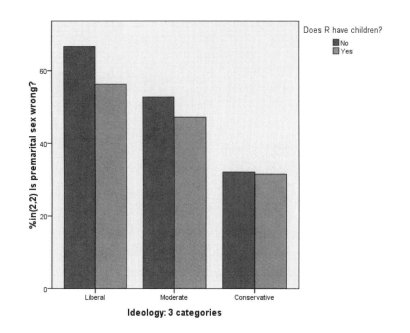

Figure 5–5 Changing the Scale Axis Title

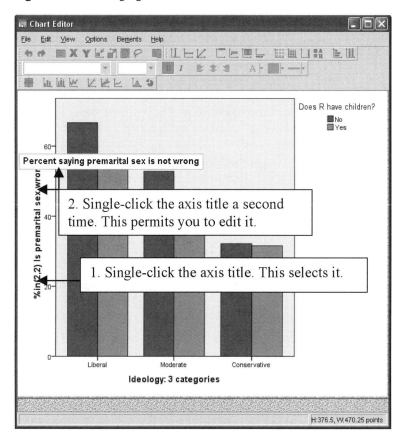

We will modify the appearance of the shorter set of bars, the bars depicting premarital sex opinions among respondents who have children. In the chart's legend area—the area that tells you which colors go with which values of the control variable—click on the color that represents the "Yes" value of kids (Figure 5-7). In the Properties window, click the Fill & Border tab. You may use the color palette to change the color

Figure 5–6 Adding Data Labels

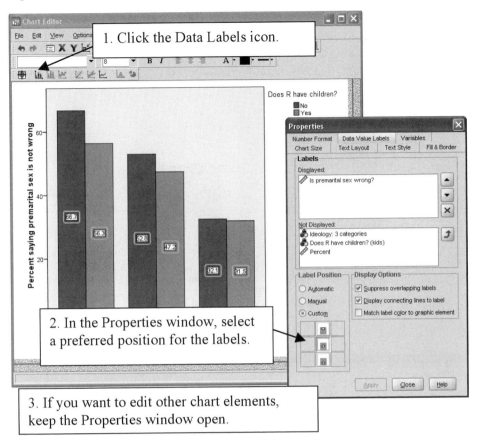

1. Click the Data Labels icon.

2. In the Properties window, select a preferred position for the labels.

3. If you want to edit other chart elements, keep the Properties window open.

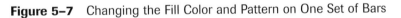

Figure 5–7 Changing the Fill Color and Pattern on One Set of Bars

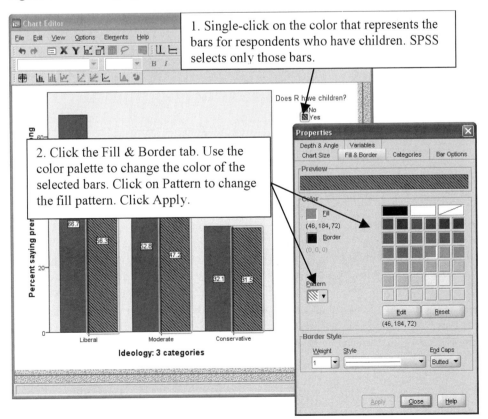

1. Single-click on the color that represents the bars for respondents who have children. SPSS selects only those bars.

2. Click the Fill & Border tab. Use the color palette to change the color of the selected bars. Click on Pattern to change the fill pattern. Click Apply.

Figure 5–8 Edited Clustered Bar Chart with Control Variable

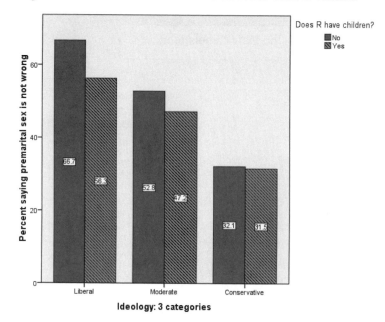

Figure 5–9 Multiple Line Chart with Control Variable

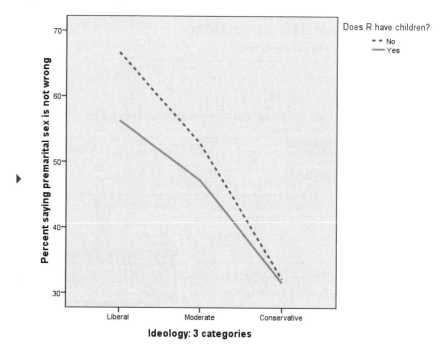

of the bars, and you can edit the fill pattern by clicking on Pattern and selecting something different. We will leave the color alone but pick a lined fill pattern. After clicking Apply, closing the Properties window, and exiting the Chart Editor, we will find our newly edited bar chart in the Viewer (Figure 5-8).

Clustered bar charts provide useful graphic support for cross-tabulation relationships. You can also use another graphic style, line charts. Sometimes line charts are better. Compared with bar charts, line charts are simpler and more elegant, and they generally have a more favorable data-ink ratio, defined as "the proportion of a graphic's ink devoted to the nonredundant display of data-information."[3] In other words, if one were to "add up" all the ink used in a graph, line charts tend to devote a larger proportion of the total ink to the essential communication of the data. For example, consider Figure 5-9, which uses Graph → Legacy

Dialogs → Line (Multiple) to graph the premarsx2-polview3 relationship, controlling for kids. Instead of showing ink-heavy bars extending upward to each percentage, the line chart simply connects the percentages for each value of the independent variable separately for each value of the control. The interaction relationship is clearly communicated in the differing slopes of the lines. Using what you have just learned about clustered bar charts, you would need to invest very little effort to recreate this graph.[4]

For depicting a set of relationships between a categorical dependent variable and categorical independent and control variables, it's your call: Clustered bar charts or multiple line charts can be used. When graphing mean comparison analyses—that is, relationships in which the dependent variable is interval level and the independent and control variables are categorical—then line charts must be used. We turn now to a discussion of mean comparison analysis with a control variable. We will then return to the topic of multiple line charts.

USING COMPARE MEANS WITH LAYERS AND OBTAINING MULTIPLE LINE CHARTS

As you know, mean comparison analysis is used when the dependent variable is interval level and the independent variable and the control variable are nominal or ordinal level. In most ways the procedure for using Compare Means with layers to obtain controlled comparisons is similar to that for using Crosstabs. However, the two procedures differ in one important way. We will work through two guided examples using NES2004. The first example shows an interesting pattern of interaction. The second example gives you a chance to identify a set of additive relationships. Open NES2004 and let's run through the first guided example.

Example of an Interaction Relationship

One of the most durable fixtures in U.S. politics is the relationship between income and partisanship. For many years, people who made less money were attracted to the Democratic Party more strongly than were more affluent individuals. Yet newer partisan divisions, such as that based on gender, may be altering the relationship between partisanship and income. How can this be? Think about this question for a moment. Suppose that, among men, traditional partisan differences persist, with lower-income males more pro-Democratic than higher-income males. But suppose that, among women, income-based partisan differences are much weaker, with lower-income females and higher-income females holding similar pro-Democratic views. This idea suggests that the relationship between income and partisanship is weaker for females than for males. If this idea is correct, then we should find a set of interaction relationships between income, partisanship, and gender. Let's investigate.

NES2004 contains a number of feeling thermometer variables, which record respondents' ratings of different political groups and personalities on a scale from 0 (negative, or "cold") to 100 (positive, or "warm"). Previously in this book, you have analyzed NES feeling thermometer scales. Another of these variables, dem_therm, which gauges feelings toward the Democratic Party, will be the dependent variable in the current example. The independent variable is income_r3, the ordinal measure of income that you created in Chapter 3. Recall that income_r3 classifies respondents into three income levels: "Low" (coded 1), "Middle" (coded 2), and "High" (coded 3). The control variable is gender, coded 1 for males and coded 2 for females. We will use Analyze → Compare Means → Means to produce mean values of dem_therm for each value of income_r3, controlling for gender.

Click Analyze → Compare Means → Means. Find dem_therm in the variable list and click it into the Dependent List box. Now we want SPSS to proceed as follows. First, we want it to separate respondents into two groups, men and women, on the basis of the control variable, gender. Second, we want SPSS to calculate mean values of dem_therm for each category of the independent variable, income_r3. SPSS handles mean comparisons by first separating cases on the variable named in the first Layer box. It then calculates means of the dependent variable for variables named in subsequent Layer boxes. For this reason, it is best to put the control variable in the first layer and to put the independent variable in the second layer. Because gender is the control variable, locate gender in the variable list and click it into the Layer 1 of 1 box (Figure 5-10). Click Next. The next Layer box, labeled "Layer 2 of 2," opens. The independent variable, income_r3, goes in this box. Click income_r3 into the Layer 2 of 2 box (Figure 5-11). One last thing: Click Options. In Cell Statistics, click Standard Deviation back into the left-hand Statistics box, and then click Continue. Ready to go. Click OK. The following SPSS control table appears in the Viewer:

dem_therm Feeling Thermometer: Democratic party			
gender ...	income_r3 ...	Mean	N
1 1 Male	1 Low	62.52	119
	2 Middle	56.07	163
	3 High	52.51	222
	Total	56.03	504
2 2 Female	1 Low	59.80	245
	2 Middle	60.81	188
	3 High	60.68	139
	Total	60.34	572
Total	1 Low	60.69	364
	2 Middle	58.61	351
	3 High	55.66	361
	Total	58.32	1076

Figure 5–10   Means Window with Dependent Variable and Control Variable

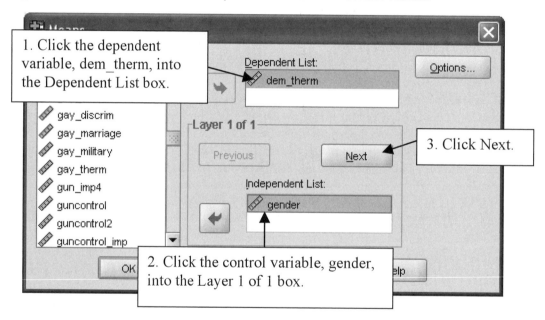

This is a highly readable table. The values of the control variable, gender, appear along the left-hand side of the table. The topmost set of mean comparisons shows the mean Democratic Party thermometer ratings of males, the next set is for females, and the bottom set (labeled "Total") shows the uncontrolled relationship between dem_therm and income_r3—for males and females combined. Let's interpret this table. What is the relationship for male respondents? As income increases, mean ratings of the Democratic Party decline. Males in the lowest income category averaged 62.52, compared with 56.07 for the males in the middle income category and 52.51 for those in the highest value of income_r3. For males, then, the mean drops about 10 degrees between the lowest income group and the highest income group.

Do we find the same pattern for females? Well, no we don't. Women in the lowest income group rate the Democratic Party at 59.80. Moving to middle income women, this rating does not budge, staying at about 60. What about the most affluent women? Again Democratic ratings stick stubbornly at a quite-positive 60 degrees. One would have to say that the relationship between income and Democratic ratings "works" for men—as income goes up, ratings go down—but it does not work for women. This is a clear case of interaction. For one value of the control (males), an increase in the independent variable is associated with a decrease in the dependent variable. For the other value of the control (females), there is no relationship

Figure 5–11 Means Window with Independent Variable in Layer Box

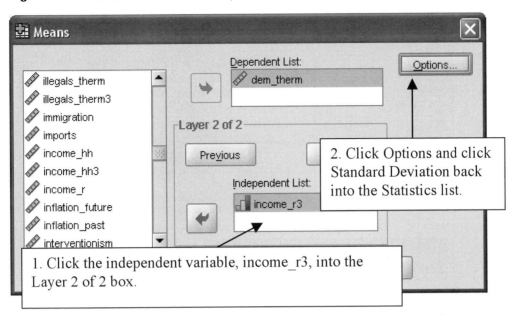

between the independent variable and the dependent variable. Thus the way we describe the relationship between the independent variable and the dependent variable depends on whether we are talking about men or women.

By examining this control table more closely, we can make some further observations about how the gender gap may be altering the long-standing relationship between income and partisanship. Consider the bottom panel of the table, which shows the overall relationship between dem_therm and income_r3 for the entire sample. Consider, too, the distribution of cases (N) across the values of income_r3 for the whole sample. The overall relationship shows the time-honored pattern: decreasing mean values of dem_therm between low-income people (60.69, or about 61 degrees) and those in the highest income category (55.66, or about 56 degrees). This is a mean difference of about 5 degrees. Which gender, male or female, makes a greater contribution to the low-income mean of 61? Well, there are 364 low-income people in the sample. From the Male panel of the table, we can see that only 119 of them are men, whereas the Female panel shows that 245 are women. So about two-thirds of the low-income group are women. Because the low-income category is populated more heavily with women than with men, women make a larger contribution to the low-income mean of dem_therm. What about the highest value of income_r3? Here the situation is reversed. There are 361 high-income individuals: 222 males and only 139 females. So males make a bigger contribution to the high-income mean, tugging it down to about 56.

Now let's speculate a bit. Imagine what the data would look like if, over time, more and more women were to enter the higher-income category, perhaps achieving equity with men. Under this scenario, women would make a larger and larger contribution to the high-income mean. Assuming, of course, that these women were as favorably disposed toward the Democratic Party as are the women in our control table, the *overall* relationship between dem_therm and income_r3 would become weaker and weaker. Speculation aside, this example illustrates why controlled comparisons are of central importance to the methodology of political research. Controlling for gender reveals the true gender-specific relationship between the independent and dependent variables.

A line chart of these relationships will permit us to make some further substantive observations about the gender gap. Click Graphs → Legacy Dialogs → Line. In the Line Charts window, click Multiple, and then click Define, as shown in Figure 5-12. In the Lines Represent panel of the Define Multiple Line window, select the Other statistic radio button. Click dem_therm into the Variable box (see Figure 5-13). SPSS moves dem_therm into the Variable box with its default designation, "MEAN(dem_therm)." This default is precisely what we want. In this case, we want to graph mean values of the dependent variable, dem_therm, for different values of income_r3. We want two lines to appear within the graph: one line showing the dem_therm-income_r3 relationship for men and a separate line showing the relationship for women. Because income_r3 is the independent variable, click income_r3 into the Category Axis box. Click gender

Figure 5–12 Opening the Define Multiple Line Window

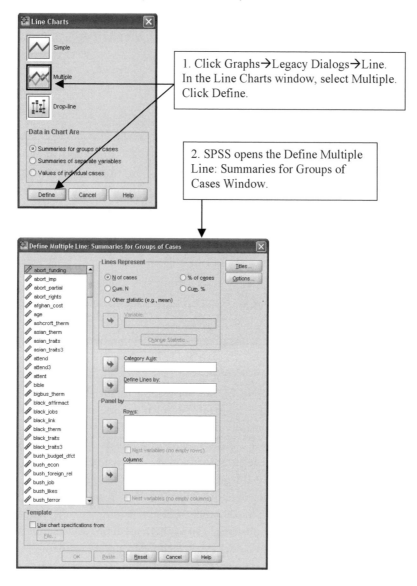

1. Click Graphs→Legacy Dialogs→Line. In the Line Charts window, select Multiple. Click Define.

2. SPSS opens the Define Multiple Line: Summaries for Groups of Cases Window.

into the Define Lines by box. (Refer to Figure 5-13.) Click OK. We now have a tailor-made line chart of the relationship (Figure 5-14).

You can see why line charts lend clarity and simplicity to complex relationships. By tracing along each line, from lower income to higher income, you can see the effect of income on thermometer ratings. Among males, the line drops sharply. The female line, by contrast, shows that the independent variable has no effect on the dependent variable. Notice, too, the relationship between gender and the dependent variable at different levels of income. For the lowest-income group, something of a reverse gender gap occurs, with males giving ratings to the Democratic Party that are slightly higher than those given by females. As income increases, however, the gender gap widens. This pattern suggests some general questions that you may wish to investigate in your own research. Do male-female differences on other political and social issues become more pronounced as income increases? Do some issues produce a larger gender gap than others?

If you planned to include the dem_therm-income_r3-gender line chart in a report or research paper, you would of course want to edit its appearance. SPSS's default line weights are weak, and a printed copy of this graphic would make it hard to distinguish the male line from the female line. Before going on to the next guided example, let's take a quick trip to the Chart Editor. Double-click the chart to get into the Chart Editor. First we will make both lines heavier, and then we will change the style of the male line, to clearly distinguish it from the female line. Click on one of the lines. SPSS selects both lines (Figure 5-15). Open the Properties window by clicking on its iconic button. In the Lines panel of the Lines tab, select a heavier

Figure 5–13 Obtaining a Multiple Line Chart

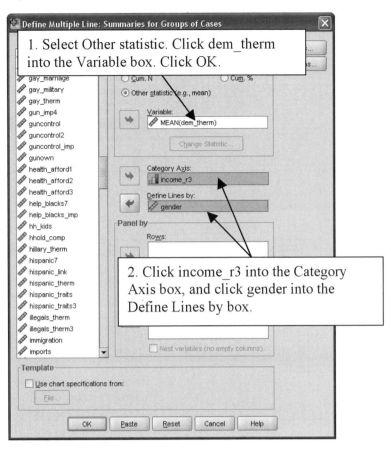

Figure 5–14 Multiple Line Chart with Control Variable: Interaction Relationship

Graph

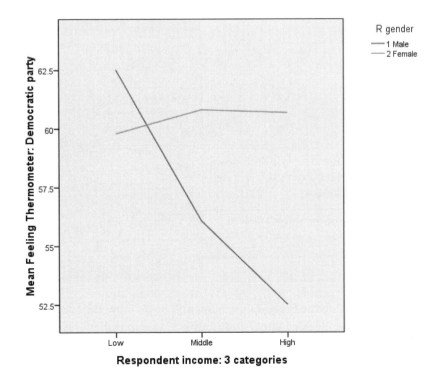

Figure 5–15 Changing the Line Weights of a Multiple Line Chart

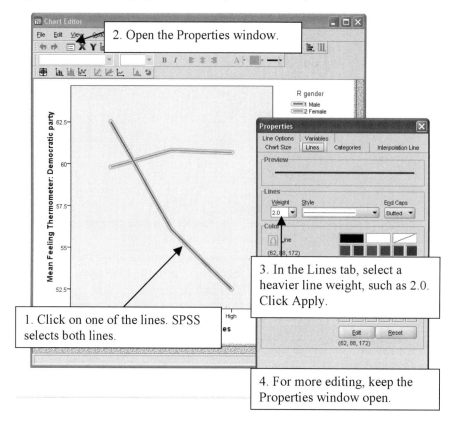

Figure 5–16 Changing Line Styles in a Multiple Line Chart

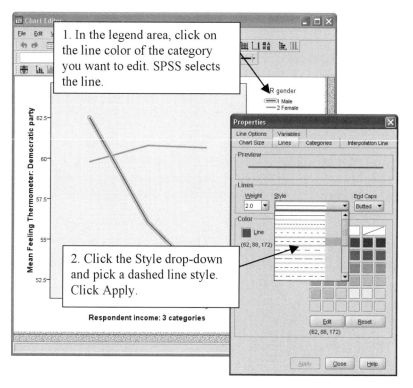

weight, such as 2.0. Click Apply. With the Properties window still open, move the cursor to the legend area and click on the color representing the male line (Figure 5-16). SPSS selects that line. Click the Style drop-down, select one of the dashed styles, and click Apply. Exit the Chart Editor. Now we have an edited line chart (Figure 5-17). Is the data-ink ratio acceptable? Well, not too bad.

Figure 5–17   Multiple Line Chart with Control Variable:
Interaction Relationship (edited)

Graph

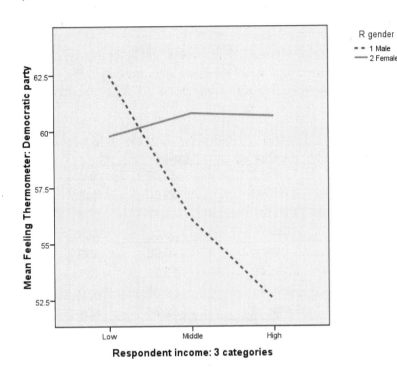

So far, this has been the "interaction chapter." In learning how to perform cross-tabulation analysis with a control variable, we found two instances of interaction: the grass-polview3-kids relationships and the premarsx2-polview3-kids relationships. In each case, the relationship between the independent and dependent variables had the same tendency for both values of the control variable, but the relationship was stronger for one value (respondents without children) than for the other value (respondents with children). In the mean comparison analysis we just performed, we saw a different pattern of interaction: a prominent relationship for one value of the control (males), but no relationship for the other value (females). That's what makes identifying interaction such a challenge—it can take on several different forms. Interaction has two fieldmarks, though, that will give it away. First, when you examine the relationship between the independent variable and the dependent variable at different values of the control variable, you may find that the relationship varies in tendency, perhaps positive for one value of the control variable, zero or negative for other control values. Second, the relationship may have the same tendency for all control values but differ in strength, negative-weak versus negative-strong or positive-weak versus positive-strong. Practice makes perfect. And, believe it or not, statistics can help (see Chapter 9).

Example of an Additive Relationship

Compared with the protean complexity of interaction, additive relationships are the soul of simplicity. In a set of additive relationships, both the independent and the control variables help to explain the dependent variable. More than this, the effect of the independent variable is the same or very similar—same tendency, same strength—for all values of the control variable. Interaction relationships assume several forms. Additive relationships assume only one.

A final example will illustrate this point. Consider this dependent variable: the NES feeling thermometer toward gays (gay_therm). Why might some people have negative feelings about homosexuals while others have more positive feelings? Social class could have an effect here, with lower-income individuals harboring more negative feelings than higher-income people. (The variable income_r3 again is the independent variable.) Plus, if you wish to explain attitudes toward gays, you need to include gender. Women are more likely than men to be positive toward gays and to be supportive of gay-oriented policies. (Gender is the control variable.) In fact, to isolate the effect of income on feelings toward homosexuals, we *must* control for gender. Why? Because, as our previous analysis showed, as income goes up, the proportion of women declines and the proportion of men increases. Thus when we compare the attitudes of lower-income people with the attitudes of higher-income people, we are also comparing the attitudes of a group dispro-portionally composed of women with those of a group disproportionally composed of men. These offset-ting tendencies might obscure the true relationship between income and gay attitudes. Let's run the analysis and find out what's going on.

Click Analyze → Compare Means → Means. Everything is still in place from our dem_therm-income_r3-gender analysis. As omens go, this is a good one, since the only thing we need to change is the dependent variable. Click dem_therm back into the variable list. Click gay_therm into the Dependent List. Click OK.

gay_therm Feeling Thermometer: Gay Men and Lesbians

gender ...	income_r3 ...	Mean	N
1 1 Male	1 Low	39.46	106
	2 Middle	42.50	148
	3 High	49.77	200
	Total	45.00	454
2 2 Female	1 Low	48.98	230
	2 Middle	53.29	161
	3 High	59.09	110
	Total	52.58	501
Total	1 Low	45.98	336
	2 Middle	48.12	309
	3 High	53.08	310
	Total	48.98	955

First consider the effect of income_r3 on gay_therm for males and for females. Low-income males have a rather chilly feeling toward gays, at 39.46 degrees. For men, moving from low income to high income boosts the temperature by about 10 degrees, to 49.77, among high-income men. Low-income women rate gays at 48.98, on average, or (interestingly) about the same as high-income men. For women, what happens as income goes up? Gay ratings increase by about 10 degrees, from 48.98 to 59.09. So the "income effect" is the same (or very nearly the same) for men and women alike, 10 degrees.

Now turn the analysis around and examine the effect of gender, controlling for income. In a comparison of low-income males with low-income females, how large is the "gender effect"? About 10 degrees, from 39.49 to 48.98. Is the gender effect the same for high-income people? Yes, pretty much. High-income males rate gays at 49.77, compared with 59.09 for high-income females, a 9-point gradient. Does the effect of income on feel-ings toward gays vary in tendency or strength between men and women? No, it is practically the same for both genders. The income effect is 10. The gender effect is 10. Thus, if we compare the mean values of the dependent variable for individuals having the least gay-friendly values on both variables (low-income males) with individuals having the most gay-friendly values on both variables (high-income females), we should find a 20-degree difference on gay_therm. Indeed, that is what we find: 39.49 versus 59.09.

All additive relationships have this sort of symmetry. But please note: It doesn't always work out, as it did here, that the independent variable and the control variable have the same magnitude of impact (for example, a 10-degree income effect and a 10-degree gender effect) on the dependent variable. But it does always work out that the effect of the independent variable is pretty much the same at all values of the control variable, and that the effect of the control variable is pretty much the same at all values of the inde-pendent variable. So, in the current example, if the income effect were 10 and the gender effect were 5, we'd still have a set of additive relationships on our hands.

There is one more thing to consider before we move on. Just as you did with the dem_therm example, examine the uncontrolled relationship between gay_therm and income_r3, shown in the bottom-most "Total" portion of the mean comparison table. Notice that the overall relationship is weaker than the controlled relationship. There is only a 7-degree difference between low-income respondents (45.98) and

Figure 5–18 Multiple Line Chart with Control Variable: Additive Relationship

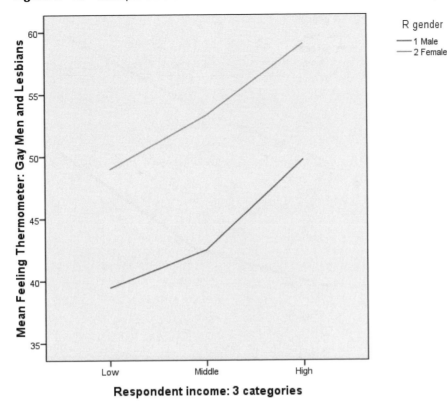

Respondent income: 3 categories

high-income respondents (53.08). But, of course, now we know what's going on. Women are overrepresented in the low-income group and pull the gay_therm average up among low-income respondents. Men are overrepresented in the high-income group and pull the gay_therm average down among high-income respondents. This results in an attenuation or shrinkage of the true relationship between income_r3 and gay_therm. As you can see, controlled comparisons are absolutely essential for revealing underlying processes that might affect what we observe in simple, uncontrolled relationships.

By now, obtaining a multiple line chart of the gay_therm-income_r3-gender relationships is a straightforward exercise. Follow these steps:

1. Click Graphs → Legacy Dialogs → Line. Click Multiple and then click Define.
2. In the Lines Represent panel of the Define Multiple Line window, click Other statistic.
3. Click gay_therm into the Variable box of the Lines Represent panel.
4. Click income_r3 into the Category Axis box.
5. Click gender into the Defines Line by box.
6. Click OK.

Consider how this line chart (Figure 5-18) communicates the additive relationship. Moving from left to right, from low income to high income, each line rises by 10 degrees. That's the income effect. The gender effect is conveyed by the distance between the lines. This effect, too, is fairly consistent. At each value of income_r3, women are about 10 degrees warmer toward gays than are men. Now, you might encounter additive relationships in which the lines slope downward, imparting a negative relationship between the independent and dependent variables. And the lines might "float" closer together, suggesting a consistent but weaker effect of the control variable on the dependent variable, controlling for the independent variable. But you will always see symmetry in the relationships. The effect of the independent variable on the dependent variable will be the same or very similar for all values of the control variable, and the effect of the control variable on the dependent variable will be the same or very similar for all values of the independent variable.

Invoke the Chart Editor and make some improvements to this multiple line chart. Experiment with the Chart Editor. Ask the Properties window to do new things. Embark on a quest to improve the data-ink ratio. Figure 5-19 may serve as inspiration. (Advanced chart editing is covered in Chapter 8.)

Figure 5–19 Multiple Line Chart with Control Variable: Additive Relationship (edited)

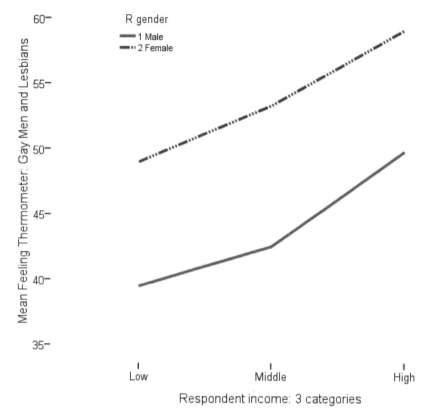

EXERCISES

1. (Dataset: World. Variables: democ_regime, frac_eth3, gdp_cap2.) Some countries have democratic regimes, and other countries do not. What factors help to explain this difference? One idea is that the type of government is shaped by the ethnic and religious diversity in a country's population. Countries that are relatively homogeneous, with most people sharing the same language and religious beliefs, are more likely to develop democratic systems than are countries having more linguistic conflicts and religious differences. Consider the ethnic heterogeneity hypothesis: Countries with lower levels of ethnic heterogeneity will be more likely to be democracies than will countries with higher levels of ethnic heterogeneity.

   A. According to the ethnic heterogeneity hypothesis, if you were to compare countries having lower heterogeneity with countries having higher heterogeneity, you should find (check one)

   ❑ a lower percentage of democracies among countries having lower heterogeneity.

   ❑ a higher percentage of democracies among countries having lower heterogeneity.

   ❑ no difference between the percentage of democracies among countries having lower heterogeneity and the percentage of democracies among countries with higher heterogeneity.

   B. World contains the variable democ_regime, which classifies each country as a democracy (coded 1) or a dictatorship (coded 0). This is the dependent variable. World also contains frac_eth3, which classifies countries according to their level of ethnic heterogeneity: low (coded 1), moderate (coded 2), or high (coded 3). This is the independent variable. Run Crosstabs, testing the ethnic heterogeneity hypothesis. Fill in the percentages of democracies:

	Ethnic heterogeneity		
	Low	Moderate	High
Percentage of democracies	?	?	?

C. Based on these results, you could say that (check one)

❏ as ethnic heterogeneity increases, the percentage of democracies increases.

❏ as ethnic heterogeneity increases, the percentage of democracies decreases.

❏ as ethnic heterogeneity increases, there is little change in the percentage of democracies.

D. A country's level of economic development also might be linked to its type of government. According to this perspective, countries with higher levels of economic development are more likely to be democracies than are countries with lower levels. The World dataset contains the variable gdp_cap2. This variable, based on gross domestic product per capita, is an indicator of economic development. Countries are classified as low (coded 1) or high (coded 2). Obtain a cross-tabulation analysis of the democ_regime-frac_eth3 relationship, controlling for gdp_cap2. Fill in the percentages of democracies:

	Ethnic heterogeneity		
	Low	Moderate	High
Low GDP per capita Percentage of democracies	?	?	?
High GDP per capita Percentage of democracies	?	?	?

E. Examine the relationship between ethnic heterogeneity and democracy in high-GDP countries and low-GDP countries.

Among low-GDP countries, as you move from the lowest value of heterogeneity ("Low") to the highest value ("High"), the percentage of democracies (circle one)

increases   decreases by (circle one)

less than 10   20   40

percentage points.

Among high-GDP countries, as you move from the lowest value of heterogeneity ("Low") to the highest value ("High"), the percentage of democracies (circle one)

increases   decreases by (circle one)

less than 10   20   40

percentage points.

F. Think about the set of relationships you just analyzed. How would you describe the relationship between ethnic heterogeneity and democracy, controlling for GDP per capita (circle one)?

Spurious   Additive   Interaction

Briefly explain your reasoning._____

_____

_____

_____

G. Obtain a clustered bar chart or a multiple line chart depicting the percentage of democracies for each value of frac_eth3, controlling for gdp_cap2. (Remember that democracies are coded 1 on democ_regime.) In the Chart Editor, give the scale axis this new title: "Percentage of democracies." Edit the chart for appearance and readability. Print the chart you created.

2. (Dataset: World. Variables: women05, pr_sys, womyear2.) In Chapter 2 you analyzed the distribution of the variable women05, the percentage of women in the lower house of the legislatures in a number of countries. In this exercise you will analyze the relationship between women05 and two variables that could have an impact on the number of women serving in national legislatures.

First consider the role of history and tradition. In some countries, women have had a long history of political empowerment. New Zealand, for example, gave women the right to vote in 1893. In other countries, such as Switzerland (where women were not enfranchised until 1971), women have had less experience in the electoral arena. Thus it seems reasonable to hypothesize that countries with longer histories of women's suffrage will have higher percentages of women in their national legislatures than will countries in which women's suffrage is a more recent development.

Now consider the role of the type of electoral system. Many democracies have proportional representation (PR) systems. PR systems foster multiple parties having diverse ideological positions—and, perhaps, having diverse demographic compositions as well. Non-PR systems, like the system used in U.S. elections, militate in favor of fewer and more homogeneous parties. Thus you might expect that non-PR countries will have fewer women in their national legislatures than will countries with PR-based electoral systems.

A. In addition to the dependent variable, women05, World contains womyear2, which measures the timing of women's suffrage by two values: 1944 or before (coded 0) and after 1944 (coded 1). Use womyear2 as the independent variable. World also has pr_sys, coded 0 for countries with non-PR systems and coded 1 for those having PR systems. Use pr_sys as the control variable. Run Compare Means to determine the mean of women05 for each value of womyear2, controlling for pr_sys. (Pr_sys is the control variable, so remember to put it in the first Layer box.) Refer to your output. In the "Mean" column of the table below, record the mean values of women05:

Mean percentage of women in lower house of parliament, 2005

PR electoral system?	Women's suffrage	Mean
No	1944 or before	?
	After 1944	?
Yes	1944 or before	?
	After 1944	?

B. Obtain a multiple line chart depicting the relationship between women05 and womyear2, controlling for pr_sys. In the Chart Editor, make both lines heavier. Also, change the style of one of the lines. Make any other desirable changes to the graphic. Print the multiple line chart you created.

C. Examine the table (part A) and the chart (part B). Among non-PR countries, those that gave women the right to vote after 1944 average about how much lower on the dependent variable than those that enfranchised women in 1944 or before? (circle one)

Between about 4.5 and 6 percentage points lower   Between about 6.5 and 8 percentage points lower

More than 8.5 percentage points lower

Among PR countries, those that gave women the right to vote after 1944 average about how much lower on the dependent variable than those that enfranchised women in 1944 or before? (circle one)

Between about 4.5 and 6 percentage points lower   Between about 6.5 and 8 percentage points lower

More than 8.5 percentage points lower

D. Based on your answers in part C, you can conclude that the effect of women's suffrage on the percentage of women in parliament is (check one)

❏ Much stronger for PR countries than for non-PR countries

❏ About the same for PR countries and non-PR countries

E. Now figure out the effect of the control variable, pr_sys, on the dependent variable separately for countries that enfranchised women in 1944 or before and for those that did so after 1944.

Among countries that gave women the right to vote in 1944 or before, non-PR countries average about how much lower on the dependent variable than PR countries? (circle one)

Between about 6.5 and 8 percentage points lower   Between about 8.5 and 10 percentage points lower

More than 10.5 percentage points lower

Among countries that gave women the right to vote after 1944, non-PR countries average about how much lower on the dependent variable than PR countries? (circle one)

Between about 6.5 and 8 percentage points lower   Between about 8.5 and 10 percentage points lower

More than 10.5 percentage points lower

F. Based on your answers in part E, you can conclude that the effect of the type of electoral system (PR or non-PR) on the percentage of women in parliament is (check one)

❏ much stronger for countries that enfranchised women in 1944 or before than for countries that enfranchised women after 1944.

❏ about the same for countries that enfranchised women in 1944 or before and for countries that enfranchised women after 1944.

G. Think about the set of relationships you just analyzed. How would you describe the relationship between the independent variable, womyear2, and the percentage of women in the legislature, controlling for the type of electoral system? (circle one)

Spurious   Additive   Interaction

Briefly explain your reasoning._____

_____

_____

_____

3. (Dataset: NES2004. Variables: bush_therm, married, attend3.) Several analyses of the 2004 presidential contest emphasized a sharpening of cultural and demographic differences between the supporters of the two major party candidates. According to one of these accounts, Bush's support "was particularly high among whites, regular churchgoers, evangelical Christians, residents of smaller communities, married people, those not in union households, those with a family income above $50,000, and southerners."[5] Of course, we know that several of these attributes are related to each other. For example, marital status and religiosity are linked rather closely: Regular churchgoers are more likely to be married than are the less observant.[6] Thus, if you sought to assess the effect of religious attendance on support for Bush, and you did not control for marital status, you would confuse the effect of religiosity with the effect of marital status.

In this exercise you will isolate the effect of religiosity on support for Bush, controlling for marital status. However, before running any analyses, you will graphically depict different possible scenarios for the relationships you might discover.

Parts A, B, and C contain graphic shells showing George W. Bush feeling thermometer ratings along the vertical axis and levels of religious attendance along the horizontal axis. (Notice that the values of attendance decline from left to right, from "High" on the left to "Low" on the right.) For each shell, you will draw two lines within the graphic space, a solid line depicting the relationship for married people and a dashed line depicting the relationship for unmarried people.

A. Draw an additive relationship fitting this description: As religious attendance declines, Bush ratings decline, and married people rate Bush higher than do unmarried people. (Hint: In additive relationships, the strength and tendency of the relationship is the same or very similar for all values of the control variable.) Remember to use a solid line to depict the relationship for married people and a dashed line to depict the relationship for unmarried people.

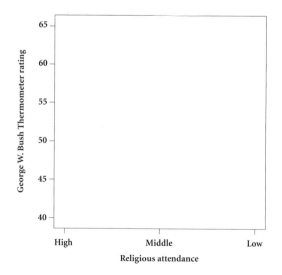

B. Draw a spurious relationship: Religious attendance has no effect on ratings of Bush; marital status has a big effect on ratings of Bush.

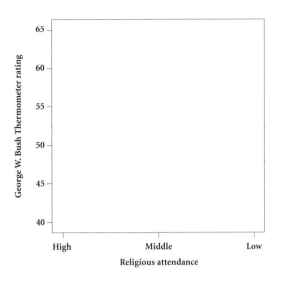

C. Draw a set of interaction relationships fitting this description: For unmarried people, as religious attendance declines, support for Bush declines. For married people, religious attendance has no effect on ratings of Bush.

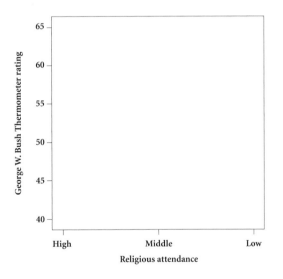

NES2004 contains bush_therm, a 100-point feeling thermometer of George W. Bush. NES2004 also has attend3, which classifies respondents' frequency of religious attendance into three ordinal categories: "High" (coded 1), "Middle" (coded 2), and "Low" (coded 3). For the control, use married, a variable you created in Chapter 3. Recall that unmarried respondents are coded 0 on married and married respondents are coded 1.

D. Run the appropriate mean comparison analysis. In the table that follows, record your results in each cell that contains a question mark (?).

Feeling Thermometer: GW Bush

Is R married?	Relig attend: 3 cats	Mean	N
Not married	High	?	?
	Middle	?	?
	Low	?	?
	Total	?	?
Married	High	?	?
	Middle	?	?
	Low	?	?
	Total	?	?
Total	High	?	?
	Middle	?	?
	Low	?	?
	Total	?	?

E. Create a line chart of this set of relationships. Use the Chart Editor to enhance the appearance of the graphic. Print the line chart.

F. Review your artistic work in parts A–C. Examine the table (part D) and the line chart (part E). Which possible scenario—the line chart you drew in A, B, or C—most closely resembles the pattern shown in the data? (circle one)

    The line chart in A   The line chart in B   The line chart in C

4. (Dataset NES2004. Variables: vote_2004, memnum2, educ3.) Two political analysts are discussing factors that affect turnout in U.S. elections.

Political analyst 1: "I think that the role of voluntary association memberships has been overlooked as an important causal factor. People who belong to community associations, interest groups, or labor unions will interact with other individuals, discuss politics, and become aware of important issues. As a result, they will be more likely to vote than will people who do not belong to any groups."

Political analyst 2: "Your idea is interesting. . .but flawed. We know that education is strongly linked to voter turnout. People with more education are more likely to vote. Plus—and here is the relationship you are overlooking—people with more education are also more likely to join voluntary organizations. So as education goes up, people will be more likely to join groups *and* to vote. But there is no causal connection between belonging to groups and voting."

A. According to political analyst 1, if you were to compare the voter turnout of people who are not members of voluntary groups with that of people who are members, you would find (check one)

☑ higher turnout among people who are members of voluntary groups.

❏ lower turnout among people who are members of voluntary groups.

❏ no difference in turnout between people who are members of voluntary groups and those who are not.

B. According to political analyst 2, if you were to compare the group memberships of people who have less education with the group memberships of people who have more education, you would find (check one)

☑ people with more education are more likely to be members of voluntary groups.

❏ people with more education are less likely to be members of voluntary groups.

❏ people with more education and people with less education are equally likely to be members of voluntary groups.

C. Think about a Crosstabs analysis showing three cross-tabulations: the relationship between group membership (independent variable) and turnout (dependent variable) for people with low education, medium education, and high education. According to political analyst 2, such an analysis would reveal (check all that apply)

❏ that at each level of education, people who are members of voluntary groups are no more likely to vote than are people who are not members of voluntary groups.

☑ that regardless of group memberships, people with more education are more likely to vote than are people with less education.

❏ that the relationship between voter turnout and group membership is spurious.

D. Run Crosstabs to analyze the relationship between turnout and group membership. For this run, you will not control for education. Dataset NES2004 contains vote_2004, coded 1 for respondents who voted and coded 0 for those who did not. The dataset also contains memnum2, which classifies people into two categories: those with no group memberships (coded 0) and those having at least one membership (coded 1). Use your analysis to fill in the blanks:

Among respondents with no memberships, (fill in the blank) _____ 71.9 _____ percent voted.

Among respondents having at least one membership, (fill in the blank) _____ 87.6 _____ percent voted.

E. Now run the analysis again, controlling for education. Use educ3 as the control. In the table the follows, record the percentages who voted.

Percentages who voted:

Education level:	Group member?	
	No	Yes
<=HS	?	?
Some coll	?	?
>=Coll	?	?

F. Obtain a line chart of the relationship between vote_2004 and memnum2, controlling for educ3. The vertical axis should depict the percentage who voted. Put memnum2 on the category axis. There should be three lines inside the chart, one for each value of educ3. Edit the chart for appearance and readability. Print the chart.

G. Examine the tabular evidence (part E) and the supporting graphic (part F). Think about the set of relationships you just analyzed. How would you describe the relationship between the independent variable, memnum2, and turnout, controlling for education? (circle one)

Spurious   Additive   Interaction

Explain your reasoning. _____

_____

_____

_____

_____

_____

5. (Dataset: NES2004. Variables: rep_therm, race_2, south.) The two political analysts are at it again. This time they are discussing reputed partisan differences between southern and non-southern states.

Political analyst 1: "Media pundits and confused academics tend to exaggerate the South's reputation as a stronghold of Republican sentiment. In fact, people who live outside the South and people who live in the South don't differ that much in their ratings of the Republican Party. Look at my latest mean comparison analysis. Non-southerners rated the Republicans at 51.88 on the feeling thermometer, compared with a slightly warmer 55.85 for southerners. That's a paltry 4-point difference on the 100-point scale!"

Political analyst 2: "Hmmm . . . that's interesting. But did you control for race? I wonder what happens to the relationship between region and Republican ratings after you take race into account. After all, blacks are less strongly attracted to the Republicans than are whites. And since southern states have a higher proportion of blacks than do non-southern states, racial differences in Republican ratings could affect regional differences in Republican ratings."

A. According to political analyst 2, why did political analyst 1's analysis find a small difference between the Republican ratings of non-southerners and the Republican ratings of southerners? (check two)

❑ Because southern respondents are more likely to be white than are non-southern respondents

❑ Because southern respondents are more likely to be black than are non-southern respondents

❑ Because blacks give the Republican Party lower ratings than do whites

❑ Because blacks give the Republican Party higher ratings than do whites

B. Perform a mean comparison analysis, using thermometer ratings of the Republican Party (rep_therm) as the dependent variable. Use south as the independent variable. Non-southern respondents are coded 0 on south and southern respondents are coded 1. Use race_2 as the control variable. Whites are coded 1 on race_2 and blacks are coded 2. In the following table, fill in the mean values (Mean) and numbers of observations (N) next to each question mark (?):

Feeling Thermometer: Republican party

Race: White/Black	Non-South/South	Mean	N
White	Non-South	?	?
	South	?	?
	Total	?	851
Black	Non-South	?	?
	South	?	?
	Total	?	177
Total	Non-South	51.97	670
	South	55.34	358
	Total	53.14	1028

C. In the table in part B, refer to the mean value you recorded in the Total row for all 851 whites and the mean value you recorded in the Total row for all 177 blacks. According to these numbers, whites are

_____ points (circle one)

cooler   warmer

toward the Republican Party than are blacks.

D. Now figure out if the group of southern respondents has proportionately more blacks than the group of non-southern respondents. According the table in part B, there are 358 southern respondents in the dataset.

About what percentage of these 358 southern respondents are black? (circle one)

About 10 percent   About 20 percent   About 30 percent

About what percentage of the 670 non-southern respondents are black? (circle one)

About 10 percent   About 20 percent   About 30 percent

E. Evaluate the rep_therm-south relationship, controlling for race_2. When you compare southern whites with non-southern whites, you find that southern whites are (fill in the blank)

_____ points (circle one)

cooler   warmer

toward the Republican Party than are non-southern whites.

When you compare southern blacks with non-southern blacks, you find that southern blacks are

(fill in the blank) _____ points (circle one)

cooler   warmer

toward the Republican Party than are non-southern blacks.

F. Obtain a clustered bar chart or a multiple line chart of the relationship between rep_therm and south, controlling for race_2. Edit the chart for appearance and readability. Print the graphic you have created.

G. Consider all of the evidence you have obtained in this exercise. How would you characterize this set of relationships? (circle one)

Spurious   Additive   Interaction

Explain your reasoning. _____

_____

_____

_____

_____

_____

6. (Dataset: GSS2006. Variables: natrace, natfare, natsci, intrace_2, race_2.) For an exercise in Chapter 4, you tested for the presence of preference falsification, the tendency for respondents to offer false opinions that they nonetheless believe to be socially desirable under the circumstances. You evaluated the hypothesis that respondents are more likely to express support for government policies aimed at helping blacks (such as "government spending to improve the conditions of blacks") when questioned by a black interviewer than when questioned by a white interviewer. But you did not control for the respondent's race. That is, you did not look to see whether whites are more (or less) likely than are blacks to misrepresent their support for racial policies, depending on the race of the interviewer.[7]

Furthermore, it may be that whites, and perhaps blacks as well, will engage in the same preference falsifying behavior for policies that do not explicitly reference race but that may symbolize race, such as "government spending for welfare." Although "welfare" does not mean "blacks," it may be that whites see "welfare" through a racially tinged lens and will respond as if the question refers to a racial policy. Of course, some policies, such as "government spending for scientific research," do not evoke such symbolic connections. Questions about these race-neutral policies should not show the same race-of-interviewer effects as questions that make explicit—or implicit—reference to race.[8]

In this exercise you will extend your Chapter 4 analysis in two ways. First, you will analyze the relationship between interviewer race (intrace_2, the independent variable) and three dependent variables: opinions on an explicitly racial policy (natrace, which measures attitudes toward spending to improve the conditions of blacks), a symbolically racial policy (natfare, opinions on spending for welfare), and a race-neutral policy (natsci, spending for scientific research). Second, you will perform these analyses while controlling for respondent's race (race_2).

Based on previous research in this area, what might you expect to find? Here are two plausible expectations:

Expectation 1: For both white and black respondents, the race-of-interviewer effect will be strongest for the explicitly racial policy (natrace), weaker for the symbolically racial policy (natfare), and nonexistent for the race-neutral policy (natsci).

Expectation 2: For the explicitly racial policy (natrace) and for the symbolically racial policy (natfare), the race-of-interviewer effect will be greater for white respondents than for black respondents. For the race-neutral policy (natsci), the race-of-interviewer effect will be the same (close to zero) for both white respondents and black respondents (see expectation 1).

A. Run the appropriate cross-tabulation analyses. In the table that follows, record the percentages of respondents saying that we are spending "too little" in each of the policy areas. For each policy, obtain the race-of-interviewer effect by subtracting the percentage of respondents saying "too little" when interviewed by a white questioner from the percentage saying "too little" when interviewed by a black questioner. (For example, if 50.0 percent of respondents said we are spending "too little" when questioned by a white and 70.0 percent said "too little" when questioned by a black, then the race-of-interview effect would be 70.0 – 50.0, or 20.0.)

Race of respondent	Percent saying we are spending "too little" on:	Race of interviewer		Race-of-interviewer effect (Black %–White %)
		White	Black	
White	Improving the conditions of blacks (natrace)	?	?	?
	Welfare (natfare)	?	?	?
	Supporting scientific research (natsci)	?	?	?
Black	Improving the conditions of blacks (natrace)	?	?	?
	Welfare (natfare)	?	?	?
	Supporting scientific research (natsci)	?	?	?

B. Examine the tabular data closely.

Among white respondents, would you say that expectation 1 is or is not supported by the evidence? (circle one)

Expectation 1 is not supported   Expectation 1 is supported

Explain your reasoning. _____

_____

_____

_____

Among black respondents, would you say that expectation 1 is or is not supported by the evidence? (circle one)

Expectation 1 is not supported   Expectation 1 is supported

Explain your reasoning. _____

_____

_____

_____

C. Now compare the race-of-interviewer effects between respondents of different races. That is, compare the race-of-interviewer effect on natrace among white respondents with the race-of-interviewer effect on natrace among black respondents. Do the same for natfare and natsci. Generally speaking, would you say that expectation 2 is supported or is not supported by the evidence? (circle one)

Expectation 2 is not supported   Expectation 2 is supported

Explain your reasoning. _____

_____

_____

_____

D. This exercise shows that the survey environment can be quite sensitive to extraneous contextual stimuli. On a more sobering note, the findings suggest how the survey environment can be manipulated by cynical, unprincipled pollsters to produce a desired result. Suppose one wanted to "prove" that no discernible racial difference exists on the explicitly racial policy of spending to improve the conditions of blacks. One would (check two)

❏ assign black interviewers to black respondents.

❏ assign black interviewers to white respondents.

❏ assign white interviewers to black respondents.

❏ assign white interviewers to white respondents.

That concludes the exercises for this chapter. Before exiting SPSS, be sure to save your output file.

NOTES

1. According to GSS2006, 78.3 percent of conservatives have children, compared with 61.8 percent of liberals—almost a 17-point difference.
2. For instance, if you were to analyze NES2004, you would find a 10-point difference in the percentages of unmarried women and men who voted for John Kerry (61.9 percent of unmarried women versus 52.0 percent of unmarried men) but virtually no gender difference among married people (42.9 percent for married women versus 41.2 percent for married men).
3. Edward R. Tufte, *The Visual Display of Quantitative Information,* 2d edition (Cheshire, CT: Graphics Press, 2001): 93.
4. The Graphs → Legacy Dialogs → Line (Multiple) interface is identical to the clustered bar chart interface in every detail. The Chart Editor presents somewhat different options for line charts. These are covered later in the text.
5. Gerald M. Pomper, "The Presidential Election: The Ills of American Politics After 9/11." In Michael Nelson, ed., *The Elections of 2004* (Washington, D.C.: CQ Press, 2005): 42–68. This quote, p. 47.
6. According to a cross-tabulation of variables contained in NES2004, 58 percent of regular churchgoers are married, compared with only 44 percent of the less observant.
7. See Darren W. Davis and Brian D. Silver, "Stereotype Threat and Race of Interviewer Effects in a Survey of Political Knowledge," *American Journal of Political Science,* 47 (January 2003), 33–45.
8. There is a large body of literature on "symbolic racism." For an excellent review and analysis, see Stanley Feldman and Leonie Huddy, "Racial Resentment and White Opposition to Race-Conscious Programs: Principles or Prejudice?" *American Journal of Political Science,* 49 (January 2005), 168–183.

6

Making Inferences about Sample Means

Procedures Covered

Analyze → Descriptive Statistics → Descriptives
Analyze → Compare Means → One-Sample T Test
Analyze → Compare Means → Independent-Samples → T Test

Political research has much to do with observing patterns, creating explanations, framing hypotheses, and analyzing relationships. In interpreting their findings, however, researchers often operate in an environment of uncertainty. This uncertainty arises, in large measure, from the complexity of the political world. As we have seen, when we infer a causal connection between an independent variable and a dependent variable, it is hard to know for sure whether the independent variable is causing the dependent variable. Other, uncontrolled variables might be affecting the relationship, too. Yet uncertainty arises, as well, from the simple fact that research findings are often based on random samples. In an ideal world, we could observe and measure the characteristics of every element in the population of interest—every voting-age adult, every student enrolled at a university, every bill introduced in every state legislature, and so on. In such an ideal situation, we would enjoy a high degree of certainty that the variables we have described and the relationships we have analyzed mirror what is really going on in the population. But of course we often do not have access to every member of a population. Instead we rely on a sample, a subset drawn at random from the population. By taking a random sample, we introduce random sampling error. In using a sample to draw inferences about a population, therefore, we never use the word *certainty*. Rather, we talk about *confidence* or *probability*. We know that the measurements we make on the sample will reflect the characteristics of the population, within the boundaries of random sampling error.

What are those boundaries? If we calculate the mean income of a random sample of adults, for example, how confident can we be that the mean income we observe in our sample is the same as the mean income in the population? The answer depends on the standard error of the sample mean, the extent to which the mean income of the sample departs by chance from the mean income of the population. If we use a sample to calculate a mean income for women and a mean income for men, how confident can we be that the difference between these two sample means reflects the true income difference between women and men in the population? Again, the answer depends on the standard error—in this case, the standard error of the *difference* between the sample means, the extent to which the difference in the sample departs from the difference in the population.

In this chapter you will use three SPSS procedures to explore and apply inferential statistics. First, you will learn to use Descriptives to obtain basic information about interval-level variables. Second, using the One-Sample T Test procedure, you will obtain confidence intervals for a sample mean. The 95 percent confidence interval will tell you the boundaries within which there is a .95 probability that the true population mean falls. You will also find the 90 percent confidence interval, which is applied in testing hypotheses at the .05 level of significance. Third, using the Independent-Samples T Test procedure, you will test for statistically significant differences between two sample means.

USING DESCRIPTIVES AND ONE-SAMPLE T TEST

To gain insight into the properties and application of inferential statistics, we will work through an example using NES2004. We begin by looking at the Descriptives procedure, which yields basic information about interval-level variables. We then demonstrate the fundamentals of inference using the One-Sample T Test procedure.

NES2004 contains spend11, a measure of individuals' opinions about federal spending. Respondents were asked whether spending should be increased, kept the same, or decreased for each of eleven federal programs. Spend11 was created by adding up the number of times the respondent said "increased."[1] We can use Descriptives to obtain summary information about spend11. Open NES2004. Click Analyze → Descriptive Statistics → Descriptives. In the main Descriptives window, scroll down the left-hand variable list until you find spend11. Click spend11 into the Variable(s) list (Figure 6-1). Click the Options button. Now you can specify which descriptive statistics you would like SPSS to produce. These defaults should already be checked: mean, standard deviation, minimum, and maximum. That's fine. Also check the box beside "S.E. mean," which stands for "standard error of the mean," as shown in Figure 6-1. Click Continue, and then click OK.

Descriptive Statistics

	N	Minimum	Maximum	Mean		Std. Deviation
	Statistic	Statistic	Statistic	Statistic	Std. Error	Statistic
spend11 IncrFed$, 11 prgms	1139	.00	11.00	5.5759	.07270	2.45347
Valid N (listwise)	1139					

SPSS has reported the requested statistics for spend11: number of cases analyzed (N), minimum and maximum observed values for spend11, mean value of spend11, standard error of the mean, and standard deviation. Among the 1,139 respondents, scores on spend11 range from 0 (increase spending on none of the programs) to 11 (increase spending on all of the programs). The mean value of spend11 is 5.5759 (which rounds to 5.58), with a standard deviation of 2.45.[2] How closely does the mean of 5.58 reflect the true mean in the population from which this sample was drawn? If we had measured spend11 for every U.S. citizen of voting age and calculated a population mean, how far off the mark would our sample estimate of 5.58 be?

Figure 6–1 Descriptives Window and Descriptives: Options Window (modified)

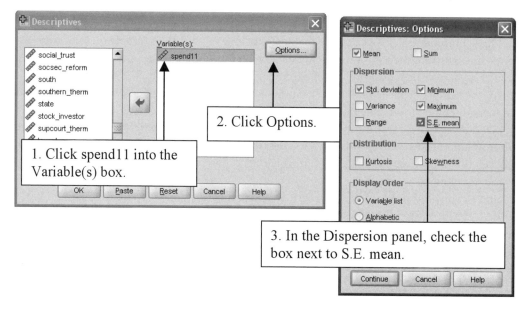

Figure 6–2 One-Sample T Test Window: Setting Up the Confidence Interval

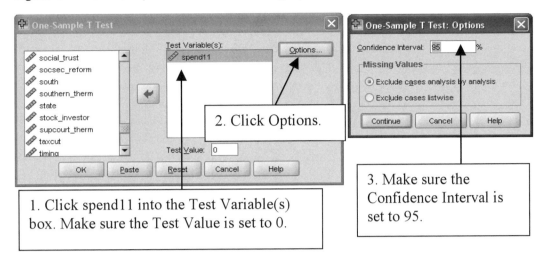

The answer depends on the standard error of the sample mean. The standard error of a sample mean is based on the standard deviation and the size of the sample. SPSS determines the standard error just as you would—by dividing the standard deviation by the square root of the sample size. For spend11, the standard error is the standard deviation, 2.45, divided by the square root of 1,139. Performed with a hand calculator: $2.45 / \sqrt{(1139)} = 2.45 / 33.749 \approx .07$.

This number, .07, tells us the extent to which the sample mean of 5.58 departs by chance from the population mean. The standard error is the essential ingredient for making inferences about the population mean. But let's get SPSS to help us make these inferences. Specifically, we will use the One-Sample T Test procedure to do three things: find the 95 percent confidence interval of the mean, calculate the 90 percent confidence interval of the mean, and test a hypothetical claim about the population mean using the .05 level of significance.

First, let's use One-Sample T Test to obtain the 95 percent confidence interval of spend11. Click Analyze → Compare Means → One-Sample T Test, causing the One-Sample T Test window to open (Figure 6-2). The user supplies SPSS with information in two places: the Test Variable(s) panel and the Test Value box, which currently contains the default value of 0. Now, One-Sample T Test is not naturally designed to report the 95 percent confidence interval for a mean. Rather, it is set up to compare the mean of a variable in the Test Variable(s) panel with a hypothetical mean (provided by the user in the Test Value box) and to see if random error could account for the difference. (We will discuss this calculation below.) However, if you run One-Sample T Test on its defaults, it will provide the 95 percent confidence interval. To obtain the 95 percent confidence interval for the mean of a variable, do three things. First, click the variable into the Test Variable(s) panel, as shown in Figure 6-2. Second, make sure that the Test Value box contains the default value of 0. Third, click Options and ensure that the confidence interval is set at 95 percent. To follow these steps for spend11, we would click it into the Test Variable(s) panel and leave the Test Value box as it is. After clicking Options to make sure that the confidence interval is set at 95 percent, we would click Continue, and then click OK to run the analysis.

One-Sample Statistics

	N	Mean	Std. Deviation	Std. Error Mean
spend11 IncrFed$, 11 prgms	1139	5.5759	2.45347	.07270

One-Sample Test

	Test Value = 0					
					95% Confidence Interval of the Difference	
	t	df	Sig. (2-tailed)	Mean Difference	Lower	Upper
spend11 IncrFed$, 11 prgms	76.701	1138	.000	5.57594	5.4333	5.7186

Figure 6–3 Bell-Shaped Curve Showing 95 Percent Confidence Interval of the NES Sample Mean

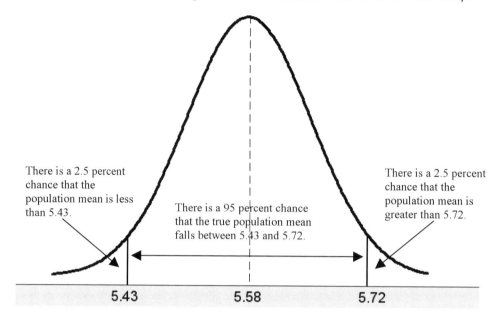

There is a 2.5 percent chance that the population mean is less than 5.43.

There is a 95 percent chance that the true population mean falls between 5.43 and 5.72.

There is a 2.5 percent chance that the population mean is greater than 5.72.

5.43   5.58   5.72

SPSS output for One-Sample T Test includes two tables. In the One-Sample Statistics table, SPSS reports summary information about spend11. This information is similar to the Descriptives output discussed earlier. Again, we can see that spend11 has a mean of 5.58, a standard deviation of 2.45, and a standard error of .07 (which, reassuringly, is the same number we calculated by hand). We are interested mainly in the second table, the One-Sample Test table. In fact, when using One-Sample T Test to obtain confidence intervals, you may safely ignore all of the information in the One-Sample Test table except for the rightmost cells. The values appearing under the label "95% Confidence Interval of the Difference," 5.43 and 5.72, define the lower and upper boundaries of the 95 percent confidence interval.

Now, we are trying to determine how much confidence to invest in our sample mean of 5.58. Is the true population mean right around 5.58? The 95 percent confidence interval provides a probabilistic—not a definitive—answer. There is a high probability, a 95 percent probability, that the true population mean lies in the region between 5.43 at the low end and 5.72 at the high end. If we were to take a very large number of random samples from the population and calculate, for each sample, the mean of spend11, 95 percent of those calculated means would fall in the interval between 5.43 and 5.72. To be sure, there is random "noise" in each random sample. Yet 95 percent of the time, that noise will give us a sample mean within the bandwidth of 5.43 to 5.72. On uncommon occasions—5 percent of the time—we would obtain a sample mean that falls outside those boundaries, below 5.43 or above 5.72. Therefore, we can infer that there is a 95 percent chance that the true population mean is in the 5.43 to 5.72 range and a 5 percent probability that the population mean lies outside these boundaries—a 2.5 percent chance that the population mean is less than 5.43 and a 2.5 percent chance that it is greater than 5.72. You might be wondering where the "2.5 percent chance" came from. Aren't we discussing the 95 percent confidence interval? Shouldn't the statement instead say "5 percent chance"? The "2.5 percent" terminology is correct. This is an understandable confusion, so let's pause and clear things up.

Figure 6-3 shows a bell-shaped curve. The 2004 NES sample mean, 5.58, is in the center of the distribution, bisecting the curve, with half of the distribution of possible sample means falling in the negative tail below 5.58 and half falling in the positive part of the tail, above 5.58. Now consider the two shorter lines, one in the negative tail, drawn at 5.43, and one in the positive tail, at 5.72. These numbers, as we have seen, define the 95 percent confidence interval. Because the 95 percent confidence interval brackets 95 percent of all possible sample means, the remaining 5 percent must fall below 5.43 *or* above 5.72. Furthermore, because the curve is symmetrical, one-half of the remaining 5 percent (2.5 percent) will fall in the negative tail, below 5.43, and one-half (the other 2.5 percent) will fall in the positive tail, above 5.72.

Suppose someone made the (weak) hypothetical claim that the true population mean was *different from* 5.58. This person didn't say whether they thought the true mean was greater than 5.58 or less than 5.58. Just different from 5.58. Such a claim—a claim that the real population mean is different from the observed sample mean—requires a two-tailed test of statistical significance. Because we don't know the direction of the hypothetical difference, we would apply the 95 percent confidence interval and use a .05 two-tailed test. We would simply say that there is a 95 percent chance that the population mean falls between 5.43 and 5.72 and a 5 percent chance that it falls above or below these numbers.

Most hypothesis-testing in political science uses a one-tailed test, not a two-tailed test. Why? Because any properly constructed hypothesis communicates the direction of a relationship. A gender-gap hypothesis, for example, would not suggest ambiguously that women and men will have different means on spend11. The hypothesis will tell you that the female mean on spend11 should be found to be higher (or lower) than the male mean on spend11. Because hypothetical relationships imply direction, only one-half of the bell-shaped curve (either the half with the negative tail or the half with the positive tail) comes into play in testing the hypothesis. For this reason, the 95 percent confidence interval provides a stringent .025 one-tailed test of statistical significance for hypothetical claims about the population mean.

Let's apply this one-tailed logic. Suppose someone were to suggest that spend11's population mean is 5.80 and that we obtained a sample mean of 5.58 by chance when we drew the sample. Notice that this hypothetical assertion, 5.80, locates the population mean above the observed sample mean of 5.58, out in the positive region of the curve. For the purposes of testing this claim, then, we ignore the negative tail and use the positive tail. Return your attention to Figure 6-3, focusing only on the upper half, the half with the line drawn at 5.72, the upper boundary of the 95 percent confidence interval. Ask yourself this question: If the unseen population mean is equal to the observed sample mean of 5.58, what is the probability that a random sample would produce a mean of 5.80? As Figure 6-3 illustrates, we know that if we were to draw a large number of random samples from a population in which the true mean is 5.58, only 2.5 percent of those samples would yield sample means of 5.72 *or higher*. Using the .025 standard, then, we would reject the claim that the population mean is 5.80. If the population mean is equal to the observed sample mean of 5.58, then random processes would yield a mean of 5.80 less than 2.5 percent of the time. Naturally, the same logic applies to any hypothetical claim that locates the population mean below the observed sample mean, down in the negative tail of the distribution.

A .025 one-tailed test of statistical significance is stringent and conservative—it makes it tougher to reject hypothetical challenges to relationships observed in a sample. Researchers tend to err on the side of caution, and you will never be criticized for using the .025 criterion. However, a somewhat less stringent test, the .05 one-tailed test of statistical significance, is perhaps a more widely applied standard. There are two ways to apply this standard—the confidence interval approach and the P-value approach. In the confidence interval approach, the researcher finds the 90 percent confidence interval of the mean. Why the 90 percent confidence interval? For the same reasons that the 95 percent confidence interval provides a .025 one-tailed test, the 90 percent confidence interval sets the limits of random sampling error in applying the .05 standard. There is a 90 percent probability that the population mean falls between the lower value of the 90 percent confidence interval and the higher value of the 90 percent confidence interval. There is a 10 percent chance that the population mean falls outside these limits—5 percent *below* the lower boundary and 5 percent *above* the upper boundary.

In the P-value approach the researcher determines the exact probability associated with a hypothetical claim about the population mean. First, let's find the 90 percent confidence interval for spend11. Then we will use the P-value approach to evaluate a hypothetical claim about the mean of spend11.

Click Analyze → Compare Means → One-Sample T Test. Make sure that spend11 is still in the Test Variable(s) panel and that the default value of 0 appears in the Test Value box. Click Options. This time, type "90" in the confidence interval box. Click Continue, and then click OK.

One-Sample Test

	Test Value = 0					
	t	df	Sig. (2-tailed)	Mean Difference	90% Confidence Interval of the Difference	
					Lower	Upper
spend11 IncrFed$, 11 prgms	76.701	1138	.000	5.57594	5.4563	5.6956

You can be confident, 90 percent confident anyway, that the population mean of spend11 lies between 5.46 at the low end and 5.70 at the high end. There is a probability of .10 that the population mean lies beyond these limits—a .05 probability that it is less than 5.46 and a .05 probability that it is greater than 5.70.

We can apply this knowledge to the task of testing a hypothetical claim. Suppose you hypothesized that political science majors will be more likely to favor government spending than will most people. To test this idea, you ask a group of political science majors a series of 11 questions about federal programs—the same set of questions that appears in the 2004 National Election Study. Whereas the 2004 NES reports a mean value of 5.58 on spend11, you find a higher mean value, 5.68, among the respondents in your study. Thus it would appear that your respondents are, on average, more supportive of government spending than are the individuals in the NES's random sample of U.S. adults. But is this difference, 5.58 versus 5.68, *statistically* significant at the .05 level? No, it is not. Why can we say this? Because the political science majors' mean, 5.68, does not exceed the NES sample's upper confidence boundary, 5.70.

Think about it this way. Imagine a population of U.S. adults in which spend11's true mean is equal to 5.58. Now suppose you were to draw a random sample from this population and calculate the mean of spend11. The upper confidence boundary tells you that such a sample would yield a mean of greater than 5.70 *less frequently* than 5 times out of 100. The upper confidence boundary also says that such a sample would produce a mean of less than 5.70 *more frequently* than 5 times out of 100. Because 5.68 is less than 5.70, you must conclude that the political science majors' mean is not significantly higher than the NES mean. Put another way, there is a probability of greater than .05 that your sample of political science majors and the NES's sample of adults were both drawn from the same population—a population in which spend11 has a mean equal to 5.58.

Confidence interval approaches to statistical significance work fine. Find the 90 percent confidence interval and compare the hypothetical mean to the appropriate interval boundary. If the hypothetical mean falls above the upper boundary (or below the lower boundary), then conclude that the two numbers are significantly different at the .05 level. If the hypothetical mean falls below the upper boundary (or above the lower boundary), then conclude that the two numbers are not significantly different at the .05 level. Thus the confidence interval approach tells you that a random sample of U.S. adults would produce a sample mean of 5.68 more frequently than 5 percent of the time. However, the P-value approach to statistical significance is more precise. The P-value will tell you *exactly* how frequently a sample mean of 5.68 would occur.

Let's run One-Sample T Test on spend11 one more time and obtain the information we need to determine the P-value associated with the political science majors' mean of 5.68. Click Analyze → Compare Means → One-Sample T Test. Spend11 should still be in the Test Variable(s) panel. Now click in the Test Value box and type "5.68" (Figure 6-4). SPSS will calculate the difference between the mean of the test variable, spend11, and the test value, 5.68. SPSS will then report the probability that the test value, 5.68, came from the same population as did the mean of the test variable, spend11. Click OK. Again, we have One-Sample T Test output (Figure 6-5).

Figure 6–4   Testing a Hypothetical Claim about a Sample Mean

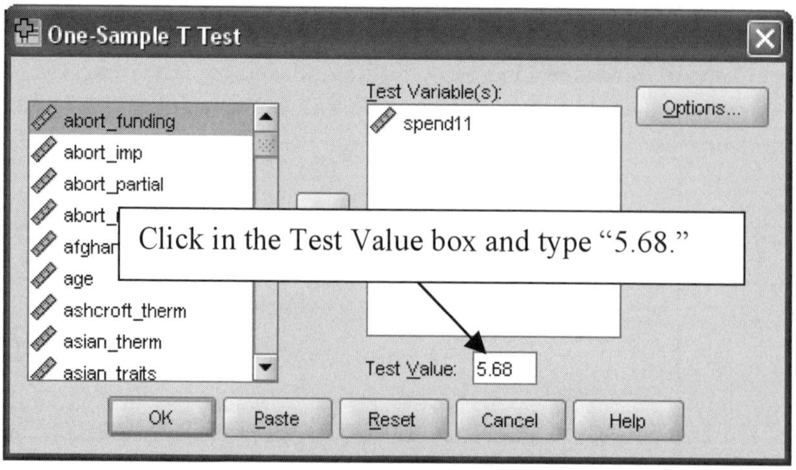

Figure 6–5 Testing for Statistical Significance

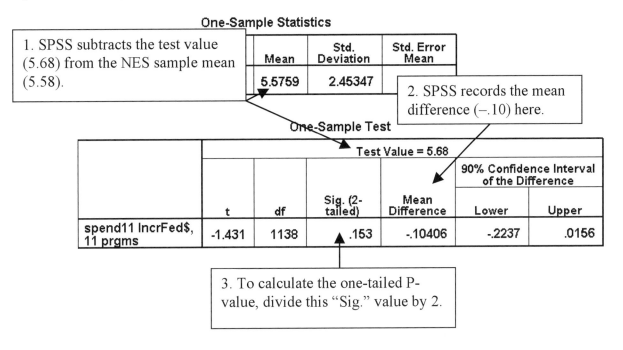

Consider the One-Sample Test table. What do these numbers mean? SPSS follows this protocol in comparing the mean of a test variable with a test value. First, it finds the difference between them: test variable mean minus test value. So SPSS calculates 5.58 – 5.68 and reports the result, –.10, in the "Mean Difference" column of the One-Sample Test table.

Obviously, if the political science majors' mean were the same as the NES mean of 5.58, then SPSS would report a mean difference of 0. But the calculated difference, –.10, is not 0. In fact, given the way that SPSS evaluates test values, the difference between the political science majors' mean and the NES mean falls in the hypothesized direction. Consider another bell-shaped curve, shown in Figure 6-6. This time the curve is centered at 0, the difference we would observe if the political science majors' mean were the same as the NES mean. The curve tapers off into a skinnier and skinnier tail above 0. This is the region of the curve you would use to test the hypothesis that the majors' mean is less than the NES mean. Because our hypothesis stipulated that the majors' mean would be greater than the NES mean, we can ignore the positive tail of the curve. The curve also tapers off into a skinnier tail below 0. This region, the half of the curve below 0, is the region we must focus on to test the hypothesis. The inferential question: If the true difference between the NES mean and the test value is equal to 0, how often would random processes produce an observed difference of –.10? If the answer is "less than 5 percent of the time," then we can infer that the two means are significantly different from one another. If the answer is "more than 5 percent of the time," then we must conclude that the two means are not significantly different.

Unhelpfully, SPSS does not report the one-tailed P-value that would permit us to settle the issue. As Figure 6-6 illustrates, SPSS does not ignore the positive half of the bell-shaped curve, the half that, in the present example, is irrelevant to testing the hypothesis. Instead, SPSS answers this inferential question: If in the population there is really no difference between the test variable mean and the test value, how often would one obtain a difference of at least .10 in *either direction*? How often would one observe a difference of less than –.10 (in the negative region) *or* more than +.10 (in the positive region)? To answer this inferential question, SPSS calculates a Student's t-test statistic, or t-ratio, the mean difference divided by spend11's standard error.[3] This value of t, –1.43, appears in the leftmost cell of the One-Sample Test table. Thus spend11's mean falls 1.43 standard errors below the political science majors' mean. Turn your attention to the cell labeled "Sig. (2-tailed)," which contains the number .153. This two-tailed value tells you the proportion of the curve that lies below t = –1.43 *and* above t = +1.43. Thus despite SPSS's willingness to report a t-test statistic whose negative sign communicates the region of the curve we are interested in, it disappoints by reporting the proportion of both tails—the proportion below t = –1.43 plus the proportion above t = +1.43.

Figure 6–6 SPSS Reports Two-Tailed P-values

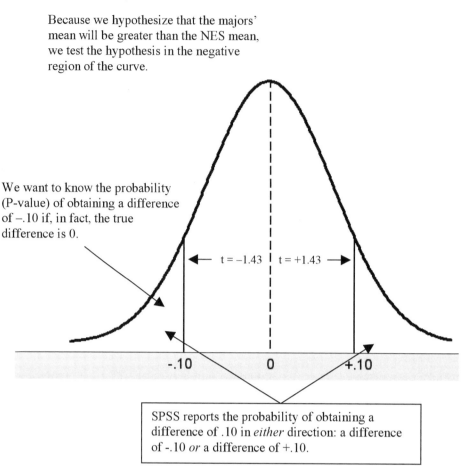

Because we hypothesize that the majors' mean will be greater than the NES mean, we test the hypothesis in the negative region of the curve.

We want to know the probability (P-value) of obtaining a difference of –.10 if, in fact, the true difference is 0.

t = –1.43   t = +1.43

-.10   0   +.10

SPSS reports the probability of obtaining a difference of .10 in *either* direction: a difference of -.10 *or* a difference of +.10.

Again, we are not testing the hypothesis that the political science majors' mean is *different from* the mean of the adult population. We are not asking SPSS to tell us if the majors' mean is less than or greater than the population mean. Rather, we are testing the hypothesis that the majors' mean is *greater than* the population mean. In testing this hypothesis, we do not want to know the proportion of the curve that lies above t = +1.43. We only want to know the P-value associated with t = –1.43, the test statistic for spend11's mean of 5.58 minus the political science majors' mean of 5.68. Because the Student's t-distribution is perfectly symmetrical, one-half of .153, or .0765 (which rounds to .077), falls below t = –1.43. From the confidence interval approach, we already know that there is a probability of greater than .05 that the political science majors' mean and the NES sample mean came from the same population, that the true difference between them is equal to 0. We can now put a finer point on this probability. We can say that, if the true difference between the majors' mean and the NES mean is equal to 0, then the observed difference between them (–.10) would occur, by chance, .077 of the time. Because .077 exceeds the .05 threshold, we would conclude again that the test subjects' mean is not significantly higher than the population mean.

Before proceeding, let's pause to review the steps for obtaining a P-value from the One-Sample T Test procedure. This is also the appropriate place to introduce a useful template for writing an interpretation of your results. First, click the variable of interest into the Test Variable(s) panel. Second, type the hypothetical value in the Test Value box and click OK. Third, obtain a P-value by dividing the number in the "Sig. (2-tailed)" cell by 2. Use the P-value to fill in the blank in the following template:

If in the population there is no difference between the mean of [the test variable] and [the test value], then the observed difference of [the mean difference] would occur _____ of the time by chance.

Of course, you can embellish the template to make it fit the hypothesis you are testing. For the spend11 example, you could complete the sentence this way: "If in the population there is no difference between the NES mean of spend11 and the political science majors' mean of 5.68, then the observed difference of −.10 would occur .077 of the time by chance." It is also acceptable to express the P-value as a percentage, as in: "would occur 7.7 percent of the time by chance." The .05 benchmark is the standard for testing your hypothesis. If the P-value is less than or equal to .05, then you can infer that the test value is significantly greater than (or less than) the mean of the test variable. If the P-value is greater than .05, then you can infer that the test value is not significantly greater than (or less than) the mean of the test variable.

USING INDEPENDENT-SAMPLES T TEST

We now turn to a common hypothesis-testing situation: comparing the sample means of a dependent variable for two groups that differ on an independent variable. Someone investigating the gender gap, for example, might test a series of hypotheses about the political differences between men and women. In the next guided example, we test two gender gap hypotheses using two feeling thermometer scales as the dependent variables:

Hypothesis 1: In a comparison of individuals, men will give gays lower feeling thermometer ratings than will women.

Hypothesis 2: In a comparison of individuals, men will give the Republican Party higher thermometer ratings than will women.

The first hypothesis suggests that when we divide the sample on the basis of the independent variable, gender, and compare mean values of the gay feeling thermometer, the male mean will be lower than the female mean. The second hypothesis suggests that a similar male-female comparison on the Republican Party feeling thermometer will show the male mean to be higher than the female mean.

The researcher always tests his or her hypotheses against a skeptical foil, the null hypothesis. The null hypothesis claims that, regardless of any group differences that a researcher observes in a random sample, no group differences exist in the population from which the sample was drawn. How does the null hypothesis explain apparently systematic patterns that might turn up in a sample, such as a mean difference between women and men on the gay feeling thermometer? It points to random sampling error. In essence the null hypothesis says, "You observed such and such a difference between two groups in your random sample. But, in reality, no difference exists in the population. When you took the sample, you introduced random sampling error. Thus random sampling error accounts for the difference you observed." For both hypotheses 1 and 2 above, the null hypothesis says that there are no real differences between men and women in the population, that men do not give gays lower ratings or the Republican Party higher ratings. The null hypothesis further asserts that any observed differences in the sample can be accounted for by random sampling error.

The null hypothesis is so central to the methodology of statistical inference that we always begin by assuming it to be correct. We then set a fairly high standard for rejecting it. The researcher's hypotheses—such as the gay thermometer hypothesis and the Republican thermometer hypothesis—are considered alternative hypotheses. The Independent-Samples T Test procedure permits us to test each alternative hypothesis against the null hypothesis and to decide whether the observed differences between males and females are too large to have occurred by random chance when the sample was drawn. For each mean comparison, the Independent-Samples T Test procedure will give us a P-value: the probability of obtaining the sample difference under the working assumption that the null hypothesis is true.

Click Analyze → Compare Means → Independent-Samples T Test. The Independent-Samples T Test window appears (Figure 6-7). SPSS wants to know two things: the name or names of the test variable(s) and the name of the grouping variable. SPSS will calculate the mean values of the variables named in the Test Variable(s) panel for each category of the variable named in the Grouping Variable box. It will then test to see if the differences between the means are significantly different from 0.

We want to compare the means for men and women on two feeling thermometers, one for gays (gay_therm) and one for the Republican Party (rep_therm). Find gay_therm and rep_therm in the variable list and click both of them into the Test Variable(s) panel. Click Options and change the confidence interval to 90 percent and click Continue (Figure 6-8). Because you want the means of these variables to be calculated separately for each sex, gender is the grouping variable. When you click gender into the Grouping Variable box, SPSS moves it into the box with the designation "gender(? ?)." The Define Groups button is activated (Figure 6-9). SPSS needs more information. It needs to know the codes of the two groups you wish to

Figure 6–7 Independent-Samples T Test Window

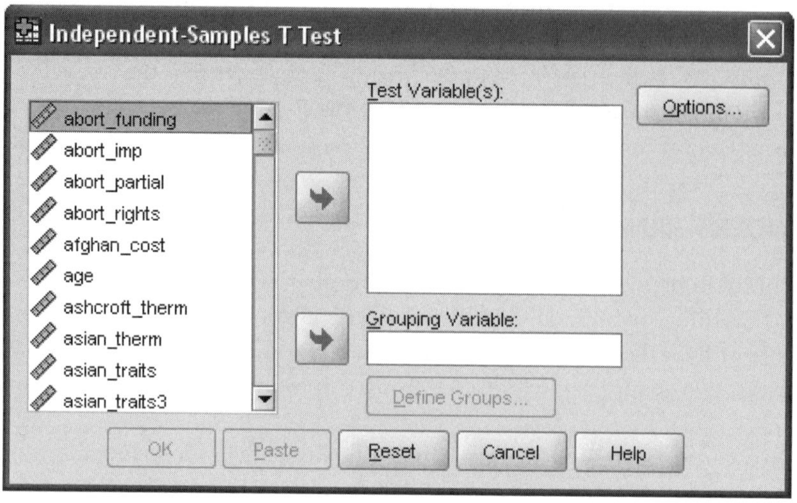

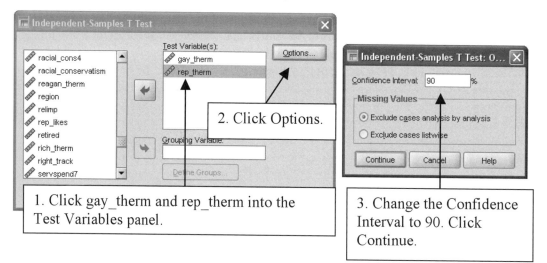

Figure 6–8 Specifying Test Variables and Setting Confidence Interval

compare. Men are coded 1 on gender and women are coded 2. (Recall that by right-clicking on a variable you can reacquaint yourself with that variable's codes.) Click Define Groups (Figure 6-9). There are two ways to define the groups you want to compare: Use specified values (the default) and Cut point. The choice, of course, depends on the situation. If you opt for Use specified values, then SPSS will divide the cases into two groups based on the codes you supply for Group 1 and Group 2. If the grouping variable has more than two categories, then you may wish to use Cut point. SPSS will divide the cases into two groups based on the code entered in the Cut point box—one group for all cases having codes equal to or greater than the Cut point code and one group having codes less than the Cut point code. (You will use Cut point in one of the exercises at the end of this chapter.) Because gender has two codes—1 for males and 2 for females—we will go with the Use specified values option in this example. Click in the Group 1 box and type "1," and then click in the Group 2 box and type "2." Click Continue, as shown in Figure 6-9. Notice that SPSS has replaced the question marks next to gender with the codes for males and females. All set. Click OK.

SPSS runs both mean comparisons and reports the results in the Viewer (Figure 6-10). The top table, labeled "Group Statistics," shows descriptive information about the means of gay_therm and rep_therm. The bottom table, "Independent Samples Test," tests for statistically significant differences between men and women on each dependent variable. There is a lot of information to digest here, so let's take it one step at a time.

We will evaluate the gender difference on the gay feeling thermometer first. From the Group Statistics table we can see that males, on average, rated gays at 44.50, whereas females had a higher mean, 52.16.

Figure 6–9 Defining the Grouping Variable

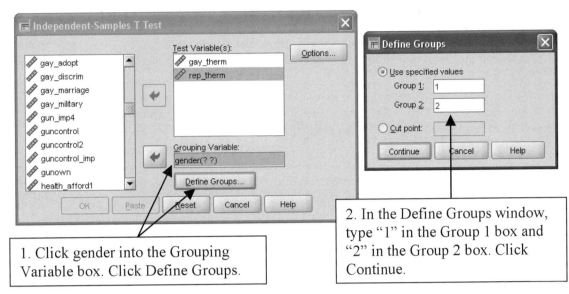

1. Click gender into the Grouping Variable box. Click Define Groups.

2. In the Define Groups window, type "1" in the Group 1 box and "2" in the Group 2 box. Click Continue.

Figure 6–10 Results of Independent-Samples T Test

Group Statistics

	gender R gender	N	Mean	Std. Deviation	Std. M...
gay_therm Feeling Thermometer: Gay Men and Lesbians	1 1 Male	496	44.50	27.543	
	2 2 Female	548	52.16	26.115	1.116
rep_therm Feeling Thermometer: Republican party	1 1 Male	552	54.69	26.476	1.127
	2 2 Female	624	51.94	27.321	1.094

1. SPSS subtracts the female mean from the male mean.

2. SPSS reports the mean difference here.

Independent Samples Test

		Levene's Test for Equality of Variances		t-test for Equality of Means					90% Confidence Interval of the Difference	
		F	Sig.	t	df	Sig. (2-tailed)	Mean Difference	Std. Error Difference	Lower	Upper
gay_therm Feeling Thermometer: Gay Men and Lesbians	Equal variances assumed	5.872	.016	-4.612	1042	.000	-7.660	1.661	-10.395	-4.926
	Equal variances not assumed			-4.599	1018.225	.000	-7.660	1.666	-10.402	-4.918
rep_therm Feeling Thermometer: Republican party	Equal variances assumed	.903	.342	1.749	1174	.081	2.752	1.573	.161	5.342
	Equal variances not assumed			1.752	1164.283	.080	2.752	1.570	.166	5.337

3. If this "Sig." Value is greater than or equal to .05, then use the "Equal variances assumed row." If it is less than .05, then use the "Equal variances not assumed" row.

It would appear that our alternative hypothesis has merit. The difference between these two sample means is 44.50 minus 52.16, or –7.660 (SPSS always calculates the difference by subtracting the Group 2 mean from the Group 1 mean. This value appears in the "Mean Difference" column of the Independent-Samples Test table.) The null hypothesis claims that this difference is the result of random sampling error and, therefore, that the true male-female difference in the population is 0. Using the information in the Independent-Samples Test table, we test the null hypothesis against the alternative hypothesis that the male mean is lower than the female mean.

Notice that there are two rows of numbers for each dependent variable. One row is labeled "Equal variances assumed" and the other "Equal variances not assumed." What is this all about? One of the statistical assumptions for comparing sample means is that the population variances of the dependent variable are the same for both groups—in this case, that the amount of variation in gay_therm among men in the population is equal to the amount of variation in gay_therm among women in the population. If this assumption holds up, then you would use the first row of numbers to test the hypothesis. If this assumption is incorrect, however, then you would use the second row of numbers. SPSS evaluates the assumption of equal variances using Levene's test, which tests the hypothesis that the two variances are equal.

Here is how to proceed. Look at the "Sig." value that appears under "Levene's Test for Equality of Variances." If this value is greater than or equal to .05, then use the "Equal variances assumed" row. If the "Sig." Value under "Levene's Test for Equality of Variances" is less than .05, then use the "Equal variances not assumed" row. Because this "Sig." value for gay_therm is less than .05, we would use the second row of numbers ("Equal variances not assumed") to evaluate the mean difference between men and women.

Let's use the 90 percent confidence interval of the mean difference to apply the .05 standard. There are, of course, two boundaries: a lower boundary, –10.40, and an upper boundary, –4.92. Here is a foolproof method for testing a hypothesis using the confidence interval approach. First, make sure that the direction of the mean difference is consistent with the alternative hypothesis. If the mean difference runs in the direction opposite from the one you hypothesized, then the game is already up—the alternative hypothesis is incorrect. Second, determine whether the confidence interval of the difference—the interval between the lower boundary and the upper boundary—includes 0. If the confidence interval includes 0, then the mean difference is not significant at the .05 level. It would occur by chance more than 5 times out of 100. If the confidence interval does not include 0, then the mean difference is statistically significant at the .05 level. The difference would occur by chance fewer than 5 times out of 100.

In the current example, we have seen that the difference between men and women on gay_therm runs in the hypothesized direction: Males have a lower mean rating than do females. So the alternative hypothesis is still alive. Does the confidence interval, the interval between –10.40 and –4.92 include 0? No, it doesn't. Conclusion: If the null hypothesis is correct, the probability of observing a mean difference of –7.660 is less than .05. Reject the null hypothesis.

Just as with the One-Sample T Test, we can arrive at an exact probability, or P-value, by dividing the "Sig. (2-tailed)" value by 2. The value in the "Sig. (2-tailed)" cell is .000. Dividing by 2, we still have .000. More precise conclusion: If the null hypothesis is correct, then random sampling error would produce a mean difference of –7.660 virtually 0 percent of the time. The alternative hypothesis is on safe inferential ground. Reject the null hypothesis.[4]

All right, so men score significantly lower on gay_therm than do women. Do men give the Republican Party higher ratings, as hypothesis 2 suggests? Again, the information on rep_therm in the Group Statistics table would appear consistent with the hypothesis. Men, with a mean of 54.69, score higher than do women, who averaged 51.94. SPSS reports the mean difference, 2.752, in the Independent-Samples Test table. Does this difference pass muster with the null hypothesis? Let's look at the Independent-Samples Test table and decide which row of numbers to use. In this case the "Sig." value under "Levene's Test for Equality of Variances" is greater than .05, so we can go ahead and use the "Equal variances assumed" row. Now pan to the right and consider the 90 percent confidence interval. Things are not looking good for the null hypothesis, which (as usual) claims that the population difference is 0. Does the 90 percent confidence interval, .161 to 5.342, include 0, the null's talisman? No, the confidence interval brackets values greater than 0. Thus we know that the mean difference would occur by chance fewer than 5 times out of 100. There is indeed a fairly small probability of observing a difference of 2.752, if in fact the null hypothesis is correct. Dividing .081 by 2 gives us a P-value of .04. If the null hypothesis is correct, then we would observe by chance a sample difference of 2.752 about 4 times out of 100. We've safely crossed the rejection frontier. Reject the null hypothesis.

EXERCISES

1. (Dataset: NES2004. Variable: egalitarianism.) The 2004 National Election Study asked people a series of questions designed to measure how egalitarian they are—that is, the extent to which they think opportunities and rewards should be distributed more equally in society. The NES2004 variable egalitarianism ranges from 0 (low egalitarianism) to 24 (high egalitarianism).[5] The 2004 NES, of course, is a random sample of U.S. adults. In this exercise you will analyze egalitarianism using One-Sample T Test. You then will draw inferences about the population mean. (Open One-Sample T Test and click Reset.)

A. Egalitarianism has a sample mean of (fill in the blank) _____.

B. There is a probability of .95 that the true population mean falls between an egalitarianism score of (fill in the blank) _____ at the low end and a score of (fill in the blank) _____ at the high end.

C. There is a probability of .90 that the true population mean falls between an egalitarianism score of (fill in the blank) _____ at the low end and a score of (fill in the blank) _____ at the high end.

D. A student researcher hypothesizes that social work majors will score significantly higher on the egalitarianism scale than the typical adult. The student researcher also hypothesizes that business majors will score significantly lower on the egalitarianism scale than the average adult. After administering the scale to a number of social work majors and a group of business majors, the researcher obtains these results: Social work majors' mean, 15.40; business majors' mean, 14.95.

Using the confidence interval approach to apply the .05 one-tailed test of significance, you can infer that (check one)

❏ social work majors probably are not more egalitarian than most adults.

❏ social work majors probably are more egalitarian than most adults.

Using the confidence interval approach to apply the .05 one-tailed test of significance, you can infer that (check one)

❏ business majors probably are not less egalitarian than most adults.

❏ business majors probably are less egalitarian than most adults.

E. Examine the t-ratio and P-value from your analysis of the business majors' mean. (fill in the blanks)

If in the population there is no difference between the mean of egalitarianism and 14.95, the observed difference of _____ would occur _____ of the time by chance.

2. (Dataset: NES2004. Variables: enviro_therm, age.) Are older people less sympathetic to the environmental movement than are younger people? Or do younger people and older people not differ significantly in their feelings toward environmentalists? In this exercise you will use Independent-Samples T Test to test this hypothesis: In a comparison of individuals, people who are 30 or older will give environmentalists lower ratings than will people who are younger than 30.

NES2004 contains enviro_therm, a feeling thermometer that gauges respondents' feelings toward environmentalists. This is the dependent variable and will go in the Test Variable(s) panel. The independent variable, age, goes in the Grouping Variable box. You want SPSS to create two groups from age: a group of respondents who are 30 or older and a group of respondents who are younger than 30. You can tell SPSS to do so by using Cut point in the Define Groups window. When you open the Define Groups window, click the radio button next to "Cut point" and type "30" in the Cut point box. Click Continue. Don't forget to click Options and ask for the 90 percent confidence interval.

A. The mean environmentalist thermometer rating for respondents who are 30 or older is (fill in the blank) _____, and the mean environmentalist thermometer rating for respondents who are under 30 years of age is (fill in the blank) _____.

B. According to the null hypothesis, in the population from which the sample was drawn, the difference between the mean for people 30 or older and the mean for people younger than 30 is equal to (fill in the blank) _____. According to your analysis of NES2004 the mean difference is equal to (fill in the blank) _____.

C. Using the confidence interval approach to apply the .05 one-tailed test of significance, you can infer that (check one)

❏ the older age group and the younger age group do not differ significantly in their ratings of environmentalists.

❏ the older age group gives environmentalists significantly lower ratings than does the younger age group.

❏ the older age group gives environmentalists significantly higher ratings than does the younger age group.

D. The P-value for the mean difference is equal to (fill in the blank) _____. Your inferential decision, therefore, is (check one)

❏ do not reject the null hypothesis.

❏ reject the null hypothesis.

3. (Dataset: GSS2006. Variables: childs, attend4.) The role of religion lies at the center of an interesting debate about the future of U.S. partisan politics. Republican presidential candidates do much better among people who frequently attend religious services than among people who are less observant. However, religious attendance has been waning. This growing secularization, according to some observers, portends a weakening of the Republican base and a growing opportunity for the Democratic Party.[6] But we also know that religious beliefs and affiliations (or the lack thereof) are strongly shaped by childhood socialization. Are less-religious people raising and socializing children at the same rate as the more religious? In this exercise you will test this hypothesis: In a comparison of individuals, those with lower levels of religiosity will have fewer children than will those with higher levels of religiosity. This hypothesis says that as religious attendance goes up, so will the average number of children.

Dataset GSS2006 contains childs, the respondent's number of children. This is the dependent variable. The independent variable is attend4, which measures religious attendance by four categories: "Never-<1/yr" (coded 1), "1/yr-sev times/yr" (coded 2), "1/mnth-nrly ev wk" (coded 3), and "Ev wk->1/wk" (coded 4).

A. Exercise a familiar skill you acquired in Chapter 4. Perform a mean comparison analysis, obtaining mean values of childs (and numbers of cases) for each value of attend4. Fill in the following table:

Religious attendance: 4 categories	Summary of Number of Children	
	Mean	N
Never-<1/yr	?	?
1/yr-sev times/yr	?	?
1/mnth-nrly ev wk	?	?
Ev wk->1/wk	?	?
Total	?	?

B. Do these findings support the childs-attend4 hypothesis? (circle one)

Yes   No

Briefly explain your reasoning. _____

_____

_____

_____

C. Focus your analysis on a comparison between respondents labeled "Never-<1/yr" (and coded 1) on attend4 and respondents labeled "Ev wk->1/wk" (and coded 4). Run Independent-Samples T Test to find out if high attenders have significantly more children than do low attenders. Apply the .05 level of significance. (fill in the blanks)

"Never-<1/yr" mean: _____

"Ev wk>1/wk" mean: _____

Mean difference: _____

90 percent confidence interval of the mean difference: Between _____ and _____.
P-value: _____.

E. Does the statistical evidence support the hypothesis that people who are more religious have significantly more children than do people who are less religious? (check one)

❏ Yes, the statistical evidence supports the hypothesis.

❏ No, the statistical evidence does not support the hypothesis.

4. (Dataset: GSS2006. Variables: relig, sibs.) Here is some conventional wisdom: Catholics have bigger families than do Protestants. Of course, this conventional wisdom is accurate. Or is it? GSS2006 contains the variable sibs, the number of siblings each respondent has. Another variable, relig, codes each respondent's religious affiliation. Protestants are coded 1, and Catholics are coded 2. In this exercise you will evaluate the conventional wisdom that the mean number of siblings among Protestants is lower than the mean number of siblings among Catholics. Run Independent-Samples T Test, using sibs as the test variable and relig as the grouping variable. Be sure to click Options and ask for the 90 percent confidence interval.

A. According to the Group Statistics table, (check one)

❏ Catholics, on average, have fewer siblings than do Protestants.

❏ Catholics, on average, have more siblings than do Protestants.

B. According to the null hypothesis, in the population from which the sample was drawn, the difference between the Protestant mean and the Catholic mean is equal to (fill in the blank) _____.

According to your analysis of GSS2006 the mean difference is equal to (fill in the blank) _____.

C. Inspect the 90 percent confidence interval. If the null hypothesis is correct, then you would obtain the observed sample difference between Protestants and Catholics (check one)

❏ less frequently than 5 times out of 100 by chance.

❏ more frequently than 5 times out of 100 by chance.

D. Refer to the "Sig. (2-tailed)" column and figure out the P-value. If the null hypothesis is correct, then you would obtain the observed sample difference between Protestants and Catholics (fill in the blank) _____ of the time by chance.

E. Suppose that someone were to read your results and conclude, "It looks like the conventional wisdom is correct." Does your statistical evidence support this inference? (check one)

❏ Yes, the statistical evidence supports this inference.

❏ No, the statistical evidence does not support this inference.

That concludes the exercises for this chapter. Before exiting SPSS, be sure to save your output file.

NOTES

1. Respondents were asked whether federal spending should be increased, decreased, or kept about the same on each of these programs or policies: building and repairing highways (2004 NES variable V043164), Social Security (V043165), public schools (V043166), science and technology (V043167), dealing with crime (V043168), welfare programs (V043169), child care (V043170), foreign aid (V043171), aid to poor people (V043172), tightening border security to prevent illegal immigration (V043173), and war on terrorism (V043174). Spend11 records the number of "increased" responses.
2. To simplify the presentation of the material in this part of the chapter, mean values will be rounded to two decimal places.
3. SPSS performs this calculation at 32-decimal precision. So if you were to check SPSS's math, using the mean difference and standard error that appear in the SPSS Viewer, you would arrive at a slightly different t-ratio than the value of t reported in the One-Sample Test table.
4. A P-value cannot exactly equal 0. There is always some chance, however remote, that random error produced the difference observed in the sample. Although the editing of SPSS tabular output is not covered in this book, it is possible to double-click on a table (SPSS refers to tabular objects as "pivot tables") to make appearance-enhancing changes or to look at the mind-numbingly precise numbers that SPSS rounds to produce the digits on display in the table. Double-clicking on the Independent-Samples T Test output shown in Figure 6-10, and then double-clicking on the "Sig. (2-tailed)" value just discussed (the value .000), reveals this number: 4.771351531527812E-6. That's scientific notation for .000004771351531527812, or about 4.8 chances in a million. Of course, this is a two-tailed probability. To arrive at a one-tailed P-value, we would divide this vanishingly small number by 2. Thus, if the null hypothesis is correct, random sampling error would produce the observed mean difference about 2.4 times in a million.
5. Egalitarianism was constructed from the following NES2004 variables: V045212 (Should do what is necessary for equal opportunity), V045213 (Have gone too far pushing equal rights), V045214 (Big problem is not giving everyone equal chance), V045215 (Better off if worried less about equality), V045216 (Not that big a problem if people have unequal chance), and V045217 (Many fewer problems if people treated equally). All variables were originally coded 1–5, from "agree strongly" to "disagree strongly." All were recoded in the more-egalitarian direction and summed. The resulting metric was then rescaled to range from 0 (low egalitarianism) to 24 (high egalitarianism).
6. John B. Judis and Ruy Teixeira, *The Emerging Democratic Majority* (New York: Scribner, 2002).

7

Chi-square and Measures of Association

Procedure Covered   Analyze → Descriptive → Crosstabs → Statistics

In the preceding chapter you learned how to test for mean differences on an interval-level dependent variable. But what if you are not dealing with interval-level variables? What if you are doing cross-tabulation analysis and are trying to figure out whether an observed relationship between two nominal or ordinal variables mirrors the true relationship in the population? Just as with mean differences, the answer depends on the boundaries of random sampling error, the extent to which your observed results "happened by chance" when you took the sample. The Crosstabs procedure can provide the information needed to test the statistical significance of nominal or ordinal relationships, and it will yield appropriate measures of association.

You are familiar with the Crosstabs procedure. For analyzing datasets that contain a preponderance of categorical variables—variables measured by nominal or ordinal categories—cross-tabulation is by far the most common mode of analysis in political research. In this section we will revisit Crosstabs and use the Statistics subroutine to obtain the oldest and most widely applied test of statistical significance in cross-tabulation analysis, the chi-square test. With rare exceptions, chi-square can always be used to determine whether an observed cross-tab relationship departs significantly from the expectations of the null hypothesis. In the first guided example, you will be introduced to the logic behind chi-square, and you will learn how to interpret SPSS's chi-square output.

In this chapter you will also learn how to obtain measures of association for the relationships you are analyzing. SPSS is programmed to produce, at the user's discretion, a large array of such measures. In doing your own analysis, you will have to tell SPSS which measure or measures you are after. If one or both variables in the cross-tabulation are nominal level, then you need to request lambda. If both are ordinal-level variables, then SPSS offers several choices: gamma, Kendall's tau-b, Kendall's tau-c, and Somers' d. Gamma tends to overstate the strength of a relationship, and so it is not recommended. The Kendall's statistics are *symmetrical* measures. Each will yield the same value, regardless of whether the independent variable is used to predict the dependent variable or the dependent variable is used to predict the independent variable. Somers' d is an *asymmetrical* measure. It will report different measures of the strength of a relationship, depending on whether the independent variable is used to predict the dependent variable or the dependent variable is used to predict the independent variable. Lambda, too, is an asymmetrical measure. Asymmetrical measures of association generally are preferred over symmetrical measures.[1] Therefore, in this book we will cover lambda for nominal relationships and Somers' d for ordinal relationships.[2]

Somers' d is a directional measure that ranges from −1 to +1. A plus (+) sign says that increasing values of the independent variable are associated with increasing values of the dependent variable. A minus (−) sign says that increasing values of the independent variable are related to decreasing values of the dependent variable. Both lambda and Somers' d are proportional reduction in error (PRE) measures of strength. A PRE measure tells you the extent to which the values of the independent variable predict the values of the dependent variable. A value close to 0 says that the independent variable provides little predictive leverage; the relationship is weak. Values close to the poles—to –1 for negative associations or to +1 for positive relationships—tell you that the independent variable provides a lot of help in predicting the dependent variable; the relationship is strong.

Lambda's PRE status stands it in good stead with political researchers because PRE measures are generally preferred over measures that do not permit a PRE interpretation. Even so, lambda tends to underestimate the strength of a relationship, especially when one of the variables has low variation. Therefore, when you are analyzing a relationship in which one or both of the variables are nominal, it is a good practice to request Cramer's V as well as lambda. Cramer's V, one of a variety of chi-square-based measures, does not measure strength by the PRE criterion. However, it is bounded by 0 (no relationship) and 1 (a perfect relationship).

ANALYZING AN ORDINAL-LEVEL RELATIONSHIP

We will begin by using NES2004 to analyze an ordinal-level relationship. (If you are running Student Version, you will need to open NES2004B_Student. For the remainder of the book, when a guided example or exercise calls for NES2004, you will use NES2004B_Student.) Consider this hypothesis: In a comparison of individuals, people with higher incomes will be more likely to oppose federal welfare spending than will people with lower incomes. NES2004 contains welfare3, an ordinal variable that measures opinions about federal welfare spending. Respondents who think spending should be "increased" are coded 1, those who say that spending should be "kept about the same" are coded 2, and those who think that spending should be "decreased" are coded 3. Welfare3 is the dependent variable. For the independent variable, we will use income_r3, a three-category measure of income that you created in Chapter 3. On income_r3, respondents are classified as "Low" (coded 1), "Middle" (coded 2), or "High" (coded 3).

Let's first test this hypothesis the old-fashioned way—by running the Crosstabs analysis and comparing column percentages. Click Analyze → Descriptive Statistics → Crosstabs. Remember to put the dependent variable, welfare3, on the rows and the independent variable, income_r3, on the columns. Request column percentages. Run the analysis and consider the output.

welfare3 Increase fed welfare$? * income_r3 Respondent income: 3 categories Crosstabulation

			income_r3 Respondent income: 3 categories			
			1 Low	2 Middle	3 High	Total
welfare3 Increase fed welfare$?	1 1 Incr	Count	53	44	40	137
		% within income_r3 Respondent income: 3 categories	14.2%	12.5%	10.9%	12.6%
	2 2 Same	Count	163	138	158	459
		% within income_r3 Respondent income: 3 categories	43.7%	39.3%	43.2%	42.1%
	3 3 Decr	Count	157	169	168	494
		% within income_r3 Respondent income: 3 categories	42.1%	48.1%	45.9%	45.3%
Total		Count	373	351	366	1090
		% within income_r3 Respondent income: 3 categories	100.0%	100.0%	100.0%	100.0%

How would you evaluate the welfare3-income_r3 hypothesis in light of this analysis? Focus on the column percentages in the "Incr" row. According to the hypothesis, as we move along this row, from lower income to higher income, the percentages of respondents favoring an increase in welfare spending should decline. Is this what happens? Sort of. The percentages decrease from 14.2 percent among low-income respondents to 10.9 percent among high-income respondents. So the difference between low-income and high-income individuals is something on the order of 3 percentage points—not a terribly robust relationship between the independent and dependent variables. Using your own interpretive skills, you would probably say that the findings are too weak to support the welfare3-income_r3 hypothesis.

Now let's reconsider the welfare3-income_r3 cross-tabulation from the standpoint of inferential statistics, the way the chi-square test of statistical significance would approach it. Chi-square begins by looking at the "Total" column, which contains the distribution of the entire sample across the values of the dependent variable, welfare3. Thus 12.6 percent of the sample favor an increase in spending, 42.1 percent think spending should be kept the same, and 45.3 percent favor a decrease in welfare spending. Chi-square then frames the null hypothesis, which claims that, in the population, welfare3 and income_r3 are not related to each other, that individuals' incomes are unrelated to their opinions about welfare. If the null hypothesis is correct, then a random sample of low-income people would produce the same distribution of welfare opinions as the total distribution. By the same token, a random sample of middle-income people would yield a distribution that looks just like the total distribution, as would a random sample of high-income individuals. If the null hypothesis is correct, then the distribution of cases down each column of the table will be the same as in the "Total" column. Of course, the null hypothesis asserts that any observed departures from this monotonous pattern resulted from random sampling error.

Now reexamine the table and make a considered judgment. Would you say that the observed distribution of cases within each category of income_r3 conforms to the expectations of the null hypothesis? Well, for the low-income category, the distribution is pretty close to the total distribution, with only slight departures—for example, a somewhat higher percentage in the "Incr" category than would be expected according to the null hypothesis and a slightly lower percentage in the "Decr" category. In fact, for each value of income_r3, there is fairly close conformity to what we would expect to find if the null hypothesis were true. The small departures from these expectations, furthermore, might easily be explained by random sampling error.

Let's rerun the analysis and find out if our considered judgment is borne out by the chi-square test. We will also obtain a measure of association for the relationship. Return to the Crosstabs window. Click Statistics. The Crosstabs: Statistics window pops up (Figure 7-1). There are many choices here, but we know what we want: We would like SPSS to perform a chi-square test on the table. Check the Chi-square box. We also know which measure of association to request. Because both welfare3 and income_r3 are ordinal-level variables, we will request Somers' d. Check the box next to "Somers' d." Click Continue, and then click OK.

SPSS runs the cross-tabulation analysis again, and this time it has produced two additional tables of statistics: Chi-Square Tests and Directional Measures (Figure 7-2). Given the parsimony of our requests, SPSS has been rather generous in its statistical output. In the Chi-Square Tests table, focus exclusively on the row labeled "Pearson Chi-Square." The first column, labeled "Value," provides the chi-square test statistic. A test statistic is a number that SPSS calculates from the observed data. Generally speaking, the larger the magnitude of a test statistic, the less likely that the observed data can be explained by random sampling error. The smaller the test statistic, the more likely that the null's favorite process—random chance—accounts for the observed data. So, if the observed data perfectly fit the expectations of the null hypothesis, then the chi-square test statistic would be 0. As the observed data depart from the null's expectations, this value grows in size, allowing the researcher to begin entertaining the idea of rejecting the null hypothesis.

For the welfare3-income_r3 cross-tabulation, SPSS calculated a chi-square test statistic equal to 4.059. Is this number, 4.059, statistically different from 0, the value we would expect to obtain if the null hypothesis were true? Put another way: If the null hypothesis is correct, how often will we obtain a test statistic of 4.059 by chance? The answer is contained in the rightmost column of the Chi-Square Tests table, under the odd label "Asymp. Sig. (2-sided)." For the chi-square test of significance, this value *is* the P-value.[3] In our example SPSS reports a P-value of .398. If the null hypothesis is correct in its assertion that no relationship exists between the independent and dependent variables, then we will obtain a test statistic of 4.059, by chance, about 40 percent of the time. Because .398 (greatly) exceeds the .05 standard, the null hypothesis is on safe ground. From our initial comparison of percentages, we suspected that the relationship might not trump the null hypothesis. The chi-square test has confirmed that suspicion. Do not reject the null hypothesis.

Turn your attention to the Directional Measures table, which reports the requested measure of Somers' d. In fact, three Somers' d statistics are displayed. Because SPSS doesn't know (or care) how we framed the hypothesis, it provided values of Somers' d for every scenario: symmetric (no hypothetical expectations

Figure 7–1 Requesting Statistics (ordinal-level relationship)

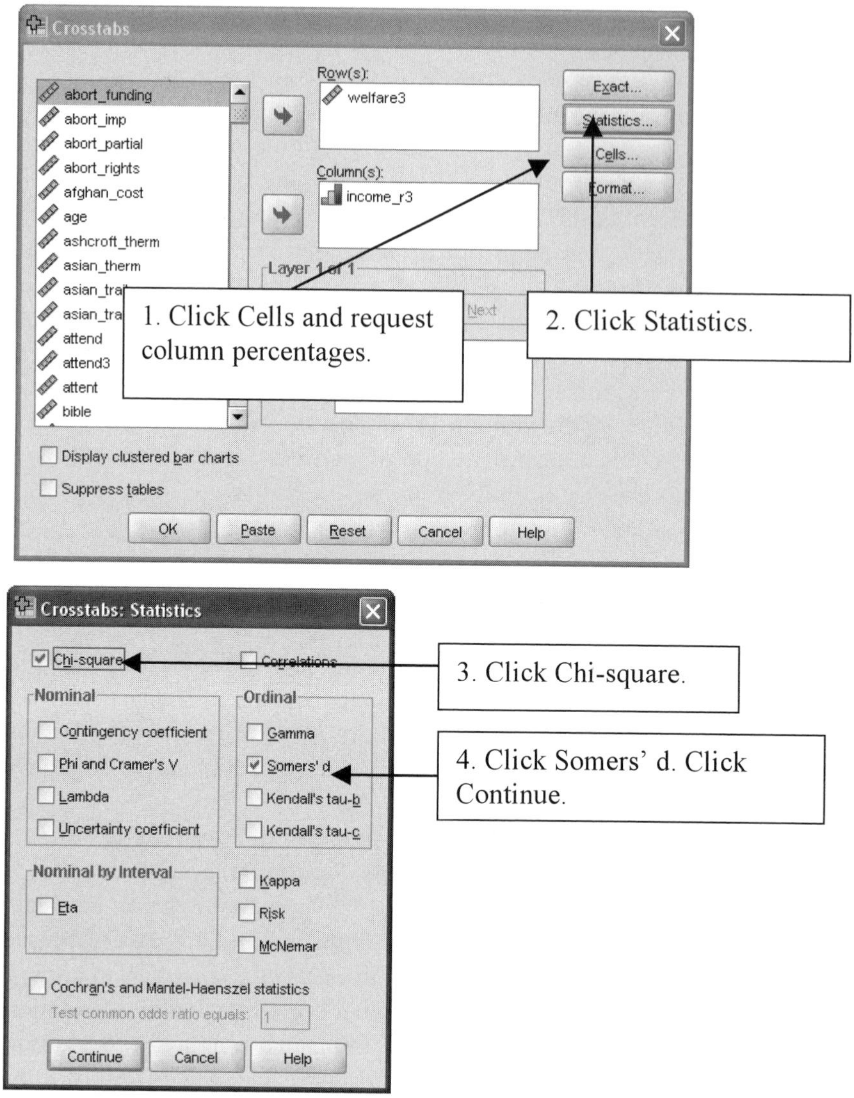

about the relationship), welfare3 dependent (welfare3 is the dependent variable and income_r3 is the independent variable), and income_r3 dependent (income_r3 is the dependent variable and welfare3 is the independent variable). Always use the dependent variable in your hypothesis to choose the correct value of Somers' d. Because welfare3 is our dependent variable, we would report the Somers's d value, .035.

What does this value, .035, tell us about the relationship? Because the statistic is positive, it tells us that increasing codes on income_r3 are associated with increasing codes on welfare3. We can discern a faint pattern: Low-income respondents (coded 1 on income_r3) are slightly more likely to support increased spending (coded 1 on welfare3), and high-income respondents (coded 3 on income_r3) are slightly more likely to favor decreased spending (coded 3 on welfare3). Because the statistic is puny—it registers .035 on a scale that runs from 0.000 to 1.000—it tells us that the relationship is weak. How weak? Because Somers' d is a PRE measure of association, we can say this: Compared to how well we can predict welfare opinions without knowing respondents' incomes, we can improve our prediction by 3.5 percent by knowing respondents' incomes.

Not much going on there. Obviously, we need to frame another hypothesis, using different variables, and see if our luck changes. But before moving on, let's quickly review the interpretation of the chi-square statistic and Somers's d. This is also a good place to introduce templates that will help you describe your findings.

Summary

SPSS reports a chi-square test statistic, labeled "Pearson Chi-Square." This test statistic is calculated from the observed tabular data. Values close to 0 are within the domain of the null hypothesis. As chi-square increases

Figure 7–2 The Chi-square Test and Directional Measures (ordinal-level relationship)

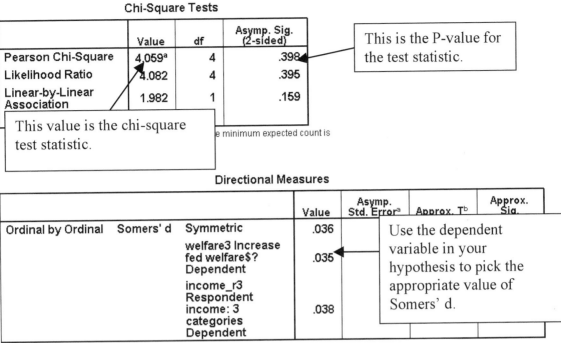

Chi-Square Tests

	Value	df	Asymp. Sig. (2-sided)
Pearson Chi-Square	4.059[a]	4	.398
Likelihood Ratio	4.082	4	.395
Linear-by-Linear Association	1.982	1	.159

This is the P-value for the test statistic.

This value is the chi-square test statistic.

e minimum expected count is

Directional Measures

			Value	Asymp. Std. Error[a]	Approx. T[b]	Approx. Sig.
Ordinal by Ordinal	Somers' d	Symmetric	.036			
		welfare3 Increase fed welfare$? Dependent	.035			
		income_r3 Respondent income: 3 categories Dependent	.038			

Use the dependent variable in your hypothesis to pick the appropriate value of Somers' d.

a. Not assuming the null hypothesis.

b. Using the asymptotic standard error assuming the null hypothesis.

in magnitude (the chi-square statistic cannot assume negative values), the null's explanation for the observed data—"it all happened by chance"—becomes increasingly implausible.

The chi-square statistic is accompanied by a P-value, which appears beneath the label "Asymp. Sig. (2-sided)." Here is a template for writing an interpretation of the P-value: "If the null hypothesis is correct that, in the population from which the sample was drawn, there is no relationship between [independent variable] and [dependent variable], then random sampling error will produce the observed data [P-value] of the time." For our example: "If the null hypothesis is correct that, in the population from which the sample was drawn, there is no relationship between income and opinions about welfare spending, then random sampling error will produce the observed data .398 of the time." (If you prefer percentages, you can make this substitution: ". . .will produce the observed data 39.8 percent of the time.") Use the .05 benchmark. If the P-value is less than or equal to .05, then reject the null hypothesis. If the P-value is greater than .05, do not reject the null hypothesis.

For ordinal-by-ordinal relationships, ask SPSS to report Somers' d. Somers' d is a directional measure, ranging from −1 to +1. Somers' d has a PRE interpretation. Here is a template for writing an interpretation of Somers' d or, for that matter, any PRE measure: "Compared to how well we can predict [dependent variable] by not knowing [independent variable], we can improve our prediction by [value of PRE measure] by knowing [independent variable]." Our example: "Compared to how well we can predict opinions on welfare spending by not knowing respondents' incomes, we can improve our prediction by .035 by knowing respondents' incomes." (Actually, percentages may sound better here: ". . .we can improve our prediction by 3.5 percent by knowing respondents' incomes.") Note that a negative sign on a PRE measure imparts the direction of the relationship, but it does not affect the PRE interpretation. Thus, if Somers' d were −.035, we would still conclude that the independent variable increases our predictive power by .035 or 3.5 percent.

ANALYZING AN ORDINAL-LEVEL RELATIONSHIP WITH A CONTROL VARIABLE

The welfare3-income_r3 hypothesis didn't fare too well against the null hypothesis. Here is a hypothesis that sounds more promising: In a comparison of individuals, younger people will be more likely than older people to think that gay marriage should be allowed. For the dependent variable we will use gay_marriage, which measures support for same-sex marriage by two ordinal response categories, "Not allow" (coded 0) and "Allow" (coded 1). For the independent variable we will use yob3, which you created in Chapter 3. Recall that yob3 measures year of birth in three ordinal categories: before 1950 (coded 1), 1950–1965 (coded 2), and after 1965 (coded 3). When testing hypotheses about political opinions in which age is an independent variable, it is often a good idea to control for education. Older people tend to have lower levels of educational attainment than do younger people. And because education might itself be related to opinions about gay marriage—people with more education may be more willing to allow the practice—we don't want our analysis to confuse the effects of age with the effects of education. Thus we will include a control variable, educ2, coded 1 for respondents with a high school education or less and coded 2 for respondents having more than a high school education.

By this point in the book, cross-tabulation analysis has become routine. Return to the Crosstabs window. Click welfare3 back into the variable list and click gay_marriage into the Row(s) panel. Click income_r3 back into the variable list and click yob3 into the Column(s) panel. We're running a controlled comparison this time, so click the control variable, educ2, into the Layer box. The other necessary choices—column percentages in Cells, Chi-square and Somers's d in Statistics—should still be in place from the previous analysis. Click OK.

gay_marriage Allow gay marriage? * yob3 Three generations * educ2 Education: 2 cats Crosstabulation

educ2 Education: 2 cats					1 Before 1950	2 1950-1965	3 After 1965	Total
					yob3 Three generations			
1 <=HS	gay_marriage Allow gay marriage?	0 Not allow	Count		153	86	77	316
			% within yob3 Three generations		84.1%	75.4%	60.6%	74.7%
		1 Allow	Count		29	28	50	107
			% within yob3 Three generations		15.9%	24.6%	39.4%	25.3%
	Total		Count		182	114	127	423
			% within yob3 Three generations		100.0%	100.0%	100.0%	100.0%
2 >HS	gay_marriage Allow gay marriage?	0 Not allow	Count		133	135	113	381
			% within yob3 Three generations		70.0%	60.3%	44.1%	56.9%
		1 Allow	Count		57	89	143	289
			% within yob3 Three generations		30.0%	39.7%	55.9%	43.1%
	Total		Count		190	224	256	670
			% within yob3 Three generations		100.0%	100.0%	100.0%	100.0%

Where people stand on gay marriage would appear to have something to do with when they were born. Consider the gay_marriage-yob3 relationship among the less educated. There are large increases in the percentages saying "Allow" as we move across the columns: from 15.9 percent among the oldest generation to 24.6 among the baby boomers to 39.4 for the youngest generation. Thus, from oldest to youngest, there is about a 25-point increase in the percentage of respondents who would allow gay marriage. In fact, we see a similar pattern among the more highly educated group. The percentage who approve of gay marriage rises from 30.0 percent among the oldest cohort to 55.9 percent among the youngest—again, about a 25-point increase. Note, too, the sizable and consistent effect of education, controlling for age. For the oldest group, only 15.9 percent of the less educated fall into the "Allow" category, compared with 30.0 percent of their better-educated cohorts, which is about a 14-point difference. Similar education effects occur for the baby boomers (a bit more than a 15-point difference between the less and more educated) and for the post-1965 generation (about a 16.5-point difference).[4] It would appear that this controlled cross-tabulation— gay_marriage-yob3, controlling for educ2—depicts a set of additive relationships. The effect of the independent variable on the dependent variable has the same tendency and about the same strength for both values of the control variable.

Reading tables and discussing patterns are familiar tasks. But do the statistics support our interpretation? First let's check for statistical significance by examining the Chi-Square Tests.

Chi-Square Tests

educ2 Education: 2 cats		Value	df	Asymp. Sig. (2-sided)
1 <=HS	Pearson Chi-Square	21.786[a]	2	.000
	Likelihood Ratio	21.444	2	.000
	Linear-by-Linear Association	21.318	1	.000
	N of Valid Cases	423		
2 >HS	Pearson Chi-Square	31.320[b]	2	.000
	Likelihood Ratio	31.632	2	.000
	Linear-by-Linear Association	30.657	1	.000
	N of Valid Cases	670		

a. 0 cells (.0%) have expected count less than 5. The minimum expected count is 28.84.

Judging from the chi-square tests at both values of the control—for the less educated (chi-square = 21.786, P-value = .000) and for the more educated (chi-square = 31.320, P-value = .000)—it is extremely unlikely that the observed patterns were produced by random sampling error. So the chi-square statistics invite us to reject the null hypothesis. How strong is the relationship between gay_marriage and yob3 for each education group? Refer to the Directional Measures table.

Directional Measures

educ2 Education: 2 cats				Value	Asymp. Std. Error[a]	Approx. T[b]	Approx. Sig.
1 <=HS	Ordinal by Ordinal	Somers' d	Symmetric	.203	.043	4.574	.000
			gay_marriage Allow gay marriage? Dependent	.160	.035	4.574	.000
			yob3 Three generations Dependent	.277	.059	4.574	.000
2 >HS	Ordinal by Ordinal	Somers' d	Symmetric	.201	.035	5.756	.000
			gay_marriage Allow gay marriage? Dependent	.175	.030	5.756	.000
			yob3 Three generations Dependent	.236	.041	5.756	.000

a. Not assuming the null hypothesis.

b. Using the asymptotic standard error assuming the null hypothesis.

For less-educated respondents we would say that, compared to how well we can predict their gay marriage opinions without knowing when they were born, we can improve our prediction by 16 percent by knowing the independent variable. Similarly, for the more-educated group, knowledge of the independent variable increases our predictive accuracy by about 17 percent. Thus, Somers' d supports our interpretation of a set of additive relationships. For both values of the control variable, the relationship is positive—as yob3 shades from lower codes (older cohorts) to higher codes (younger cohorts), respondents tend to shift from the lower code of gay_marriage (Not allow) to the higher code of gay_marriage (Allow)—and the relationships are of similar strength at both levels of education.

ANALYZING A NOMINAL-LEVEL RELATIONSHIP WITH A CONTROL VARIABLE

All of the variables analyzed thus far have been ordinal level. Many social and political characteristics, however, are measured by nominal categories—gender, race, region, or religious denomination, to name a few. In this example we will use gender to help frame the following hypothesis: In a comparison of individuals, men

Figure 7–3 Requesting Chi-square, Lambda, and Cramer's V (nominal-level relationship)

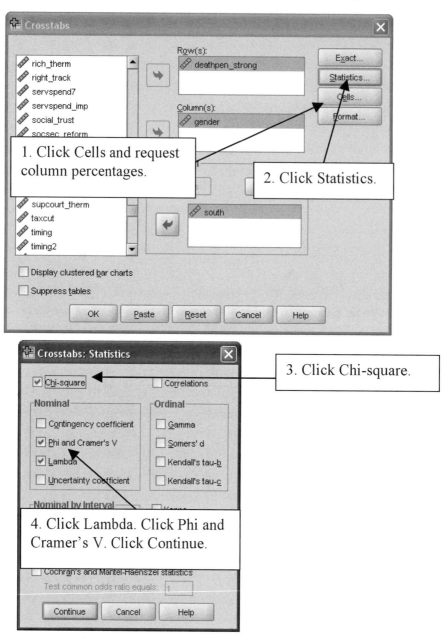

are more likely than women to favor the death penalty. To make things more interesting, we will control for another variable that might affect attitudes toward capital punishment, whether the respondent resides in the South. Would the gender gap on the death penalty be the same for southerners and non-southerners? Or might the gender gap be weaker in the South than in the non-South? Let's investigate.

NES2004 contains deathpen_strong, coded 1 for respondents who "strongly favor" the death penalty and coded 0 for respondents who are not strong supporters of the death penalty. Deathpen_strong is the dependent variable. The independent variable, of course, is gender. For the control variable we will use south, coded 0 for non-southerners and 1 for southerners. Click Analyze → Descriptive Statistics → Crosstabs. Click Reset to clear the panels. Now set up our new analysis: deathpen_strong on the rows, gender on the columns, and south in the Layer panel (Figure 7-3). Click Cells and request column percentages. Click Statistics. Again we want chi-square, so make sure that the Chi-square box is checked. Because gender is a nominal-level variable, we can't use any of the ordinal statistics. But we can use lambda and Cramer's V. In the Nominal panel, check the box next to "Phi and Cramer's V" and check the box next to "Lambda." Click Continue, and then click OK.

deathpen_strong Strongly favor death penalty? * gender R gender * south Non-South/South Crosstabulation

south Non-South/South					gender R gender		Total
					1 1 Male	2 2 Female	
0 Non-South	deathpen_strong Strongly favor death penalty?	0 No	Count		166	248	414
			% within gender R gender		46.0%	61.2%	54.0%
		1 Yes	Count		195	157	352
			% within gender R gender		54.0%	38.8%	46.0%
	Total		Count		361	405	766
			% within gender R gender		100.0%	100.0%	100.0%
1 South	deathpen_strong Strongly favor death penalty?	0 No	Count		81	104	185
			% within gender R gender		42.6%	50.2%	46.6%
		1 Yes	Count		109	103	212
			% within gender R gender		57.4%	49.8%	53.4%
	Total		Count		190	207	397
			% within gender R gender		100.0%	100.0%	100.0%

Before examining the statistics, consider the substantive relationships depicted in the cross-tabulations. Among non-southerners, 54.0 percent of the males strongly favor capital punishment, compared with 38.8 percent of the females—more than a 15-point gap. What happens when we switch to respondents who reside in the South? Among southerners, the percentage of males who strongly support the death penalty, 57.4 percent, is only slightly higher than that of non-southern males. But among women there is a large increase, from 38.8 percent for non-southern women to 49.8 percent for women in the south. Thus, the gender gap is narrower inside the South than outside the South. Clearly, interaction is going on here: The deathpen_strong-gender relationship is much stronger for one value of the control (when south = 0) than for the other value of the control (when south = 1). Now let's see what the statistics have to say.

Chi-Square Tests

south Non-South/South		Value	df	Asymp. Sig. (2-sided)	Exact Sig. (2-sided)	Exact Sig. (1-sided)
0 Non-South	Pearson Chi-Square	17.875[a]	1	.000		
	Continuity Correction[b]	17.267	1	.000		
	Likelihood Ratio	17.931	1	.000		
	Fisher's Exact Test				.000	.000
	Linear-by-Linear Association	17.852	1	.000		
	N of Valid Cases[b]	766				
1 South	Pearson Chi-Square	2.306[c]	1	.129		
	Continuity Correction[b]	2.010	1	.156		
	Likelihood Ratio	2.309	1	.129		
	Fisher's Exact Test				.132	.078
	Linear-by-Linear Association	2.300	1	.129		
	N of Valid Cases[b]	397				

According to the Chi-Square Tests table, the deathpen_strong-gender relationship defeats the null hypothesis in the non-South cross-tab (chi-square = 17.875, P-value = .000) but not in the South cross-tab (chi-square = 2.306, P-value = .129). Thus, the probability that the observed gender gap among non-southern respondents (about a 15-point difference) was produced by random sampling error is very low. The roughly 8-point gap among southerners, however, could occur 12.9 percent of the time, if in fact the independent and dependent variables are not related to each other. So the null hypothesis loses one (when south = 0) and wins one (when south = 1). This mixed record again points toward interaction. Recall the earlier gay_marriage-yob3-educ2 analysis, which revealed a set of additive relationships. In that analysis, we rejected the null hypothesis for both values of the control. But in the present analysis, the statistics bespeak an interaction pattern: significant for one value of the control but not significant for the other value of the control. Let's see what the other tables can add to our interpretation.

Scroll down to the next two tables, Directional Measures and Symmetric Measures.

Directional Measures

south Non-South/South				Value	Asymp. Std. Error[a]	Approx. T[b]	Approx. Sig.
0 Non-South	Nominal by Nominal	Lambda	Symmetric	.094	.045	2.023	.043
			deathpen_strong Strongly favor death penalty? Dependent	.082	.052	1.529	.126
			gender R gender Dependent	.105	.049	2.031	.042
		Goodman and Kruskal tau	deathpen_strong Strongly favor death penalty? Dependent	.023	.011		.000[c]
			gender R gender Dependent	.023	.011		.000[c]
1 South	Nominal by Nominal	Lambda	Symmetric	.019	.066	.280	.779
			deathpen_strong Strongly favor death penalty? Dependent	.005	.078	.070	.945
			gender R gender Dependent	.032	.075	.412	.680
		Goodman and Kruskal tau	deathpen_strong Strongly favor death penalty? Dependent	.006	.008		.129[c]
			gender R gender Dependent	.006	.008		.129[c]

a. Not assuming the null hypothesis.
b. Using the asymptotic standard error assuming the null hypothesis.
c. Based on chi-square approximation

Symmetric Measures

south Non-South/South			Value	Approx. Sig.
0 Non-South	Nominal by Nominal	Phi	-.153	.000
		Cramer's V	.153	.000
	N of Valid Cases		766	
1 South	Nominal by Nominal	Phi	-.076	.129
		Cramer's V	.076	.129
	N of Valid Cases		397	

A comparison of the Cramer's V values, .153 versus .076, supports the interaction interpretation. Although both relationships "run in the same direction," the deathpen_strong-gender relationship is stronger among non-southern respondents than among southern respondents. The lambda statistics, as well, suggest that our interpretation is correct. Notice that, just as it did with Somers's d, SPSS has reported three lambda statistics: "Symmetric," "deathpen_strong Dependent", and "gender Dependent." Always use the lambda statistic that names the dependent variable in your hypothesis. Because our dependent variable is labeled "deathpen_strong," we would use that statistic. For the non-south relationship, lambda is equal to .082. Compared with how accurately we can predict the death penalty opinions of non-southerners by not knowing gender, using gender as a predictive tool improves our prediction by about 8 percent—not earth-shattering, to be sure. But consider the magnitude of lambda for southerners, .005, which suggests virtually no gain in predictive leverage.

A Problem with Lambda

One final example will emphasize an important limitation of lambda. Here is one of the most heavily researched gender gap hypotheses: In a comparison of individuals, women are more likely to identify with the Democratic Party than are men. NES2004 has democrat, coded 1 for Democrats and coded 0 for Independents and Republicans. Return to the Crosstabs window for one more analysis, but don't reset the panels. Click deathpen_strong back into the variable list and click democrat into Row(s). Click south back into the variable list, leaving the Layer box empty. Because we just ran a nominal-level analysis—Chi-square, Lambda, Phi and Cramer's V are still selected in Statistics—we should be all set. Click OK. Take a look at the cross-tab and the Chi-Square Tests table.

democrat Is R Democrat? * gender R gender Crosstabulation

			gender R gender		Total
			1 1 Male	2 2 Female	
democrat Is R Democrat?	0 0 Non-Dem	Count	416	397	813
		% within gender R gender	74.3%	62.5%	68.0%
	1 1 Dem	Count	144	238	382
		% within gender R gender	25.7%	37.5%	32.0%
Total		Count	560	635	1195
		% within gender R gender	100.0%	100.0%	100.0%

Chi-Square Tests

	Value	df	Asymp. Sig. (2-sided)	Exact Sig. (2-sided)	Exact Sig. (1-sided)
Pearson Chi-Square	18.942[a]	1	.000		
Continuity Correction[b]	18.405	1	.000		
Likelihood Ratio	19.104	1	.000		
Fisher's Exact Test				.000	.000
Linear-by-Linear Association	18.927	1	.000		
N of Valid Cases[b]	1195				

Based on the cross-tab, the hypothesis clearly has merit. Among females 37.5 percent are Democrats, compared with 25.7 percent of males, about a 12-point gender difference. And according to chi-square (18.942, P-value = .000), we can reject the null hypothesis and infer that, in the population from which the sample was drawn, women probably are more likely than men to identify with the Democratic Party. Now look at the Directional Measures table.

Directional Measures

			Value	Asymp. Std. Error[a]	Approx. T[b]	Approx. Sig.
Nominal by Nominal	Lambda	Symmetric	.020	.030	.666	.505
		democrat Is R Democrat? Dependent	.000	.000	[c]	[c]
		gender R gender Dependent	.034	.050	.666	.505
	Goodman and Kruskal tau	democrat Is R Democrat? Dependent	.016	.007		.000[d]
		gender R gender Dependent	.016	.007		.000[d]

a. Not assuming the null hypothesis.

b. Using the asymptotic standard error assuming the null hypothesis.

c. Cannot be computed because the asymptotic standard error equals zero.

d. Based on chi-square approximation

Consider the relevant value of lambda, .000, which suggests that gender plays no role in predicting partisanship. What is going on here? Because of the way lambda is computed, it will sometimes fail to detect a relationship that, by all other evidence, clearly exists. Lambda looks at the modal value of the dependent variable for each category of the independent variable. Lambda can detect a relationship only if the mode is different between categories. In our earlier analysis of the deathpen_strong-gender relationship among non-southerners, for example, men and women had different modes on the dependent variable—the modal value for men was code 1 ("favor strongly," with 54.0 percent of the males), and the mode for women was code 0 (not "favor strongly," with 61.2 percent of the females). So lambda picked up the relationship. In the democrat-gender cross-tab, however, both genders have the same mode—code 0, "Non-Dem"—the category of 74.3 percent of the males and 62.5 percent of the females. Because men and women have the same mode on the dependent variable, lambda returned a value of .000: no relationship detected. In situations like this, which are not uncommon, rely on Cramer's V as a useful gauge of strength. In the current example, Cramer's V registers a value of .126, indicating a weak to moderate relationship between the independent and dependent variables.

EXERCISES

1. (Dataset: NES2004. Variables: egalit3, moralism3.) An interesting scholarly debate exists about the relationship between two different cultural values: moralism and egalitarianism. Moralism is the extent to which people believe that society should not accept or adjust to new moral values. People with higher moralism are more resistant to new values than are people with lower moralism. Egalitarianism is the extent to which people think the government should extend legal protections to different groups in society. People with higher egalitarianism are more likely to think that different groups require legal protections than are people with lower egalitarianism. Conceptually, these two beliefs are distinct. For example, a person might accept homosexuality as a lifestyle choice yet might also believe that homosexuals should not be legally protected by antidiscrimination laws. As an empirical matter, however, moralism and egalitarianism may be related. In this exercise you will test this hypothesis: In a comparison of individuals, people who have higher moralism will be less egalitarian than will people who have lower moralism. According to this hypothesis, as moralism increases, egalitarianism decreases.

   NES2004 contains moralism3, which classifies respondents into three ordinal categories: "Low" (coded 1), "Moderate" (coded 2), or "High" moralism (coded 3). So higher codes indicate more strongly moralistic beliefs. Moralism3 is the independent variable. The dependent variable, egalit3, is a three-category ordinal variable that gauges the extent to which respondents endorse the ideas of equal opportunity and equal rights: "Low" (coded 1), "Moderate" (coded 2), and "High" (coded 3). Higher codes on egalit3 indicate higher egalitarianism.

   A. Think about how SPSS calculates Somers' d. If the hypothesis is correct, then SPSS should report a Somers' d that shows (check one)

      ❑ a negative relationship between egalit3 and moralism3.

      ❑ a positive relationship between egalit3 and moralism3.

   B. Run a Crosstabs analysis of the relationship between egalit3 (dependent variable) and moralism3 (independent variable). Obtain Chi-square. Request Somers' d. Examine the cross-tabulation. Based on your substantive interpretation of the cross-tabulation, it would appear that (check one)

      ❑ no relationship exists between moralism and egalitarianism.

      ❑ the relationship is positive: Respondents who have higher moralism are more egalitarian than are respondents who have lower moralism.

      ❑ the relationship is negative: Respondents who have higher moralism are less egalitarian than are respondents who have lower moralism.

C. The chi-square test statistic for this relationship is (fill in the blank) _____ and has a P-value of (fill in the blank) _____.

D. Based on these results, your inferential decision is (check one)

❑ do not reject the null hypothesis.

❑ reject the null hypothesis.

E. Somers' d for this relationship is equal to (fill in the blank) _____. Thus compared with how well we can predict egalit3 by not knowing the independent variable, using moralism3 as a predictive tool increases our predictive accuracy by about (fill in the blank) _____ percent.

2. (Dataset: NES2004. Variables: hispanic_traits3, illegals_therm3, patriotism_nonH.) Are patriots bigots? Are people who profess a stronger sense of national pride more likely to hold negative attitudes toward immigrants than are people who profess a weaker sense of national pride? "Does pride imply prejudice?"[5] This question gains special relevance in times of war or economic insecurity, when in-groups may seek to strengthen their dominance by expressing hostility toward out-groups, such as immigrants or religious minorities.

In this exercise you will use NES2004 to investigate the relationship between patriotism and two dependent variables: hispanic_traits3 and illegals_therm3. Hispanic_traits3 is a three-category ordinal variable that measures prejudicial or stereotypical beliefs about Hispanics. Negative views are coded −1, neutral views are coded 0, and positive views are coded 1. Illegals_therm3, which is based on feeling thermometer ratings of illegal immigrants, has three ordinal categories: "Low" (coded 1), "Medium" (coded 2), and "High" (coded 3). For both dependent variables, then, higher values denote more positive beliefs and feelings. Dataset NES2004 also contains patriotism_nonH, a four-category ordinal measure of the respondent's identity with (and affection for) national symbols. Codes range from 0 (a weak sense of patriotism) to 3 (a strong sense of patriotism). Patriotism_nonH is the independent variable. (Because you will be investigating the relationship between patriotism and attitudes toward Hispanics, patriotism_nonH confines the measure of the independent variable to non-Hispanic respondents.)

A. Obtain cross-tabulations for the hispanic_traits3-patriotism_nonH and the illegals_therm3-patriotism_nonH relationships. For each relationship, request column percentages and Chi-square. Obtain Somers' d for each relationship. (Since both relationships have the same independent variable, only one Crosstabs run is needed.) Refer to the results to fill in the following tables. In the first table, record the percentages of respondents holding negative stereotypes of Hispanics and the percentages of respondents giving illegal immigrants "Low" ratings. In the second table, record chi-square and its P-value. Record Somers' d.

	Patriotism scale			
Dependent variable:	0 Low	1.00	2.00	3 High
hispanic_traits3, % "−1 Negative"	?	?	?	?
illegals_therm3, % "Low"	?	?	?	?

Dependent variable:	Chi-square	P-value	Somers' d
hispanic_traits3	?	?	?
illegals_therm3	?	?	?

B. Consider the illegals_therm3-patriotism relationship. If the null hypothesis is correct, then the observed relationship would occur by chance (fill in the blank) _____ of the time. Therefore, you should (circle one)

not reject the null hypothesis.   reject the null hypothesis.

C. Consider the hispanic_traits3-patriotism relationship. Compared to how well you can predict individuals' levels of prejudice toward Hispanics without knowing how patriotic they are, knowing their levels of patriotism increases your predictive accuracy by (fill in the blank) _____ percent.

D. What do you think? Does the evidence support the idea that patriots are bigots? (check one)

❏ Yes, the evidence supports the idea that patriots are bigots.

❏ No, the evidence does not support the idea that patriots are bigots.

Briefly explain your reasoning. _____

_____

_____

_____

3. (Dataset: GSS2006. Variables: abhealth, femrole2, sex.) Interested student and pedantic pontificator are discussing the gender gap in U.S. politics.

Interested student: "On what sorts of issues or opinions are men and women most likely to be at odds? What defines the gender gap, anyway?"

Pedantic pontificator: "That's easy. A couple of points seem obvious, to me anyway. First, we know that the conflict over abortion rights is the defining gender issue of our time. Women will be more likely than men to take a strong pro-choice position on this issue. Second—and pay close attention here—on more mundane cultural questions, such as whether women should play a role in work and politics, men and women will not differ at all."

A. Pedantic pontificator has suggested the following two hypotheses about the gender gap: (check two)

❏ In a comparison of individuals, women will be less likely than men to think that abortion should "always" be allowed.

❏ In a comparison of individuals, women will be no more likely than men to think that abortion should "always" be allowed.

❏ In a comparison of individuals, women will be more likely than men to think that abortion should "always" be allowed.

❏ In a comparison of individuals, women will be less likely than men to think that women should play a role in work and politics.

❏ In a comparison of individuals, women will be no more likely than men to think that women should play a role in work and politics.

❏ In a comparison of individuals, women will be more likely than men to think that women should play a role in work and politics.

B. Test pedantic pontificator's hypotheses using Crosstabs. GSS2006 contains two variables that will serve as dependent variables. Abhealth, which measures the number of conditions under which respondents think an abortion should be allowed, ranges from 0 ("Never allow") to 3 ("Allow, 3 conditions"). The variable femrole2, which measures attitudes toward the role of women in work and politics, is coded 1 ("Women domestic") or 2 ("Women in work, politics"). The independent variable is sex. Obtain Chi-square. Sex is a nominal variable, so be sure to request lambda. Because lambda may underestimate these relationships, you will need to request Cramer's V. In the abhealth-sex cross-tabulation, focus on the percentage saying "Allow, 3 conditions." In the femrole2-sex cross-tabulation, focus on the "Work, politics" category. Record your results in the table that follows.

	Sex		Chi-square		Measures of association	
Dependent variable	Male	Female	Test Statistic	P-value	Lambda	Cramer's V
Percent "Allow, 3 conds" (abhealth)	?	?	?	?	?	?
Percent "Work politics" (femrole2)	?	?	?	?	?	?

C. Based on these results, you may conclude that (check all that apply)

❏ a statistically significant gender gap exists on abortion opinions.

❏ pedantic pontificator's hypothesis about the femrole2-sex relationship appears to be incorrect.

❏ the abhealth-sex relationship you obtained could have occurred by chance more frequently than 5 times out of 100.

❏ pedantic pontificator's hypothesis about the abhealth-sex relationship appears to be correct.

❏ a higher percentage of females than males think that women belong in work and politics.

D. The P-value of the chi-square statistic in the abhealth-sex cross-tabulation tells you that, if the null

hypothesis is correct (complete the sentence) _____

_____.

4. (Dataset: GSS2006. Variables: polview3, racial_liberal3, social_cons3, spend3.) While having lunch together, three researchers are discussing what the terms *liberal, moderate,* and *conservative* mean to most people. Each researcher is touting a favorite independent variable that may explain the way survey respondents describe themselves ideologically.

Researcher 1: "When people are asked a question about their ideological views, they think about their attitudes toward government spending. If people think the government should spend more on important programs, they will respond that they are 'liberal.' If they don't want too much spending, they will say that they are 'conservative.' "

Researcher 2: "Well, that's fine. But let's not forget about social policies, such as abortion and pornography. These issues must influence how people describe themselves ideologically. People with more permissive views on these sorts of issues will call themselves 'liberal.' People who favor government restrictions will label themselves as 'conservative.' "

Researcher 3: "Okay, you both make good points. But you're ignoring the importance of racial issues in American politics. When asked whether they are liberal or conservative, people probably think about their opinions on racial policies, such as affirmative action. Stronger proponents of racial equality will say they are 'liberal,' and weaker proponents will say they are 'conservative.'"

In Chapter 3 you created an ordinal measure of ideology, polview3, which is coded 1 for "Liberal," 2 for "Moderate," and 3 for "Conservative." This is the dependent variable. GSS2006 also contains researcher 1's favorite independent variable, spend3, a three-category ordinal measure of attitudes toward government spending. Higher codes denote more supportive opinions toward spending. Researcher 2's favorite independent variable is social_cons3, a three-category ordinal measure of attitudes on social issues. Higher codes denote less permissive views. Researcher 3's favorite independent variable is racial_liberal3, also a three-category ordinal variable. Higher codes denote more strongly egalitarian opinions on racial issues.

A. Think about how SPSS calculates Somers' d. Assuming that each researcher is correct, SPSS should report (check all that apply)

❑ a negative relationship between polview3 and spend3.

❑ a positive relationship between polview3 and social_cons3.

❑ a negative relationship between polview3 and racial_liberal3.

B. Run a Crosstabs analysis of each of the relationships, using polview3 as the dependent variable and spend3, social_cons3, and racial_liberal3 as independent variables. Obtain Chi-square and Somers' d. Examine the cross-tabulations and chi-square statistics. Based on these results, you may conclude that (check all that apply)

❑ as values of spend3 increase, the percentage of respondents describing themselves as liberal increases.

❑ as values of spend3 increase, the percentage of respondents describing themselves as liberal decreases.

❑ the polview3-social_cons3 relationship is not statistically significant.

❑ if the null hypothesis is correct, you will obtain the polview3-racial_liberal3 relationship more frequently than 5 times out of 100.

❑ if the null hypothesis is correct, you will obtain the polview3-racial_liberal3 relationship less frequently than 5 times out of 100.

C. Somers' d for the polview3-social_cons3 relationship is equal to (fill in the blank) _____.
Thus, compared with how well we can predict polview3 by not knowing (complete the sentence)

_____.

_____.

D. The three researchers make a friendly wager. The researcher whose favorite independent variable does the worst job predicting values of the dependent variable has to buy lunch for the other two. Who pays for lunch? (circle one)

Researcher 1   Researcher 2   Researcher 3

5. (Dataset: GSS2006. Variables: partyid_3, trade_union2, pro_secure, genX.) Certainly you would expect partisanship and attitudes toward trade unions to be related. Unions have been mainstays of Democratic support since the New Deal of the 1930s, so you could hypothesize that people holding pro-union opinions are more likely to be Democrats than are people less favorably disposed toward unions. Yet it also seems reasonable to hypothesize that the relationship between union opinions (independent variable) and party identification (dependent variable) will not be the same for all age groups. It may be that, among an older cohort—those born from the New Deal era through the union heyday of the 1950s—attitudes toward unions will play a stronger role in shaping partisanship than they will among "Generation X," those born during and after the 1960s. Assuming that union opinions do indeed play a weaker role for Generation X, what sorts of issues might play a larger role in defining partisanship among the younger cohort? Perhaps the renewed debate between those wishing to protect civil liberties and those wanting greater security from terrorist threats figures more prominently in affecting the partisan loyalties of the younger generation.

Let's call this idea the "generational change perspective." According to the generational change perspective: (a) the relationship between union opinions and party identification will be stronger for older people than for younger people, and (b) the relationship between civil liberties opinions and party identification will be stronger for younger people than for older people. In this exercise you will test the generational change perspective.

GSS2006 contains partyid_3, which measures party identification (Democrat, coded 1; Independent, coded 2; Republican, coded 3). This is the dependent variable. (For this exercise, treat partyid_3 as an ordinal-level variable, with higher codes denoting stronger Republican identification.) One independent variable is trade_union2, a two-category ordinal measure that gauges opinions toward labor unions. Respondents could "Agree" (coded 1) or "Disagree" (coded 2) with the following statement: "Workers need strong unions to protect their interests." A second independent variable is pro_secure, coded 0 for respondents who favor protecting civil liberties over security, and coded 1 for respondents who favor security over civil liberties. The control variable is genX, which assigns respondents to an older generation (born in 1962 or before, coded 0) or younger generation (born after 1962, coded 1). Run a Crosstabs analysis using partyid_3 as the dependent variable, trade_union2 and pro_secure as independent variables, and genX as the control. Request Chi-square and Somers' d.

A. In the controlled comparison cross-tabulations, focus on the percentages of Democrats across the values of trade_union2 and pro_secure. Browse the statistical tables for the appropriate chi-square tests and measures of association. Fill in the table that follows.

trade_union2 Workers need strong unions

Generation X	Agree (% Dem)	Disagree (% Dem)	Chi-square	P-value	Somers' d
Born 1962 or before	?	?	?	?	?
Born after 1962	?	?	?	?	?

pro_secure Civil liberties or security?

Generation X	Pro-civil liberties (% Dem)	Pro-security (% Dem)	Chi-square	P-value	Somers' d
Born 1962 or before	?	?	?	?	?
Born after 1962	?	?	?	?	?

B. Which of the following inferences are supported by your analysis? (check all that apply)

❏ For both generations, people with pro-union opinions are more likely to be Democrats than are people with anti-union opinions.

❑ For both generations, people with pro–civil liberties opinions are more likely to be Democrats than are people with pro-security opinions.

❑ For the older generation, random sampling error would produce the observed relationship between partisanship and civil liberties opinions less frequently than 5 times out of 100.

❑ For the younger generation, the partyid_3-trade_union2 relationship is weaker than the partyid_3-pro_secure relationship.

C. Focus on the value of Somers' d for the younger cohort in the pro_secure cross-tabulation. This value

of Somers' d says that, compared to how well you can predict (complete the sentence)_____

_____.

D. Based on your analysis of these relationships, you can conclude that (check one)

❑ The generational change perspective is incorrect.

❑ The generational change perspective is correct.

Explain your reasoning. _____

_____

_____

_____

_____

_____.

6. (Dataset: World. Variables: protact3, gender_equal3, vi_rel3, pmat12_3.) Ronald Inglehart offers a particularly elegant and compelling idea about the future of economically advanced societies. According to Inglehart, the cultures of many postindustrial societies have been going through a value shift—the waning importance of materialist values and a growing pursuit of postmaterialist values. In postmaterialist societies, economically based conflicts—unions versus big business, rich versus poor—are increasingly supplanted by an emphasis on self-expression and social equality. Postmaterialist societies also are marked by rising secularism and elite-challenging behaviors, such as boycotts and demonstrations. In this exercise you will investigate Inglehart's theory.[6]

The World variable pmat12_3 measures the level of postmaterial values by a three-category ordinal measure: low postmaterialism (coded 1), moderate postmaterialism (coded 2), and high postmaterialism (coded 3). Higher codes denote a greater prevalence of postmaterial values. Use pmat12_3 as the independent variable. Here are three dependent variables, all of which are three-category ordinals: gender_equal3, which captures gender equality (1 = low equality, 2 = medium equality, 3 = high equality); protact3, which measures citizen participation in protests (1 = low, 2 = moderate, 3 = high); and vi_rel3, which gauges religiosity by the percentage of the public saying that religion is "very important" (1 = less than 20 percent, 2 = 20–50 percent, 3 = more than 50 percent). Higher codes on the dependent variables denote greater gender equality (gender_equal3), more protest activity (protact3), and higher levels of religiosity (vi_rel3).

A. Using pmat12_3 as the independent variable, three postmaterialist hypotheses can be framed:

Gender equality hypothesis (fill in the blanks): In a comparison of countries, those with higher levels of postmaterialism will have _____ levels of gender equality than will countries having lower levels of postmaterialism.

Protest activity hypothesis (fill in the blanks): In a comparison of countries, those with _____ levels of postmaterialism will have _____ levels of protest activity than will countries having _____ levels of postmaterialism.

Religiosity hypothesis (complete the sentence): In a comparison of countries, those with _____

_____.

B. Which of the following measures of association is most appropriate for all three relationships? (circle one)

Lambda   Somers' d

C. Consider how the independent variable is coded and how each dependent variable is coded. In the way that SPSS calculates the appropriate measure of association, which one of the three hypotheses implies a negative sign on the measure of association? (check one)

❏ The gender equality hypothesis

❏ The protest activity hypothesis

❏ The religiosity hypothesis

D. Test each hypothesis using cross-tabulation analysis. Obtain Chi-square and the appropriate measure of association. In the table that follows, record the percentages of countries falling into the highest category of each dependent variable. Also, report chi-square statistics, P-values, and measures of association.

Dependent variable	Level of postmaterialism			Chi-square	P-value	Measure of association
	Low	Moderate	High			
Percentage high gender equality	?	?	?	?	?	?
Percentage high protest activity	?	?	?	?	?	?
Percentage high religiosity	?	?	?	?	?	?

E. Which of the following inferences are supported by your analysis? (check all that apply)

❏ The gender equality hypothesis is supported.

❏ Compared with how well we can predict gender equality by not knowing the level of postmaterialism, we can improve our prediction by 20.03 percent by knowing the level of postmaterialism.

❏ The protest activity hypothesis is supported.

❏ If the null hypothesis is correct, the postmaterialism-protest activity relationship would occur, by chance, less frequently than 5 times out of 100.

❏ The religiosity hypothesis is supported.

❏ If the null hypothesis is correct, the postmaterialism-religiosity relationship would occur, by chance, less frequently than 5 times out of 100.

That concludes the exercises for this chapter. Before exiting SPSS, be sure to save your output file.

NOTES

1. Asymmetry is the essence of hypothetical relationships. Thus we would hypothesize that income causes opinions on welfare policies, but we would not hypothesize that welfare opinions cause income. We would prefer a measure of association that tells us how well income (independent variable) predicts welfare opinions (dependent variable), not how well welfare opinions predict income. Or, to cite Warner's tongue-in-cheek example: "There are some situations where the ability to make predictions is asymmetrical; for example, consider a study about gender and pregnancy. If you know that an individual is pregnant, you can predict gender perfectly (the person must be female). However, if you know that an individual is female, you cannot assume that she is pregnant." Rebecca M. Warner, *Applied Statistics* (Los Angeles: Sage, 2008): 316.

2. Somers' d may be used for square tables (in which the independent and dependent variables have the same number of categories) and for nonsquare tables (in which the independent and dependent variables have different numbers of categories). Because of its other attractive properties, some methodologists prefer Somers' d to Kendall's tau-b or tau-c. See George W. Bohrnstedt and David Knoke, *Statistics for Social Data Analysis*, 2d ed. (Itasca, Ill.: Peacock, 1988): 325.

3. Although the "2-sided" part of the P-value's label suggests a two-tailed probability, this number is indeed a one-tailed probability and so permits a one-tailed test of statistical significance.

4. You might also notice that the independent variable, age, and the control variable, education, are themselves related. Eye-balling the distribution of the oldest group, for example, reveals that 182 respondents have a high school education (or less) and 190 have some college (or more). So only about half fall into the more-educated category. This compares with about two-thirds for each of the two younger cohorts.

5. Rui J. P. de Figueiredo Jr. and Zachary Elkins, "Are Patriots Bigots? An Inquiry into the Vices of In-Group Pride," *American Journal of Political Science*, 47 (January 2003): 171–188. This quote, p. 171. These researchers identify two conceptually distinct dimensions of national pride, nationalism and patriotism. Nationalism is "a belief in national superiority and dominance" (p. 175). Patriotism, on the other hand, "refers to an attachment to the nation, its institutions, and its founding principles" (p. 175). The authors find that nationalists evince strong prejudice toward immigrants but that patriots are "no more antagonistic toward immigrants" than are average citizens (p. 186). The NES2004 variable patriotism is closer to de Figueiredo and Elkins's measure of patriotism than to their measure of nationalism.

6. Inglehart has written extensively about cultural change in postindustrial societies. For example, see his *Culture Shift in Advanced Industrial Society* (Princeton: Princeton University Press, 1990).

8

Correlation and Linear Regression

Procedures Covered

Analyze → Correlate → Bivariate
Analyze → Regression → Linear
Graphs → Legacy Dialogs → Scatter/Dot

Correlation and regression are powerful and flexible techniques used to analyze interval-level relationships. Pearson's correlation coefficient (Pearson's r) measures the strength and direction of the relationship between two interval-level variables. Pearson's r is not a proportional reduction in error (PRE) measure, but it does gauge strength by an easily understood scale—from –1, a perfectly negative association between the variables, to +1, a perfectly positive relationship. A correlation of 0 indicates no relationship. Researchers often use correlation techniques in the beginning stages of analysis to get an overall picture of the relationships between interesting variables.

Regression analysis produces a statistic, the regression coefficient, that estimates the effect of an independent variable on a dependent variable. Regression also produces a PRE measure of association, R-square, which indicates how completely the independent variable (or variables) explains the dependent variable. In regression analysis the dependent variable is measured at the interval level, but the independent variable can be of any variety—nominal, ordinal, or interval. Regression is more specialized than correlation. Researchers use regression analysis to model causal relationships between one or more independent variables and a dependent variable.

In the first part of this chapter, you will learn to perform correlation analysis using the Correlate procedure, and you will learn to perform and interpret bivariate regression using Regression → Linear. Bivariate regression uses one independent variable to predict a dependent variable. We then will turn to Scatter/Dot, an SPSS graphic routine that yields a scatterplot, a visual depiction of the relationship between two interval-level variables. With generous use of the Chart Editor, you will learn how to add a regression line to the scatterplot and how to edit the graph for elegance and clarity. Finally, you will use Regression → Linear to perform multiple regression analysis. Multiple regression, which uses two or more independent variables to predict a dependent variable, is an essential tool for analyzing complex relationships.

USING CORRELATE AND REGRESSION → LINEAR

Suppose that a student of state politics is interested in the gender composition of state legislatures. Using Descriptives to analyze the States dataset, this student finds that state legislatures range from 9 percent female to 38 percent female. (This student could be you, since you now know how to use Descriptives.) Why is there such variation in the female composition of state legislatures? The student researcher begins to formulate an explanation. Perhaps states with lower percentages of college graduates have lower percentages of women legislators than do states with more college-educated residents. And maybe a cultural variable, the percentage of Christian adherents, plays a role. Perhaps states with higher percentages of Christian residents

Figure 8–1 Bivariate Correlations Window (modified)

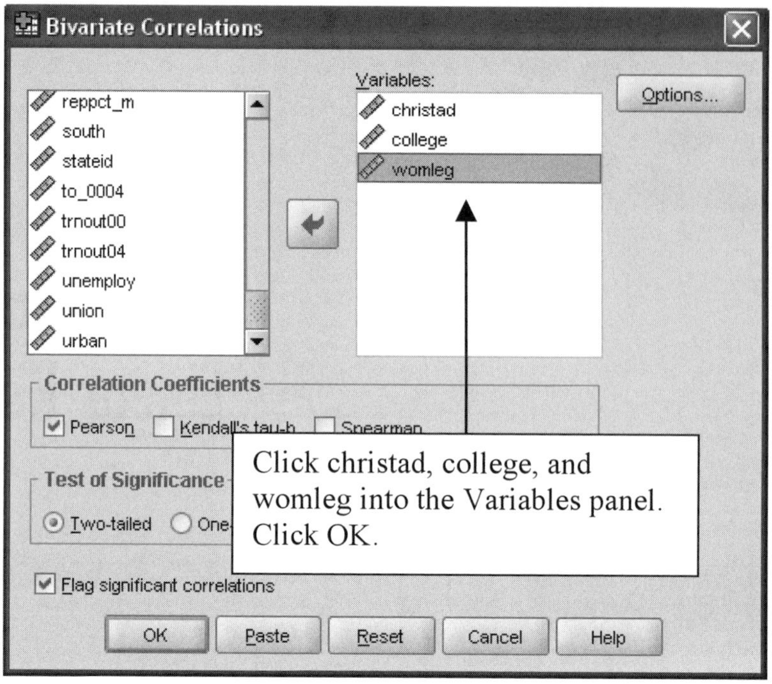

have lower percentages of female lawmakers. Correlation analysis would give this researcher an overview of the relationships among these variables. Let's use Correlate and Regression → Linear to investigate.

Open the States dataset. Click Analyze → Correlate → Bivariate. The Bivariate Correlations window is a no-frills interface (Figure 8-1). We are interested in three variables: the percentage of Christian adherents (christad), the percentage of college graduates (college), and the percentage of female state legislators (womleg). Click each of these variables into the Variables panel, as shown in Figure 8-1.

By default, SPSS will return Pearson's correlation coefficients. So the Pearson box, which is already checked, suits our purpose. Click OK. SPSS reports the results in the Viewer.

Correlations

		christad Percent of pop who are Christian adherents	college Percent of pop w/college or higher	womleg Percent of state legislators who are women
christad Percent of pop who are Christian adherents	Pearson Correlation	1	-.125	-.446**
	Sig. (2-tailed)		.387	.001
	N	50	50	50
college Percent of pop w/college or higher	Pearson Correlation	-.125	1	.580**
	Sig. (2-tailed)	.387		.000
	N	50	50	50
womleg Percent of state legislators who are women	Pearson Correlation	-.446**	.580**	1
	Sig. (2-tailed)	.001	.000	
	N	50	50	50

**. Correlation is significant at the 0.01 level (2-tailed).

The Correlations table, called a correlation matrix, shows the correlation of each variable with each of the other variables—it even shows the correlation between each variable and itself. Each of the correlations in which we are interested appears twice in the table: once above the upper-left-to-lower-right diagonal of 1's, and again below the diagonal. The correlation between christad and womleg is −.446, which tells us that

Figure 8–2 Linear Regression Window

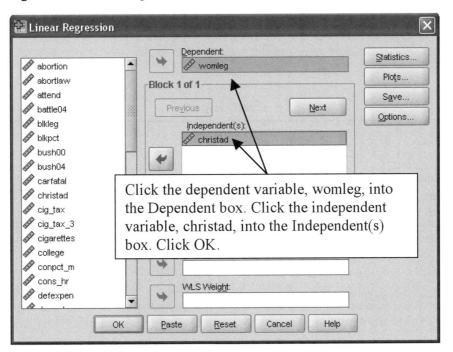

increasing values of one of the variables is associated with decreasing values of the other variable. So as the percentage of Christian adherents goes up, the percentage of female legislators goes down. How strong is the relationship? We know that Pearson's r is bracketed by −1 and +1, so we could say that this relationship is a fairly strong negative association. The correlation between college and womleg, .580, indicates a positive relationship: As states' percentages of college graduates increase, so do their percentages of women legislators. Again, this is a fairly strong association—a bit stronger than the womenleg-christad relationship. Finally, christad and college, with a Pearson's r of −.125, are weakly related. This value for Pearson's correlation coefficient suggests, of course, that the relationship has no clear, systematic pattern.

Correlation analysis is a good place to start when analyzing interval-level relationships. Even so, a correlation coefficient is agnostic on the question of which variable is the cause and which the effect. Does an increase in the percentage of Christian adherents somehow cause lower percentages of women in state legislatures? Or do increasing percentages of women in state legislatures somehow cause states to have lower percentages of Christian adherents? Either way, correlation analysis reports the same measure of association, a Pearson's r of −.446.

Regression is more powerful than correlation, in part because it helps us investigate causal relationships—relationships in which an independent variable is thought to affect a dependent variable. Regression analysis will (1) reveal the precise nature of the relationship between an independent variable and a dependent variable; (2) test the null hypothesis that the observed relationship occurred by chance, and (3) provide a PRE measure of association between the independent variable and the dependent variable. To illustrate these and other points, we will run two separate bivariate regressions. First we will examine the relationship between christad and womleg, and then we will analyze the relationship between college and womleg.

Click Analyze → Regression → Linear. The Linear Regression window appears (Figure 8-2). Click womleg into the Dependent box. Find christad in the variable list and click it into the Independent(s) box. Click OK.

SPSS regression output includes four tables: Variables Entered/Removed, Model Summary, ANOVA (which stands for analysis of variance), and Coefficients. For the regression analyses you will perform in this book, the Model Summary table and the Coefficients table contain the most important information. Let's examine them.

Model Summary

Model	R[a]	R Square	Adjusted R Square	Std. Error of the Estimate
1	.446	.199	.182	6.6235

a. Predictors: (Constant), christad Percent of pop who are Christian adherents

Coefficients[a]

Model		Unstandardized Coefficients		Standardized Coefficients	t	Sig.
		B	Std. Error	Beta		
1	(Constant)	37.929	4.357		8.705	.000
	christad Percent of pop who are Christian adherents	-.304	.088	-.446	-3.450	.001

a. Dependent Variable: womleg Percent of state legislators who are women

First, consider the Coefficients table. The leftmost column, under the heading "Model," contains the names of the key elements in the regression equation. "Constant" is the Y-intercept of the regression line, and "Percent of pop who are Christian adherents" is the label of the independent variable. The numbers along the "Constant" row report statistics about the Y-intercept, and the numbers along the "Percent of pop who are Christian adherents" row report statistics about the independent variable. Now look at the first column of numbers, which shows the regression coefficient for each parameter. According to these values, the Y-intercept is equal to 37.929, and the regression coefficient is –.304. The regression equation for estimating the effect of christad on womleg, therefore, is as follows (to make the numbers a bit simpler, we will round to two decimal places):

Percent of state legislators who are women = 37.93 – 0.30*christad.

The constant, 37.93, is the estimated value of Y when X equals 0. If you were using this equation to estimate the percentage of women legislators for a state, you would start with 37.93 percent and then subtract .30 for each percentage of the state's population who are Christian adherents. So your estimate for a state with, say, 50 percent Christian adherents would be 37.93 – .30*(50) = 37.93 – 15.00 ≈ 23 percent female legislators. The main statistic of interest, then, is the regression coefficient, –.30, which estimates the average change in the dependent variable for each unit change in the independent variable. A regression coefficient of –.30 tells us that, for each one-unit increase in the percentage of Christian adherents, there is a .30-unit decrease in the percentage of female legislators. So a 1-percentage-point increase in christad is associated with a .30-percentage-point decrease in womleg.[1]

What would the null hypothesis have to say about all this? Of course, we are not analyzing a random sample here, since we have information on the entire population of 50 states. But let's assume, for illustrative purposes, that we have just analyzed a random sample and that we have obtained a sample estimate of the effect of christad on womleg. The null hypothesis would say what it always says: In the population from which the sample was drawn, there is no relationship between the independent variable (in this case, the percentage of Christian adherents) and the dependent variable (the percentage of female legislators). In the population the true regression coefficient is equal to 0. Furthermore, the regression coefficient that we obtained, –.30, occurred by chance.

In SPSS regression results, you test the null hypothesis by examining two columns in the Coefficients table—the column labeled "t," which reports t-ratios, and the column labeled "Sig.," which reports P-values. Informally, to safely reject the null hypothesis, you generally look for t-ratios with magnitudes (absolute values) of 2 or greater. According to the results of our analysis, the regression coefficient for christad has a t-ratio of –3.45, well above the informal 2-or-greater rule. A P-value, which tells you the probability of obtaining the results if the null hypothesis is correct, helps you to make more precise inferences about the relationship between the independent variable and the dependent variable. If "Sig." is greater than .05, then the observed results would occur too frequently by chance, and you must not reject the null hypothesis. By contrast, if "Sig." is equal to or less than .05, then the null hypothesis represents an unlikely occurrence and may be rejected. The t-ratio for christad has a corresponding P-value of .001. If the null is correct, then random sampling error would have produced the observed results about one time in a thousand, a highly unlikely random event.[2] Reject the null hypothesis. It depends on the research problem at hand, of course, but for most applications you can ignore the t-ratio and P-value for the constant.[3]

How strong is the relationship between christad and womleg? The answer is provided by the R-square statistics, which appear in the Model Summary table. SPSS reports two values, one labeled "R Square," and one labeled "Adjusted R Square." Which one should you use? Most research articles report the adjusted value, so let's rely on adjusted R-square to provide the best overall measure of the strength of the relationship.[4] Adjusted R-square is equal to .182. What does this mean? R-square communicates the proportion of the variation in the dependent variable that is explained by the independent variable. Like any proportion, R-square can assume any value between 0 and 1. Thus, of all the variation in womleg between states, .18, or 18 percent, is explained by christad. The rest of the variation in womleg, 82 percent, remains unexplained by the independent variable.

So that you can become comfortable with bivariate regression—and to address a potential source of confusion—let's do another run, this time using college as the independent variable. Click Analyze → Regression → Linear. Leave womleg in the Dependent box, but click christad back into the Variables list. Click college into the Independent(s) box and click OK. Examine the Coefficients table and the Model Summary table.

Model Summary

Model	R[a]	R Square	Adjusted R Square	Std. Error of the Estimate
1	.580	.336	.322	6.0281

a. Predictors: (Constant), college Percent of pop w/college or higher

Coefficients[a]

Model		Unstandardized Coefficients B	Unstandardized Coefficients Std. Error	Standardized Coefficients Beta	t	Sig.
1	(Constant)	-1.003	4.991		-.201	.842
	college Percent of pop w/college or higher	.938	.190	.580	4.932	.000

a. Dependent Variable: womleg Percent of state legislators who are women

The regression equation for the effect of college on womleg is as follows:

$$\text{Percent of state legislators who are women} = -1.00 + 0.94 * \text{college}.$$

As is sometimes the case with regression, the constant, −1.00, represents an "unreal" situation. For states in which 0 percent of residents have college degrees, the estimated percentage of female legislators is a *minus* 1 percent. Of course, the smallest value of college in the actual data is substantially higher than 0.[5] However, for the regression line to produce the best estimates for real data, SPSS Regression has anchored the line at a Y-intercept. The regression coefficient, .94, says that for each percentage-point increase in college, there is an average increase of .94 of a percentage point in the percentage of female legislators. Again, increase the percentage of college graduates by 1, and the percentage of women legislators goes up by almost 1, on average. In the population, could the true value of the regression coefficient be 0? Probably not, according to the t-ratio (4.932) and the P-value (Sig. = .000). And, according to adjusted R-square, the independent variable does a fair amount of work in explaining the dependent variable. About 32 percent of the variation in womleg is explained by college. As bivariate regressions go, that's not too bad.

PRODUCING AND EDITING A SCATTERPLOT

An SPSS graphic routine, Scatter/Dot, adds a visual dimension to correlation and regression and thus can help you paint a richer portrait of a relationship. Consider Figure 8-3, created using Graphs → Legacy Dialogs → Scatter/Dot and edited in the Chart Editor. This graph, generically referred to as a scatterplot, displays the cases in a two-dimensional space according to their values on the two variables. The horizontal axis (X-axis) is defined by the independent variable, college, and the vertical axis (Y-axis) is defined by the dependent variable, womleg. We know from our correlation analysis that Pearson's r for this relationship is .58. We can now see what the correlation "looks like." Based on the figure, states with lower percentages of college graduates tend to have lower percentages of women legislators, with values on the Y-axis that range from 10 percent to about 25 percent or so. The percentages of women legislators for states at the higher end of the X-axis, furthermore, fall between 15–20 percent to around 35 percent. So as you move from left to right along the X-axis, values on the Y-axis generally increase, just as the positive correlation coefficient suggested.

The scatterplot has other interesting features. Notice that the dots have been overlaid by the linear regression line obtained from the analysis we just performed:

$$\text{Estimated percentage of women legislators} = -1.00 + .94 * \text{college}.$$

Thanks to this visual depiction, we can see that the linear summary of the relationship, while reasonably coherent, is far from perfect. Helpfully, to the lower right-hand side SPSS has supplied the value of R-square, one measure of the perfection (or the completeness, at least) of the relationship. Also, notice the sparse clarity of the graph. The dots are solid, but the background is white, both inside and outside the data space. The axis lines have been removed, leaving only the tick marks, which are labeled in round numbers, without decimal points. The two key data elements—the dots representing each case and the regression line summarizing the relationship—do not compete for our eye with any other lines, colors, or text. The scatterplot in Figure 8-3 comes close to what Edward R. Tufte calls an "erased" graph, a graph in which nonessential elements have been deleted. In Chapter 5, we touched on Tufte's definition of the data-ink ratio, the proportion of a graph's total ink devoted to depicting the information contained in the data. Tufte, a leading expert on the visual communication of quantitative information, recommends that the greatest share of a graph's

Figure 8–3 Scatterplot with Regression Line

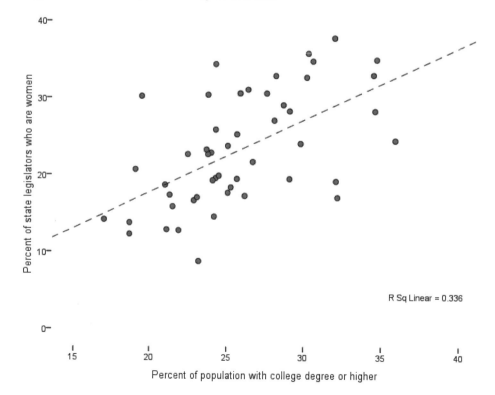

Figure 8–4 Opening Scatter/Dot

elements should be devoted to data ink—graphic features that convey the essence of the relationship.[6] Let's recreate this graphic.

Click Graphs → Legacy Dialogs → Scatter/Dot. In the Scatter/Dot window, select Simple Scatter and click Define, opening the Simple Scatterplot dialog (Figure 8-4). Nothing mysterious in this window. Click the independent variable (college) into the X Axis box, click the dependent variable (womleg) into the Y Axis box, and click OK. SPSS summons its defaults and cranks out a scatterplot (Figure 8-5). This is a good start, but improvement is always possible. Double-click on the image, opening the Chart Editor (Figure 8-6). First we will complement or enhance the data elements—add the regression line, make the dots more prominent— and then we will deemphasize the graph's nondata features by whiting out the scatterplot's fill, border, and axes. Also, we will need to un-bold the axis titles. (In SPSS's default rendition, the bolded axis titles are the first thing one looks at.) Perhaps we also will want to modify the X Axis title to make it more presentable.

To add the regression line, simply click the Add Fit Line at Total button (see Figure 8-7). SPSS superim-poses the line, selects it, and automatically opens the Properties window.[7] The Fit Line tab (the opening tab) does not require our attention. Click the Lines tab. In its solid-black attire, the regression line looks more like a sure thing than a probabilistic estimate. Click the Style drop-down and pick one of the dashed-line options. Click Apply, but be sure to keep the Properties window open.[8] Now click on any one of the hollow circles in the cloud of points or "markers." SPSS selects all of the markers (Figure 8-8). In the Color panel of the Marker tab, click Fill. The default setting, a diagonal line through a white background, means "transpar-ent." This won't do. Make a color choice in the palette, and then click Apply. SPSS fills the dots with your selected choice.

Figure 8–5 Unedited Scatterplot in the Viewer

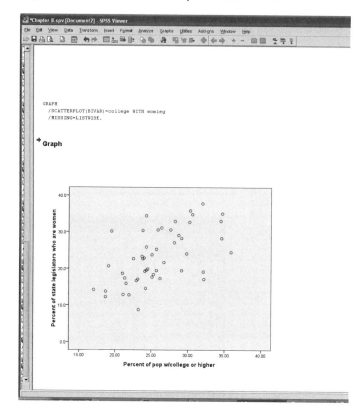

Figure 8–6 Scatterplot Ready for Editing

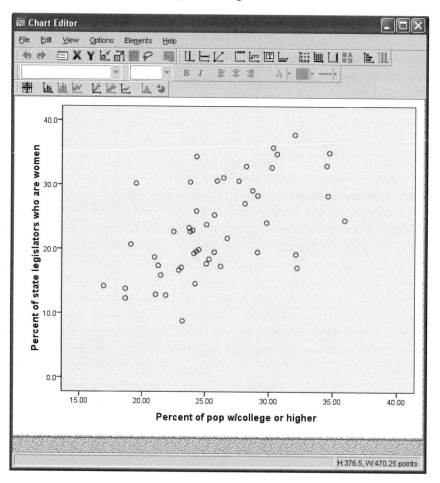

Figure 8–7 Adding a Regression Line to the Scatterplot

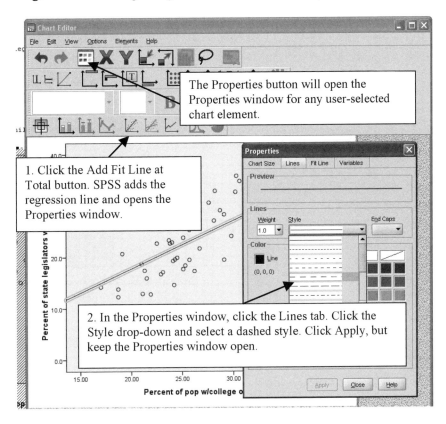

Note: To enhance readability, large buttons are shown.

Figure 8–8 Adding a Fill Color to Scatterplot Markers

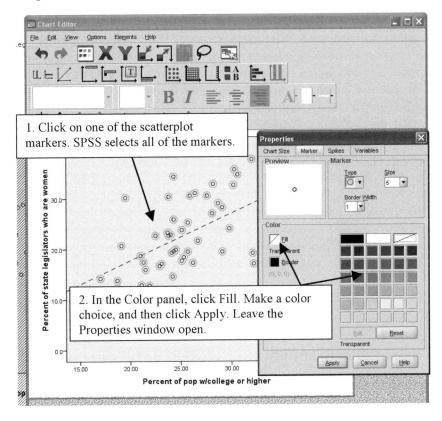

Figure 8–9 Whiting Out the Border and Fill

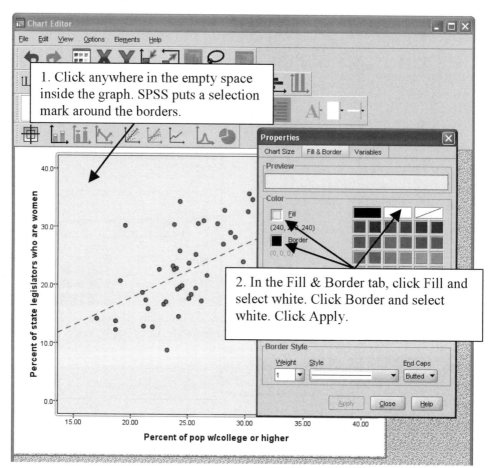

Compared with the effort involved in enhancing the scatterplot's data-related information, the task of erasing the nonessential elements of the graph is rather more labor-intensive. But with a little practice it becomes second nature. To whiten the graphic background (and dim but not whiten the axes), click anywhere on the empty gray space inside the graph (Figure 8-9). The Properties window adapts, telling us which elements are editable. In the Color panel of the Fill & Border tab, click Fill and select white, click Border and select white, and click Apply. Now we will blank out the newly dimmed axes. Carefully click on the X-axis, but make sure not to click on one of the tick labels (Figure 8-10). Depress the Control key and keep it depressed. Click on the Y-axis. Okay, now both axes are selected. In the Color panel of the Lines tab, click Line, select white, and click Apply. While we are here, we also will delete the unnecessary and distracting digits to the right of the decimal points in the axes tick-mark labels. Click the Number Format tab (refer to Figure 8-10). Click in the Decimal Places box (which may be empty) and type 0. Click Apply.

By default, SPSS bolds its graphic axis titles. This draws us away from the data and directs us toward a less important text element. Click on the X-axis title. Hold down the Control key and click on the Y-axis title (Figure 8-11). Both axis titles should now be selected. Click the Preferred Size drop-down and select 10. Click the Style drop-down and select Normal. Click Apply. At long last, you can close the Properties window. One more thing. Click on the X-axis title and modify it to make the graph more presentable (Figure 8-12). Before exiting the Chart Editor, you may want to save your chart preferences as a template, which can be opened and applied to future scatterplot-editing tasks.[9] In any event, if you want a chart that reflects current professional standards of elegant and informative graphic display, the scatterplot you have just created comes very close indeed.[10] Nicely done.

EXPLORING MULTIVARIATE RELATIONSHIPS WITH REGRESSION → LINEAR

Suppose a policy researcher is investigating factors causally related to motor vehicle deaths in the states. One such factor might be a simple characteristic of states: how densely populated they are. Residents of sparsely

Figure 8-10 Editing the Axes of a Scatterplot

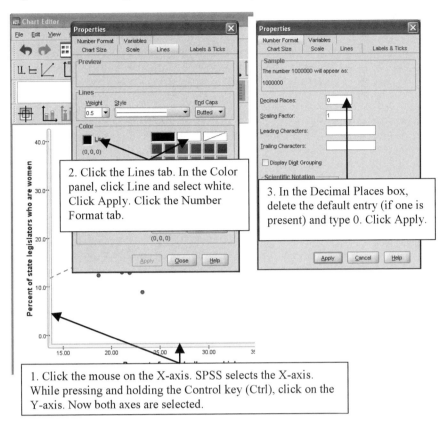

2. Click the Lines tab. In the Color panel, click Line and select white. Click Apply. Click the Number Format tab.

3. In the Decimal Places box, delete the default entry (if one is present) and type 0. Click Apply.

1. Click the mouse on the X-axis. SPSS selects the X-axis. While pressing and holding the Control key (Ctrl), click on the Y-axis. Now both axes are selected.

Figure 8-11 Editing the Axis Titles

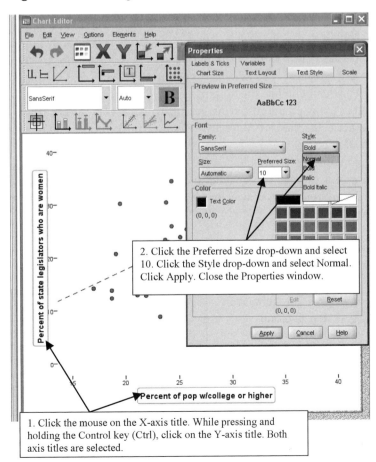

2. Click the Preferred Size drop-down and select 10. Click the Style drop-down and select Normal. Click Apply. Close the Properties window.

1. Click the mouse on the X-axis title. While pressing and holding the Control key (Ctrl), click on the Y-axis title. Both axis titles are selected.

Figure 8–12   Changing an Axis Title

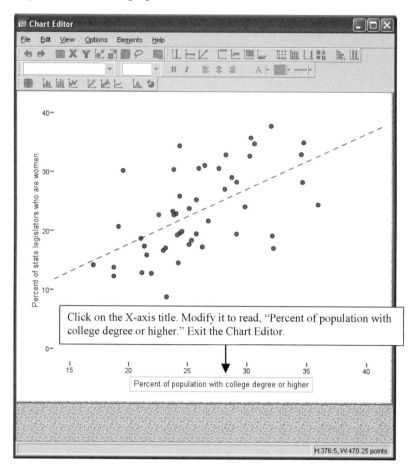

populated states, the policy researcher reasons, would typically drive longer distances at higher speeds than would residents of more densely populated states. Plus, a car accident in a thinly populated state would be more likely to be fatal, because "both Good Samaritans and hospitals are more scattered in thinly populated states compared to the denser states."[11] So as density goes up, we should find that fatalities go down. Another variable might be demographic: the proportion of young people in the population. As every insurance agent knows—and as many premium-paying parents will attest—younger drivers are more likely to be involved in automobile accidents than are older drivers. Thus, as the proportion of younger people goes up, fatalities should be found to go up, too.

The States dataset contains three variables: carfatal, the number of motor vehicle deaths per 100,000 residents; density, state population per square mile; and pop_18_24, the percentage of the population between eighteen and twenty-four years of age. Run Analyze → Correlate → Bivariate to obtain a correlation matrix of these three variables.

Correlations

		carfatal Motor vehicle fatalities (per 100,000 pop)	density Population per square mile	pop_18_24 Percent age 18-24
carfatal Motor vehicle fatalities (per 100,000 pop)	Pearson Correlation	1	-.574**	.388**
	Sig. (2-tailed)		.000	.005
	N	50	50	50
density Population per square mile	Pearson Correlation	-.574**	1	-.478**
	Sig. (2-tailed)	.000		.000
	N	50	50	50
pop_18_24 Percent age 18-24	Pearson Correlation	.388**	-.478**	1
	Sig. (2-tailed)	.005	.000	
	N	50	50	50

**. Correlation is significant at the 0.01 level (2-tailed).

Note the correlation between each independent variable and the dependent variable. The correlation between density and carfatal is negative, indicating that as density increases, motor vehicle fatalities decrease ($r = -.57$). The relationship between pop_18_24 and carfatal, as the policy researcher suspected, is positive: As the percentage of young people increases, fatalities also increase ($r = .39$). But notice, too, that the two independent variables are themselves moderately related ($r = -.48$). This correlation is negative, suggesting that densely populated states have lower percentages of young people than do sparsely populated states. (This relationship becomes important later on.)

First let's run a simple regression, using carfatal as the dependent variable and density as the independent variable. Click Analyze → Regression → Linear. Put carfatal in the Dependent box and density in the Independent(s) box. Click OK and examine the output.

Model Summary

Model	R^a	R Square	Adjusted R Square	Std. Error of the Estimate
1	.574	.329	.315	4.74790

a. Predictors: (Constant), density Population per square mile

Coefficients^a

Model		Unstandardized Coefficients		Standardized Coefficients	t	Sig.
		B	Std. Error	Beta		
1	(Constant)	19.939	.836		23.847	.000
	density Population per square mile	-.013	.003	-.574	-4.855	.000

a. Dependent Variable: carfatal Motor vehicle fatalities (per 100,000 pop)

Consider the Y-intercept (Constant) and the regression coefficient on density. According to these values, the Y-intercept is equal to 19.94, and the regression coefficient is –.013, which rounds to –.01. The regression equation for the effect of density on carfatal, therefore, is

Motor vehicle fatalities per 100,000 pop. = 19.94 – .01*Population per square mile.

What do these coefficients mean? In terms of its magnitude, for example, the regression coefficient seems to be an incredibly small number, and its meaning is not intuitively obvious. Remember to keep the substantive relationship in mind—and focus on the units of measurement. Very thinly populated states will have an estimated fatality rate close to the intercept, or about 20 fatalities per 100,000 population. Alaska, for example, has a population density of just more than 1 person per square mile. So its estimated fatality rate would be close to the intercept of about 20. The regression coefficient tells us that, for each additional person per square mile, the motor vehicle fatality rate drops by .01. New Jersey, for instance, is a very densely populated state, with a density of about 1,200 people per square mile. So New Jersey's estimated fatality rate would be 20 – .01*1200, which is equal to 20 – 12, or about 8 fatalities per 100,000 population. Thus, as density increases, by one person at a time, fatalities decrease by .01 of a fatality per 100,000 population. According to the t-ratio (–4.86) and accompanying P-value (.000), we can safely reject the null hypothesis. SPSS reports an adjusted R-square of .315. Thus, of all the variation among states in automobile fatality rates, about 32 percent is explained by population density.

Now let's run another bivariate regression. We will keep carfatal as the dependent variable, but this time we'll use the percentage of the population aged 18 to 24 (pop_18_24) as the independent variable.

Model Summary

Model	R	R Square	Adjusted R Square	Std. Error of the Estimate
1	.388[a]	.151	.133	5.34276

a. Predictors: (Constant), pop_18_24 Percent age 18-24

Coefficients[a]

Model		Unstandardized Coefficients		Standardized Coefficients	t	Sig.
		B	Std. Error	Beta		
1	(Constant)	-10.102	9.494		-1.064	.293
	pop_18_24 Percent age 18-24	2.729	.935	.388	2.919	.005

a. Dependent Variable: carfatal Motor vehicle fatalities (per 100,000 pop)

According to the estimated coefficients, the regression line for the effect of pop_18_24 on carfatal is as follows:

Motor vehicle fatalities per 100,000 pop. = −10.10 + 2.73*Percent age 18-24.

Again we have a Y-intercept depicting an unreal situation, so let's focus on the regression coefficient, 2.73. This says that for each percentage point increase in pop_18_24, there is a 2.73-unit increase in the motor vehicle fatality rate—2.73 additional fatalities per 100,000 population. As the percentage of younger people in the population increases, so too does the fatality rate. In the population, could the true value of the regression coefficient be 0? Probably not, according to the t-ratio (2.92) and the P-value (.005). And, according to the adjusted R-square of .133, about 13 percent of the variation in carfatal is explained by pop_18_24.

Let's review our analysis so far. In the first bivariate regression, we found that population density has a statistically significant negative effect on motor vehicle fatalities. Low-density states have higher fatality rates than do high-density states. In the second bivariate regression, we found that the percentage of younger people has a significant positive effect on motor vehicle fatalities. States with lower percentages of young people have lower fatality rates than do states with higher percentages of young people. But recall the initial correlation matrix. There we found that the two independent variables are related: As density goes up, the percentage of younger people goes down. So when we compare states with lower percentages of young people with states with higher percentages of young people, we are also comparing high-density states with low-density states. Perhaps states with higher percentages of young people have higher fatality rates not because they have more young people, but because they have lower population densities. Thus the relationship between pop_18_24 and carfatal might be spurious. Then again, it might not be. Unless we reexamine the pop_18_24-carfatal relationship, controlling for density, there is no way to tell.

Multiple regression is designed to estimate the partial effect of an independent variable on a dependent variable, controlling for the effects of other independent variables. Regression → Linear easily allows us to run multiple regression analysis. Let's do such an analysis, again using carfatal as the dependent variable and entering *both* density and pop_18_24 as independent variables. Return to the Regression window. Leaving pop_18_24 in place, click density into the Independent(s) box and click OK.

Model Summary

Model	Rᵃ	R Square	Adjusted R Square	Std. Error of the Estimate
1	.588	.346	.318	4.73747

a. Predictors: (Constant), density Population per square mile, pop_18_24 Percent age 18-24

Coefficientsᵃ

Model		Unstandardized Coefficients		Standardized Coefficients	t	Sig.
		B	Std. Error	Beta		
1	(Constant)	9.126	9.858		.926	.359
	pop_18_24 Percent age 18-24	1.039	.944	.148	1.101	.277
	density Population per square mile	-.011	.003	-.503	-3.748	.000

a. Dependent Variable: carfatal Motor vehicle fatalities (per 100,000 pop)

This analysis provides the information we need to isolate the partial effect of each independent variable on the dependent variable. The multiple regression equation is as follows:

$$\text{Motor vehicle fatalities per 100,000 pop.} =$$
$$9.13 + 1.04*(\text{Percent age 18-24}) - .01*(\text{Population per square mile}).$$

Let's focus on the regression coefficients for each of the independent variables. The coefficient for pop_18_24, 1.04, tells us the effect of pop_18_24 on carfatal, controlling for density. Recall that in the bivariate analysis, a 1-percentage-point increase in pop_18_24 was associated with about a 2.73-unit increase in the fatality rate. When we control for density, however, we find a substantial reduction in this effect—to about a 1-unit increase in the fatality rate. What is more, the regression coefficient for pop_18_24, with a P-value of .277, is not statistically significant. Density, on the other hand, retains much of its predictive power. The regression coefficient, –.011, is essentially the same effect we found earlier when we investigated the bivariate relationship between carfatal and density. With a P-value of .000, we can say that, controlling for the percentage of young residents, population density is significantly related to motor vehicle fatalities. It would appear, then, that the carfatal-pop_18_24 relationship is a spurious artifact of differences between states in population density.

In multiple regression, adjusted R-square communicates how well all of the independent variables explain the dependent variable. So by knowing two things about states—the percentage of younger people and the population density—we can account for about 32 percent of the variation in motor vehicle death rates. But notice that this value of adjusted R-square is practically the same as the adjusted R-square we found before, using density by itself to explain carfatal. Clearly, density does the lion's share of explanatory work in accounting for the dependent variable.

EXERCISES

1. (Dataset: States. Variables: demnat, demstate, union.) Consider a plausible scenario for the relationships between three variables: the percentages of a state's U.S. House and U.S. Senate delegations who are Democrats, the percentage of state legislators who are Democrats, and the percentage of workers in the state who are unionized. We could hypothesize that, compared with states with fewer Democrats in their state legislatures, states having larger percentages of Democratic legislators would also have greater proportions of Democrats in their U.S. congressional delegations. Furthermore, because unions tend to

support Democratic candidates, we would also expect more heavily unionized states to have higher percentages of Democratic legislators at the state and national levels. States contains three variables: demnat, the percentage of House and Senate members who are Democrats; demstate, the percentage of state legislators who are Democrats; and union, the percentage of workers who are union members.

A. Run Correlate to find the Pearson correlation coefficients among demnat, demstate, and union. Fill in the four empty cells of this correlation matrix:

		Percent U.S. House and Senate Democratic	Percent state legislators Democratic	Percent workers who are union members
Percent U.S. House and Senate Democratic	Pearson Correlation	1	.641	
Percent state legislators Democratic	Pearson Correlation	.641	1	
Percent workers who are union members	Pearson Correlation			1

B. According to the correlation coefficient, as the percentage of unionized workers increases, the percentage of Democratic U.S. representatives and U.S. senators (circle one)

increases.   decreases.

C. According to the correlation coefficient, as the percentage of unionized workers decreases, the percentage of Democratic U.S. representatives and U.S. senators (circle one)

increases.   decreases.

D. Which two of the following statements describe the relationship between the percentage of unionized workers and the percentage of state legislators who are Democrats? (check two)

❑ The relationship is negative.

❑ The relationship is positive.

❑ The relationship is stronger than the relationship between the percentage of unionized workers and the percentage of Democratic U.S. representatives and U.S. senators.

❑ The relationship is weaker than the relationship between the percentage of unionized workers and the percentage of Democratic U.S. representatives and U.S. senators.

2. (Dataset: States. Variables: cons_hr, conpct_m.) Two congressional scholars are discussing the extent to which members of the U.S. House of Representatives stay in touch with the voters in their states.

Scholar 1: "When members of Congress vote on important public policies, they are closely attuned to the ideological make-ups of their states. Members from states having lots of liberals will tend to cast votes in the liberal direction. Representatives from states with mostly conservative constituencies, by contrast, will take conservative positions on important policies."

Scholar 2: "You certainly have a naïve view of congressional behavior. Once they get elected, members of congress adopt a 'Washington, D.C., state of mind,' perhaps voting in the liberal direction on one policy and in the conservative direction on another. One thing is certain: The way members vote has little to do with the ideological composition of their states."

Think about an independent variable that measures the percentage of self-described conservatives among the mass public in a state, with low values denoting low percentages of conservatives and high values denoting high percentages of conservatives. And consider a dependent variable that gauges the degree to which the state's House delegation votes in a conservative direction on public policies. Low scores on this dependent variable tell you that the delegation tends to vote in a liberal direction, and high scores say that the delegation votes in a conservative direction.

A. Below is an empty graphic shell showing the relationship between the independent variable and the dependent variable. Draw a regression line inside the shell that depicts what the relationship should look like if scholar 1 is correct.

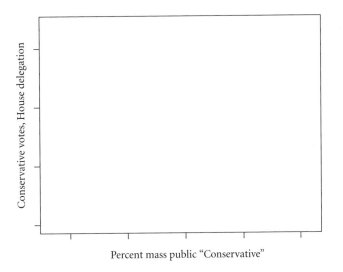

B. Below is another graphic shell showing the relationship between the independent variable and the dependent variable. Draw a regression line inside the shell that depicts what the relationship should look like if scholar 2 is correct.

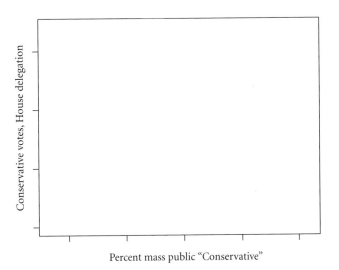

C. States contains the variable conpct_m, the percentage of the mass public calling themselves conservative. This is the independent variable. States also contains cons_hr, a measure of conservative votes by states' House members. Scores on this variable can range from 0 (low conservatism) to 100 (high conservatism). This is the dependent variable. Run Regression to analyze the relationship between cons_hr and conpct_m.

According to the regression equation, a 1-percentage-point increase in conservatives in the mass public is associated with (check one)

❏ about a 42-point decrease in House conservatism scores.

❏ about a 3-point increase in House conservatism scores.

❏ about a 7-point increase in House conservatism scores.

D. If you were to use this regression to estimate the mean House conservatism score for states having 30 percent conservatives, your estimate would be (circle one)

a score of about 45.   a score of about 65.   a score of about 85.

E. The adjusted R-square for this relationship is equal to (fill in the blank) _____. This tells you that about (fill in the blank) _____ percent of the variation in cons_hr is explained by conpct_m.

F. Use Graphs → Legacy Dialogs → Scatter/Dot to create a scatterplot of the relationship between conpct_m (X-axis) and cons_hr (Y-axis). Enhance the graph's data-ink ratio by following the procedures described in this chapter for creating an erased graph. Print the graph.

G. Based on your inspection of the graph, the regression line, and adjusted R-square, which congressional scholar is more correct? (fill in the appropriate blank)

Scholar 1 is more correct because _____

_____.

Scholar 2 is more correct because _____

_____.

3. (Dataset: States. Variables: kerry04, to_0004.) An article of faith among Democratic Party strategists (and a source of apprehension among Republican strategists) is that high voter turnouts help Democratic candidates. Why should this be the case? According to the conventional wisdom, Democratic electorates are less likely to vote than are Republican voters. Thus low turnouts naturally favor Republican candidates. As turnouts push higher, the reasoning goes, a larger number of potential Democratic voters will go to the polls, creating a better opportunity for Democratic candidates. Therefore, as turnouts go up, so should the Democratic percentage of the vote.[12]

A. Use Regression → Linear to test this conventional wisdom. States contains to_0004, the percentage-point change in presidential election turnout between 2000 and 2004. States in which turnout declined between 2000 and 2004 have negative values on to_0004; states in which turnout increased have positive values on to_0004. (For example, Florida's turnout increased from 53.4 percent to 61.8 percent, giving Florida a value of 8.4 on to_0004. Arizona's turnout dropped from 44.6 percent to 42.3 percent, giving it a value of −2.3 on to_0004.) In this exercise, to_0004 is the independent variable. Another variable, kerry04, the percentage of the vote cast for Democratic candidate John Kerry, is the dependent variable.

Based on your results, the regression equation for estimating the percentage votes cast for Kerry is (fill in the blank)

44.561 + _____*to_0004.

B. The P-value for the regression coefficient on to_0004 is (fill in the blank) _____ , and the value of adjusted R-square is (fill in the blank) _____ .

C. Consider your findings in parts A and B. You may conclude that (fill in the appropriate blank)

the conventional wisdom is correct because _____

_____ .

the conventional wisdom is incorrect because _____

_____ .

4. (Dataset: States. Variables: abortlaw, permit.) As you are no doubt aware, in its momentous decision in *Roe v. Wade* (1973) the U.S. Supreme Court declared that states may not outlaw abortion. Even so, many state legislatures have enacted restrictions and regulations that, while not banning abortion, make an abortion more difficult to obtain. Other states, however, have few or no restrictions. What factors might explain these differences in abortion laws among the states? We know that the mass public remains divided on this issue. Public opinion in some states is more favorable toward permitting abortion and in other states is less favorable. Does public opinion guide state policy on this issue?

States contains abortlaw, which measures the number of abortion restrictions a state has enacted into law. Values on abortlaw range from 0 (least restrictive) to 10 (most restrictive). This is the dependent variable. States also has the variable permit, the percentage of the mass public saying that abortion should "always" be permitted. This is the independent variable.

A. If you were to use regression analysis to test the idea that public opinion on abortion affects state abortion policy, then you would expect to find (check one)

❏ a negative sign on permit's regression coefficient.

❏ a positive sign on permit's regression coefficient.

B. Analyze the abortlaw-permit relationship using Regression → Linear. According to the results, the regression equation for estimating the number of abortion restrictions is (fill in the blanks)

_____   _____ * permit
　　(constant)　　　　　　　　(regression coefficient)

C. The P-value for the regression coefficient is (fill in the blank) _____ . The value of adjusted R-square is (fill in the blank) _____ .

D. According to States, about 44 percent of Colorado residents believe that abortion should "always" be permitted. In Louisiana, by contrast, only about 14 percent of the public holds this view. Based on the regression equation (fill in the blanks)

you would estimate that Colorado would have about _____ abortion restrictions.

you would estimate that Louisiana would have about _____ abortion restrictions.

E. Use Graphs → Legacy Dialogs → Scatter/Dot to create a scatterplot of the relationship between permit (X-axis) and abortlaw (Y-axis). Enhance the graph's data-ink ratio by following the procedures described in this chapter for creating an erased graph. Print the graph.

5. (Dataset: States. Variables: demstate, dempct_m, libpct_m.) In Exercise 2 you analyzed the connection between mass political attitudes and congressional voting, and in Exercise 4 you examined the link between public opinion and public policy. In this exercise you will use correlation and multiple regression to examine a set of relationships between mass attitudes and the partisan make-up of state legislatures. State legislatures are remarkably varied in this regard—ranging in partisan composition from about 25 percent Democratic to over 88 percent Democratic. What accounts for this variation? Consider two plausible independent variables: the percentage of a state's citizens who are self-identified Democrats, and the percentage of citizens who are self-described liberals. Each of these variables should have a positive relationship with the percentage of Democrats in the state legislature.

A. States contains these three variables: demstate, the percentage of state legislators who are Democrats; dempct_m, the percentage of Democrats in the mass electorate; and libpct_m, the percentage of self-described liberals in the mass public. Run Correlate to find the Pearson correlation coefficients among demstate, dempct_m, and libpct_m. Fill in the four empty cells of this correlation matrix:

		Percent state legislators Democratic	Percent mass public Democratic	Percent mass public liberal
Percent state legislators Democratic	Pearson Correlation	1	.437	
Percent mass public Democratic	Pearson Correlation	.437	1	
Percent mass public liberal	Pearson Correlation			1

B. According to the correlation coefficient, as the percentage of liberals in the mass public increases, the percentage of Democratic state legislators (circle one)

increases.   decreases.

C. Suppose someone were to make this claim: "Being a Democrat and being a liberal are practically synonymous. The relationship between the percentage of Democratic identifiers and the percentage of liberals, therefore, will be positive and strong." According to the correlation coefficient, this claim is (fill in the appropriate blank)

correct because_____

_____.

incorrect because_____

_____.

D. Run Regression → Linear to obtain multiple regression estimates for the partial effects of dempct_m and libpct_m on demstate. Demstate is the dependent variable, and dempct_m and libpct_m are the independent variables. Based on your results, the multiple regression for estimating the percentage of Democratic state legislators is (fill in the blanks)

−24.938 + _____ *dempct_m + _____ *libpct_m.

E. The P-value for the regression coefficient on dempct_m is (fill in the blank) _____, and the P-value for the regression coefficient on libpct_m is (fill in the blank) _____.

F. As you may know, Nebraska's state legislature is unique in two ways: It is unicameral (all other state legislatures are bicameral), and it is nonpartisan. Candidates do not run for the state legislature using party labels, and the legislature is not organized on the basis of party. Thus Nebraska has a missing value on the variable demstate, and it was not included in the regression analysis you just performed. However, if you were to peruse States, you would find that 29.03 percent of Nebraskans are Democrats and 16.44 percent are self-described liberals.

For the sake of speculation, assume that Nebraska decided that all members of the state legislature should declare a partisan allegiance. Based on your regression model, about what percentage of state legislators would be Democrats? (circle one)

About 25 percent   About 40 percent   About 55 percent

G. Based on your interpretation of the multiple regression output, you can conclude that (check all that apply)

❏ controlling for the percentage of the mass public who are liberal, a 1-percentage-point increase in the percentage of Democrats in the mass public is associated with about a 4.3-percentage-point increase in the percentage of Democratic state legislators.

❏ controlling for the percentage of the mass public who are Democratic, a 1-percentage-point increase in the percentage of liberals in the mass public is associated with about a 2.3-percentage-point increase in the percentage of Democratic state legislators.

❏ both independent variables are significantly related to the dependent variable.

❏ the relationship between dempct_m and demstate is spurious.

❏ taken together, both independent variables explain about one-half of the variation in the dependent variable.

6. (Dataset: GSS2006. Variables: civlibs, educ, polviews, pol_trust, exploit3.) In recent years there has been much public debate about the "civil liberties tradeoff." To what extent is the public's belief in free speech and association counterbalanced by a desire for more security against terrorist activity? What factors determine whether a person is pro–civil liberties or pro-security? Certainly such attitudes have more than one source. According to Darren W. Davis and Brian D. Silver, people with higher levels of education will be more reluctant than the less educated to trade civil liberties for security, and liberals will be more pro–civil liberties than will conservatives. Interestingly, Davis and Silver theorize that two sorts of trust—political trust and interpersonal trust—will pull in opposite directions. People harboring greater trust in political authorities will be more pro-security. However, as faith in fellow citizens goes up—as interpersonal trust increases—individuals will be more pro–civil liberties. Thus "higher interpersonal trust might partly compensate for the effect of higher trust in government."[13] In this exercise you will test the Davis-Silver model of support for civil liberties.

In Chapter 3 you created civlibs, a scale that ranges from 0 (the strongest pro-security position) to 9 (the strongest pro–civil liberties position). The GSS2006 variable educ measures education by the number of years of formal schooling (from 0 to 20 years). Self-described ideology is captured by polviews, which runs from 1 ("extremely liberal") to 7 ("extremely conservative"). The variable pol_trust uses a five-point scale, "strongly disagree" (coded 1) through "strongly agree" (coded 5), to elicit the level of agreement with the statement "Most government administrators can be trusted." Higher scores denote higher political trust. Finally, exploit3 measures interpersonal trust by asking respondents whether they agree (coded 0), neither agree nor disagree (coded 1), or disagree (coded 2) with the statement "If you are not careful, other people will take advantage of you." So, higher codes mean higher interpersonal trust.

A. Imagine running a multiple regression using civlibs as the dependent variable and educ, exploit3, pol_trust, and polviews as independent variables. According to the Davis-Silver idea, which of the following variables will have a positive relationship with civlibs? (circle all that apply)

educ   exploit3   pol_trust   polviews

Which of the following variables will have a negative relationship with civlibs? (circle all that apply)

educ   exploit3   pol_trust   polviews

B. Run the regression analysis. Which of the following variables have a positive and statistically significant relationship with civlibs? (circle all that apply)

educ   exploit3   pol_trust   polviews

Which of the following variables have a negative and statistically significant relationship with civlibs? (circle all that apply)

educ   exploit3   pol_trust   polviews

C. Use the regression equation to estimate the civil liberties scale score for the typical respondent, which we will define as a person having the median values of all the independent variables. Run Frequencies to obtain the median values for each independent variable. Write the medians in the table that follows (the median of educ already appears in the table):

	educ	exploit3	pol_trust	polviews
Median	13	?	?	?

D. When you use the median values to estimate the civil liberties score for the typical person, you obtain an estimate equal to (fill in the blank) _____ .

E. Now let's get an idea of the importance of political trust in shaping civil liberties beliefs, controlling for educ, exploit3, and polviews.

Assuming median values for educ, exploit3, and polviews, the estimated civil liberties score for people having the lowest level of political trust is equal to (fill in the blank) _____ .

Again assuming median values for educ, exploit3, and polviews, the estimated civil liberties score for people having the highest level of political trust is equal to (fill in the blank) _____ .

F. Suppose you witnessed two nonviolent but vociferous groups of demonstrators outside the halls of Congress. One side carries signs indicating that they strongly favor an expansion of civil liberties. The other side's signs advocate just the opposite: more government power to detain people and gather information about citizens. You administer a questionnaire to individuals from both sides, obtaining information on educational attainment, social trust, political trust, and ideology. Before you can administer the civlibs questions, however, the demonstrators disperse. When you examine your incomplete dataset, you find that both groups average 13 years of schooling. On the other variables, though, the groups are starkly different. The pro–civil liberties demonstrators have the highest level of interpersonal trust and the lowest level of political trust, and they are all "extremely liberal." The pro-security crowd has the lowest level of interpersonal trust and the highest level of political trust, and they are all "extremely conservative."

Based on your regression analysis, the pro–civil liberties demonstrators would have a score of (fill in the blank) _____ on the civlibs scale.

Based on your regression analysis, the pro-security demonstrators would have a score of (fill in the blank) _____ on the civlibs scale.

That concludes the exercises for this chapter. Before exiting SPSS, be sure to save your output file.

NOTES

1. Regression analysis on variables measured by percentages can be confusing. Always stay focused on the exact units of measurement. One percentage point would be 1.00. So if christad increases by 1.00, then womleg decreases, on average, by .30, or .30 of a percentage point.
2. SPSS Regression → Linear reports two-tailed P-values, not one-tailed P-values. Strictly speaking, then, you may correctly apply the .05 standard by rejecting the null hypothesis for any reported P-value of equal to or less than .10. However, in this book we follow the more conservative practice of rejecting the null hypothesis for P-values of equal to or less than .05.
3. The t-ratio for the Y-intercept permits you to test the null hypothesis that, in the population, the Y-intercept is 0. In this case we have no interest in testing the hypothesis that states having 0 Christian adherents have 0 percent women in their state legislatures.
4. Most data analysis programs, SPSS included, provide two values of R square—a plain version, which SPSS labels "R Square," and an adjusted version, "Adjusted R Square." Adjusted R Square is often about the same as (but is always less than) plain R Square. What is the difference? Just like a sample mean, which provides an estimate of the unseen population mean, a sample R-square provides an estimate of the true value of R-square in the population. And just like a sample mean, the sample R-square is equal to the population R-square, give or take random sampling error. However, unlike the random error associated with a sample mean, R-square's errors can assume only positive values—squaring any negative error, after all, produces a positive number—introducing upward bias into the estimated value of R-square. This problem, which is more troublesome for small samples and for models with many independent variables, can be corrected by adjusting plain R-square "downward." For a sample of size N and a regression model with k predictors, adjusted R-square is equal to: $1 - (1 - \text{R-square})[(N - 1)/(N - k - 1)]$. See Barbara G. Tabachnick and Linda S. Fidell, *Using Multivariate Statistics,* 3d ed. (New York: HarperCollins, 1996), 164–165.
5. If you do a quick Descriptives run, you will find that the lowest value of college is 17 percent.
6. Edward R. Tufte, *The Visual Display of Quantitative Information,* 2d ed. (Cheshire, Conn.: Graphics Press, 2001). Tufte's work has inspired other excellent treatments of visual communication. For example, see Stephen Few, *Show Me the Numbers: Designing Tables and Graphs to Enlighten* (Oakland, Calif.: Analytics Press, 2004); Howard Wainer, *Graphic Discovery: A Trout in the Milk and Other Visual Adventures* (Princeton: Princeton University Press, 2005).
7. When SPSS adds a regression line to a chart, it also adds the text box, "R Sq Linear = ." The Chart Editor always adds this box in the same place, the graph's lower right-hand corner, even if the box obstructs the data points or the fit line. This annoyance is easily resolved. Click on the box and drag it to a different location inside the graphic space. Alternatively, you might want to click on the box and delete it.
8. You will want to keep the Properties window open for your entire excursion into the Chart Editor. Each time you select a different part of the graph for editing, SPSS automatically adjusts the Properties window to reflect the editable features of the graphic element you have selected. Naturally, you can open the Properties window upon entering the Chart Editor by clicking the Properties button.
9. With the Chart Editor still open, click File → Save Chart Template. In the Save Chart Template window, click in the All Settings box, which selects all chart features. Now uncheck the box next to Text Content. (You don't want SPSS to apply the same axis titles to all of your scatterplots.) Click Continue. Find a good place to save the template (and concoct a descriptive name for the file), which SPSS saves with the .sgt extension. To apply the template to future editing projects: In the Chart Editor, click File → Apply Chart Template, find the .sgt file, and click Open. Experience teaches that SPSS will apply most of the template's features to the new graphic, although some minor editing may still be required.
10. You might want to make two additional Tufte-recommended changes to the scatterplot, one of which SPSS can accommodate and one of which it can't. The accommodated edit: Instead of the default sans serif font (Arial), which SPSS applies to all text elements, use a serif font, such as Times New Roman. Tufte and other graphics experts argue that serif fonts, although busier and more detailed, are easier to read. On the unaccommodating side, SPSS

will not permit the user to edit the displayed regression line so that it falls only within the range of observed values of the independent variable. SPSS anchors the line to the Y-axis and extrapolates beyond the highest observed value of X. Properly displayed, an estimation line should be restricted to the *range frame,* defined by the lowest and highest observed values of the independent variable. Currently, SPSS permits this sort of editing only in the case of loess fit lines, for which the user may control the range of display. The separate graphics editor for the classic SPSS Interactive → Scatterplot—an editor that SPSS no longer supports—did it correctly. This capability has not been carried forward into the current SPSS Chart Editor, which edits images created by all SPSS graphics protocols, including Interactive → Scatterplot.

11. Edward R. Tufte, *Data Analysis for Politics and Policy* (Englewood Cliffs, N.J.: Prentice Hall, 1974), 21. Tufte uses regression analysis to evaluate the effectiveness of motor vehicle inspections, controlling for population density.

12. See Michael D. Martinez and Jeff Gill, "The Effects of Turnout on Partisan Outcomes in U.S. Presidential Elections 1960–2000," *Journal of Politics* 67 (4), November 2005: 1248–1274. Martinez and Gill find that the Democratic advantage from higher turnouts has declined over time.

13. Darren W. Davis and Brian D. Silver, "Civil Liberties vs. Security: Public Opinion in the Context of the Terrorist Attacks on America," *American Journal of Political Science,* 48 (January 2004), 28–46. This quote, p. 31.

9

Dummy Variables and Interaction Effects

Procedures Covered

Transform → Recode into Different Variables (dummy variables)
Analyze → Regression → Linear (with dummy variables)
Transform → Compute → If (optional case selection condition)
Analyze → Regression → Linear (with interaction variable)

You can adapt regression analysis to different research situations. In one situation you might have nominal or ordinal independent variables. Provided that these variables are dummy variables, you can run a regression analysis, using categorical variables to predict values of an interval-level dependent variable. In this chapter you will learn how to construct dummy variables and how to use them in regression analysis. In a second research situation you might suspect that the effect of one independent variable on the dependent variable is not the same for all values of another independent variable—in other words, that interaction is going on in the data. Provided that you have created an interaction variable, you can use multiple regression to estimate the size and statistical significance of interaction effects. In this chapter you will learn how to create an interaction variable and how to perform and interpret multiple regression with interaction effects.

REGRESSION WITH DUMMY VARIABLES

A dummy variable can take on only two values, 1 or 0. Each case being analyzed either has the characteristic being measured (a code of 1) or does not have it (a code of 0). For example, a dummy variable for gender might code females as 1 and males as 0. Everybody who is coded 1 has the characteristic of being female, and everybody who is coded 0 does not have that characteristic. To appreciate why this 0 or 1 coding is the essential feature of dummy variables, consider the following regression model, which is designed to test the hypothesis that women will give Democratic presidential candidate John Kerry higher feeling thermometer ratings than will men:

$$\text{Kerry feeling thermometer} = a + b(\text{female}).$$

In this formulation, gender is measured by a dummy variable, female, which is coded 0 for males and 1 for females. Since males are scored 0 on the dummy, the constant or intercept, a, will tell us the average Kerry rating among men. Why so? Substituting 0 for the dummy yields: $a + b*0 = a$. In the language of dummy variable regression, males are the "omitted" category, the category whose mean value on the dependent variable is captured by the intercept, a. The regression coefficient, b, will tell us how much to adjust the intercept for women—that is, when the dummy switches from 0 to 1. Thus, just as in any regression, b will estimate the average change in the dependent variable for a unit change in the independent variable. Since in this case a unit change in the independent variable is the difference between men (coded 0 on female) and women (coded 1 on female), the regression coefficient will reflect the mean difference in Kerry thermometer

Figure 9–1 Recoding to Create a Dummy Variable

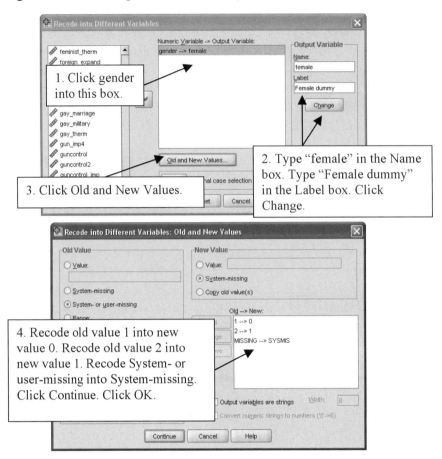

ratings between males and females. It is important to be clear on this point: The coefficient, b, does not communicate the mean Kerry rating among females. Rather, it estimates the mean difference between males and females. (Of course, an estimated value of the dependent variable among females can be arrived at easily by summing a and b: a + b*1 = a + b.) As with any regression coefficient, we can rely on the coefficient's t-ratio and P-value to test the null hypothesis that there is no statistically meaningful gender difference in thermometer ratings of Kerry.

Let's open NES2004 and figure out how to use gender as an independent variable in a regression analysis of Kerry thermometer ratings. We'll use kerry_therm as the dependent variable. The independent variable, gender, is a nominal-level measure, coded 1 for males and 2 for females. Because of the way it is currently coded, gender could not be used in regression analysis. How can we create a dummy variable, female, coded 0 for males and 1 for females? We could assign these values by using Transform → Recode into Different Variables and applying this recoding scheme:

Respondent's gender	Old value (gender)	New value (female)
Male	1	0
Female	2	1
	Missing	Missing

You know how to use Transform → Recode into Different Variables, so go ahead and create female, which you can label "Female dummy." (Figure 9-1 helps to reacquaint you with the Recode procedure.)

To check your work, run Frequencies on gender and female to ensure that the distributions are the same.

gender R gender

		Frequency	Percent	Valid Percent	Cumulative Percent
Valid	1 1 Male	566	46.7	46.7	46.7
	2 2 Female	646	53.3	53.3	100.0
	Total	1212	100.0	100.0	

female Female dummy

		Frequency	Percent	Valid Percent	Cumulative Percent
Valid	0	566	46.7	46.7	46.7
	1	646	53.3	53.3	100.0
	Total	1212	100.0	100.0	

The 646 respondents coded 2 on gender are coded 1 on female, and the 566 respondents coded 1 on gender are coded 0 on female. Return to the Variable View and assign value labels to the dummy variable you have created ("Male" for value 0, "Female" for value 1).

Now let's run linear regression, using the John Kerry feeling thermometer (kerry_therm) as the dependent variable and female as the independent variable. Click Analyze → Regression → Linear. Click kerry_therm into the Dependent box, and click female into the Independent(s) box. Click OK.

Model Summary

Model	R	R Square	Adjusted R Square	Std. Error of the Estimate
1	.082[a]	.007	.006	26.283

a. Predictors: (Constant), female Female dummy

Coefficients[a]

Model		Unstandardized Coefficients		Standardized Coefficients	t	Sig.
		B	Std. Error	Beta		
1	(Constant)	50.730	1.111		45.676	.000
	female Female dummy	4.320	1.526	.082	2.831	.005

a. Dependent Variable: kerry_therm Feeling Thermometer: John Kerry

According to the Coefficients table, the regression equation is as follows:

Kerry feeling thermometer = 50.730 + 4.320*Female dummy.

How would we interpret these estimates? As always, the constant estimates the value of the dependent variable when the independent variable is 0. Because males have a value of 0 on female, the mean thermometer rating of John Kerry for males is 50.730, the intercept. The regression coefficient on female communicates the mean change in the dependent variable for each unit change in the independent variable. So when the dummy switches from 0 to 1, the Kerry rating goes up, on average, about 4.3 degrees. We can use this value to estimate the mean rating for females: 50.730 + 4.320 = 55.05. So men rated Kerry at about 51 and women rated him at about 55. Was this gender difference produced by random sampling error? Not according to the P-value, .005. Do gender differences account for a big chunk of the variation in John Kerry thermometer ratings? Not exactly. According to the adjusted R-square, gender alone accounts for less than 1 percent of the variation in the dependent variable. There must be other variables that contribute to the explanation of Kerry's ratings. Let's expand the model.

We would expect partisanship to have a big effect on the Kerry thermometer scale. Democrats should score higher on the dependent variable than do Independents or Republicans. Plus, we know that women are more likely than men to be Democrats, so the kerry_therm-female relationship might be the spurious result of partisan differences, not gender differences. NES2004 contains partyid3, which codes Democrats as 1, Independents as 2, and Republicans as 3. Because partyid3 is a categorical variable, we cannot use it in a regression—not in its present form, anyway. But we can use partyid3 to create a dummy variable for partisanship.

Actually, we need to create not one but two dummy variables from partyid3. Why two? Here is a general rule about dummy variables: If the variable you want to "dummy-ize" has k categories, then you need k–1 dummies to measure the variable. Because partyid3 has three categories, we need two dummy variables. One of these variables, which we will call demdum, is equal to 1 for Democrats and 0 for Independents and Republicans. The second dummy variable, repdum, is equal to 1 for Republicans and 0 for Democrats and Independents. Independents, then, are uniquely identified by their exclusion from both dummies. Independents have values of 0 on demdum and 0 on repdum. Consider this recoding protocol:

Party ID: 3 Categories	Old value (partyid3)	New value (demdum)	New value (repdum)
Democrat	1	1	0
Independent	2	0	0
Republican	3	0	1
	Missing	Missing	Missing

We will create demdum and repdum one at a time. To create demdum, click Transform → Recode into Different Variables. (The gender recode is still in the window, so click Reset.) Follow these steps:

1. Click partyid3 into the Numeric Variable → Output Variable panel.
2. Click in the Name box and type "demdum."
3. Click in the Label box and type "Democrat dummy." Click Change.
4. Click Old and New Values.

In the Recode Into Different Variables: Old and New Values window, recode old value 1, the partyid3 code for Democrats, into new value 1, the code for Democrats on demdum. Old values 2 and 3, the partyid3 codes for Independents and Republicans, are equal to new value 0, the code for Independents and Republicans on demdum. You can use the Range boxes in the Old Value panel to accomplish this change. Make sure that you recode missing values on partyid3 into missing values on demdum. The Old and New Values window should now look like the upper panel of Figure 9-2.

Repeat the recoding procedure, using partyid3 to create repdum. Return to the Transform → Recode into Different Variables window. To avoid confusion, click the Reset button first and then follow these steps:

1. Click partyid3 into the Numeric Variable → Output Variable panel.
2. Type "repdum" in the Name box.
3. Type "Republican dummy" in the Label box and click Change.
4. Click Old and New Values.

This time recode old value 3 on partyid3 into new value 1 on repdum. Old values 1 and 2 become 0 in the new values of repdum. Again, make sure to recode missing values on partyid3 into missing values on repdum (see the lower panel of Figure 9-2).

Before analyzing these new variables, it would be prudent to check your work. Run a quick Frequencies on partyid3, demdum, and repdum.

partyid3 R Party ID: 3 cats

		Frequency	Percent	Valid Percent	Cumulative Percent
Valid	1 Democrat	382	31.5	32.0	32.0
	2 Independent	466	38.4	39.0	71.0
	3 Republican	347	28.6	29.0	100.0
	Total	1195	98.6	100.0	
Missing	System	17	1.4		
Total		1212	100.0		

demdum Democrat dummy

		Frequency	Percent	Valid Percent	Cumulative Percent
Valid	0	813	67.1	68.0	68.0
	1	382	31.5	32.0	100.0
	Total	1195	98.6	100.0	
Missing	System	17	1.4		
Total		1212	100.0		

repdum Republican dummy

		Frequency	Percent	Valid Percent	Cumulative Percent
Valid	0	848	70.0	71.0	71.0
	1	347	28.6	29.0	100.0
	Total	1195	98.6	100.0	
Missing	System	17	1.4		
Total		1212	100.0		

Figure 9–2   Creating Two Dummy Variables from a Three-Category Ordinal

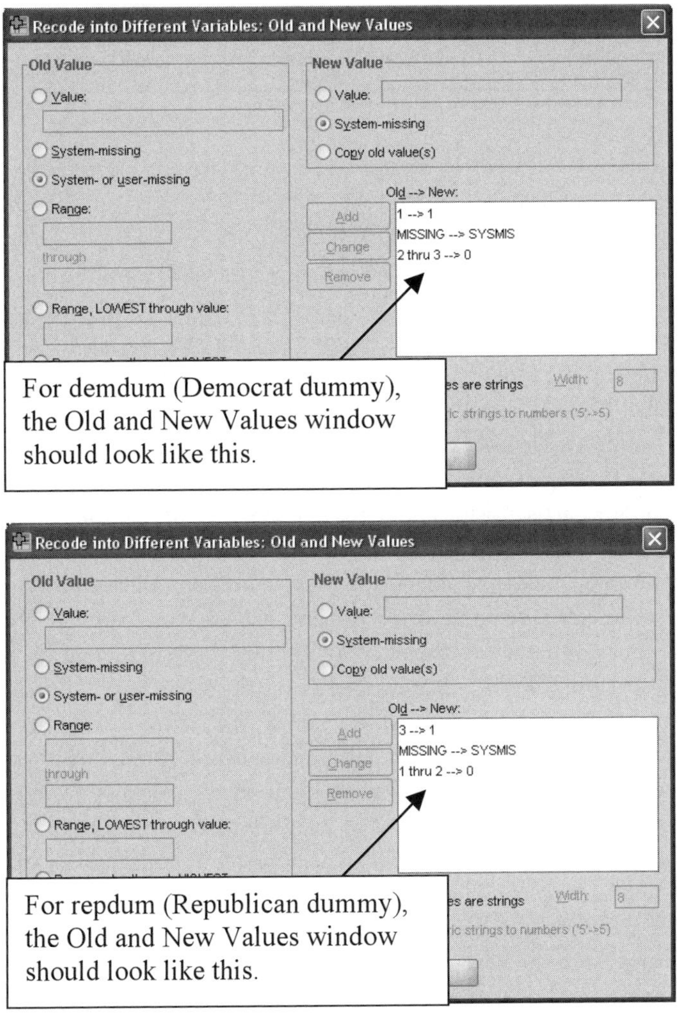

For demdum (Democrat dummy), the Old and New Values window should look like this.

For repdum (Republican dummy), the Old and New Values window should look like this.

According to the distribution of partyid3, NES2004 has 382 Democrats, 347 Republicans, and 466 Independents. According to the distribution of demdum, 382 respondents are coded 1 on the Democrat dummy. And according to the distribution of repdum, 347 respondents are coded 1 on the Republican dummy. So these codes check out. Notice, too, that the number of people coded 0 on the Democrat dummy (813) is equal to the number of Republicans (347) plus the number of Independents (466); and the number of respondents coded 0 on the Republican dummy (848) is equal to the number of Democrats (382) plus the number of Independents (466). Everything checks. Before proceeding, return to the Variable View and label the values for each of these new variables.

At last we are ready to run a multiple regression analysis of kerry_therm, using female, demdum, and repdum as independent variables. Click Analyze → Regression → Linear. Kerry_therm should still be in the Dependent panel and female in the Independent(s) panel. Fine. Click both of the partisanship dummies, demdum and repdum, into the Independent(s) panel. Click OK. Let's see what we have.

Model Summary

Model	R	R Square	Adjusted R Square	Std. Error of the Estimate
1	.602[a]	.362	.361	21.011

a. Predictors: (Constant), repdum Republican dummy, female Female dummy, demdum Democrat dummy

Coefficients[a]

Model		Unstandardized Coefficients		Standardized Coefficients	t	Sig.
		B	Std. Error	Beta		
1	(Constant)	53.425	1.144		46.709	.000
	female Female dummy	1.922	1.239	.037	1.551	.121
	demdum Democrat dummy	16.888	1.478	.300	11.425	.000
	repdum Republican dummy	-23.099	1.503	-.401	-15.367	.000

a. Dependent Variable: kerry_therm Feeling Thermometer: John Kerry

The regression equation is as follows (to enhance readability, we will shorten the variable names to "Female," "Democrat," and "Republican"):

Kerry thermometer rating = 53.425 + 1.922*Female + 16.888*Democrat − 23.099*Republican.

First, get oriented by using the constant, 53.425, as a point of reference. Again, because this value estimates the dependent variable when all the independent variables are 0, 53.425 is the mean Kerry rating for males who are Independents. Why so? Because all the dummies are switched to 0: Female is 0 (that's the "male" part of the intercept) and both the Democrat dummy and the Republican dummy are 0 (that's the "Independent" part of the intercept). The regression coefficient on Female tells us how much to adjust the "male" part of the intercept, controlling for partisanship. The regression coefficients on the partisanship dummies tell us how much to adjust the "Independent" part of the intercept, controlling for gender. Thus, compared with Independents, Democrats average nearly 17 degrees higher—and Republicans score more than 23 degrees lower—on the Kerry thermometer. The partisan coefficients are large and statistically significant, with huge t-ratios and miniscule P-values. What about the effect of gender? The coefficient on Female, 1.922, tells us that women, on average, score only about 2 degrees higher on the Kerry scale, controlling for partisanship. This weak effect fails to trump the null hypothesis (t = 1.55 with P-value = .121). In the earlier regression, using the female dummy alone, we found a gender difference of more than 4 degrees. That regression, of course, didn't account for the fact that women are more likely than men to be Democrats. After taking party differences into account, the gender difference fades to insignificance.

Overall, however, the model performs fairly well. The adjusted R-square value of .361 tells us that all of the independent variables, taken together, account for about 36 percent of the variation in the dependent variable. So the "glass is 36 percent full." A skeptic would point out, of course, that the "glass is still 64 percent empty." Before going on to the next section, you may want to exercise your new skills by creating new dummies and further expanding the model. In any event, before proceeding be sure to save the dataset.

INTERACTION EFFECTS IN MULTIPLE REGRESSION

Multiple regression is a linear and additive technique. It assumes a linear relationship between the independent variables and the dependent variable. It also assumes that the effect of one independent variable on the dependent variable is the same for all values of the other independent variables in the model. In the regression we just estimated, for example, multiple regression assumed that the effect of being female is the same for all values of partisanship—that Democratic females are about 2 degrees warmer toward John Kerry than are Democratic males and that Republican females and independent females also are 2 degrees warmer than are their male counterparts. This assumption works fine for additive relationships. However, if interaction is taking place—if, for example, the gap between male and female ratings is significantly larger among Republicans than among Democrats or independents—then multiple regression will not capture this effect. Before researchers attempt to model interaction effects by using multiple regression, they have usually performed preliminary analyses that suggest such effects are occurring in the data.

Consider an interesting theory in American public opinion. According to this perspective, which we will call the "polarization perspective," political disagreements are often more intense among people who are more interested in and knowledgeable about public affairs than they are among people who are disengaged or who lack political knowledge.[1] For example, it could be reasonably hypothesized that individuals who favor a bigger role for government in solving problems would give the Democratic Party higher ratings than would individuals who advocate a smaller role for government. So if we were to compare ratings on a Democratic Party feeling thermometer for anti–big government and pro–big government respondents, we should find a higher mean among those who support bigger government. According to the polarization perspective, however, this relationship will be weaker for people with low political knowledge than for people with higher political knowledge. Among people with lower political knowledge, the mean difference in Democratic ratings may be modest, with pro–big government respondents giving the Democratic Party somewhat higher average ratings than do anti–big government respondents. As political knowledge increases, however, this mean difference should increase, reflecting greater polarization between the opponents and supporters of bigger government. Thus the strength of the relationship between attitudes toward big government and evaluations of the Democratic Party will depend on the level of political knowledge.

NES2004 contains dem_therm, which records respondents' feeling thermometer ratings of the Democratic Party. Another variable, progovmnt, is a 4-point scale that gauges respondents' attitudes toward the role of government, from 0 (anti–big government) to 3 (pro–big government).[2] A third variable, polknow3, measures each respondent's political knowledge by three values: 0 (low knowledge), 1 (medium knowledge), or 2 (high knowledge). First, let's do a preliminary analysis and find out if the polarization perspective has merit. We want to examine the mean of dem_therm (dependent variable) for each value of progovmnt (independent variable), controlling for polknow3. Figure 9-3 provides a review of how to accomplish this. Let's look at the results of this analysis.

dem_therm Feeling Thermometer: Democratic party

polknow3 ...	progovmnt ...	Mean	N
0 Low	0 Less govt	52.07	46
	1	58.62	58
	2	63.17	101
	3 More govt	63.53	187
	Total	61.36	392
1 Med	0 Less govt	40.40	72
	1	59.47	76
	2	61.39	83
	3 More govt	65.66	139
	Total	58.52	370
2 High	0 Less govt	38.57	87
	1	45.93	41
	2	60.24	37
	3 More govt	68.16	68
	Total	51.94	233
Total	0 Less govt	42.24	205
	1	56.02	175
	2	62.01	221
	3 More govt	65.08	394
	Total	58.10	995

Figure 9–3 Obtaining Mean Comparisons with a Control Variable

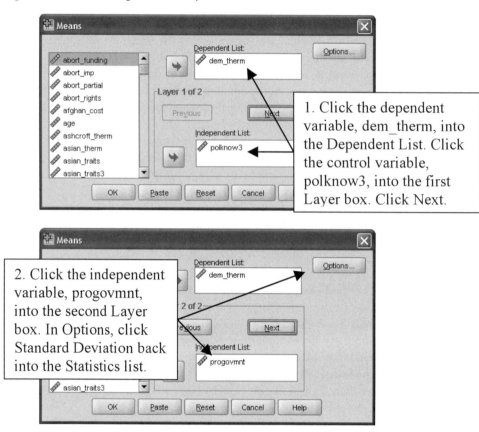

In the way that the table is set up, we would assess the effect of the pro-government scale, at each level of political knowledge, by reading down the column within each value of the control. Consider low-knowledge respondents. As we move from low values of progovmnt to higher values of the variable, do mean Democratic ratings increase? Yes, they do—from about 52 degrees for individuals favoring less government to about 64 degrees for individuals favoring more government, about a 12-point increase. But notice that this relationship is substantially stronger at medium and high levels of political knowledge. For people with medium knowledge, Democratic ratings rise from around 40 to nearly 66, a 26-point jump. And the relationship is somewhat stronger still for individuals at the highest knowledge level, for whom the data show about a 30-point difference in ratings of the Democratic Party. So it looks like the dem_therm-progovmnt relationship does indeed strengthen as political knowledge increases. And notice that the control variable, polknow3, is also related to dem_therm. According to the "Total" row in each panel of the table, thermometer ratings decline by about 10 degrees across the values of political knowledge, from about 61 among the low-knowledge group to nearly 52 among the high-knowledge group. How would we use regression analysis to estimate the size and statistical significance of these relationships?

We want to specify a regression model that does three things. First, we want to estimate the effect of progovmnt on dem_therm. Second, we want to estimate the effect of political knowledge on the dependent variable. So our regression model would begin in a familiar way:

$$\text{Democratic thermometer rating} = a + b1*progovmnt + b2*polknow3.$$

These parameters—a, b1, and b2—are the additive building blocks of the model. We can use them to estimate the mean of the dependent variable for different combinations of the independent variables. We will be able to add the effect of, say, scoring 2 on the progovmnt variable to the effect of scoring a 2 on the knowledge variable. The constant will estimate dem_therm for respondents who have a value of 0 on both independent variables—anti–big government respondents who have low political knowledge. Based on our inspection of the mean comparison table, we can expect the sign on b2 to be negative—as knowledge goes up, ratings of the Democratic Party decline. And we can expect the sign on b1 to be positive. The table shows that, at all levels of political knowledge, as people become more favorably disposed toward big government, their ratings of the Democratic Party increase. Third, however, we want to adjust the effect of progovmnt on dem_therm, depending on "where we are" on the political knowledge variable. For low values of knowledge, we want a modest adjustment to the additive effect. But as knowledge goes up, we want a larger adjustment, since the mean difference between anti–big government and pro–big government respondents increases as political knowledge increases.

In multiple regression we accomplish the third goal by including an interaction variable as an independent variable. To create an interaction variable, we multiply one independent variable by the other independent variable. Consider how we would create an interaction variable for the problem at hand: progovmnt*polknow3. All respondents who are coded 0 on progovmnt will, of course, have a value of 0 on the interaction variable. For respondents coded 1 or higher on progovmnt, however, the magnitude of the interaction variable will increase as political knowledge increases. Let's include this term in the model just discussed and see what it looks like:

Democratic thermometer rating = a + b1*progovmnt + b2*polknow3 + b3*(progovmnt*polknow3).

We would estimate the mean Democratic rating for respondents having a value of 0 on progovmnt by using the additive building blocks. We would start with the intercept, a, and add the political knowledge effect, which would be b2 times a respondent's political knowledge score. What happens as progovmnt increases? We would start with the intercept, add the pro-government effect, b1 times the respondent's progovmnt score, and add the political knowledge effect, b2 times the respondent's political knowledge score. But we also would add the interaction effect, b3 times the value of the interaction variable (progovmnt*polknow3). This coefficient, b3, tells us how much to adjust our additive estimate for each one-unit increase in political knowledge.

Let's work through the research problem and get SPSS to estimate the model for us. First, we will use Compute to create an interaction variable. Then we will run Regression to estimate the additive effects and the interaction effect.

USING COMPUTE FOR INTERACTION VARIABLES

Because NES2004 does not have the interaction variable we need for our model, we will use Compute to calculate it. Click Transform → Compute. What to name the variable? In naming interaction variables, it's a good idea to choose a name that implies higher codes on the variables whose product you wish to compute. Higher codes on progovmnt denote more favorable attitudes toward government. And higher codes on polknow3 denote greater knowledge. So a name such as "govyes_knowhi" will tell you that respondents who like big government (that's the "govyes" part of the name) and who have higher levels of knowledge (the "knowhi" part) will have higher values on the interaction variable. Type "govyes_knowhi" in the Target Variable box (Figure 9-4). In the Numeric Expression box, type "progovmnt*polknow3." Next, click Type & Label. In the Compute Variable: Type and Label window, type "progovmnt*polknow3" in the Label box. Click Continue, returning to the Compute Variable window.

Before clicking OK and computing the variable, there is one more thing to do. Whenever you create a new variable by multiplying one variable by another (as we are doing), and at least one of the variables can take on the value of 0 (as is the case here), you need to make sure that the computation is restricted to cases that have nonmissing values on both variables.[3] In the Compute Variable window, click the button labeled "If (optional case selection condition)," as shown in Figure 9-5. The grayed out Compute Variable: If Cases window appears. Select the radio button next to "Include if case satisfies condition." Doing so wakes up the window. Click in the box and type "not missing(progovmnt) & not missing(polknow3)." Click Continue. Finally set. Click OK. Click Analyze → Regression → Linear. Click Reset to clear the panels for our new analysis. Click dem_therm into the Dependent box. Click progovmnt, polknow3, and govyes_knowhi into the Independent(s) box, as shown in Figure 9-6. Click OK.

Figure 9–4 Computing an Interaction Variable

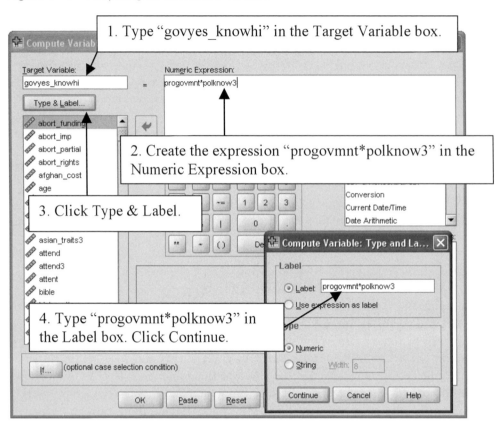

1. Type "govyes_knowhi" in the Target Variable box.

2. Create the expression "progovmnt*polknow3" in the Numeric Expression box.

3. Click Type & Label.

4. Type "progovmnt*polknow3" in the Label box. Click Continue.

Figure 9–5 Restricting Compute to Non-missing Cases

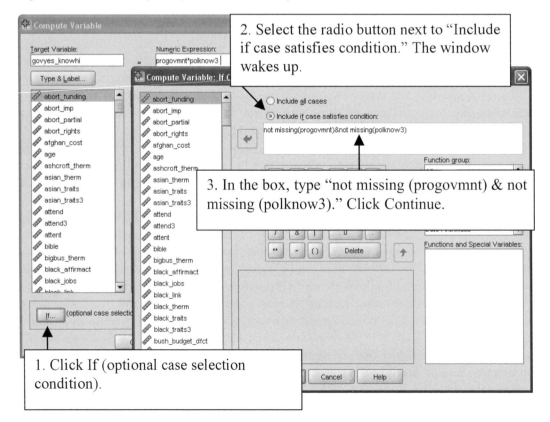

2. Select the radio button next to "Include if case satisfies condition." The window wakes up.

3. In the box, type "not missing (progovmnt) & not missing (polknow3)." Click Continue.

1. Click If (optional case selection condition).

Figure 9–6 Requesting Multiple Regression with Interaction Variable

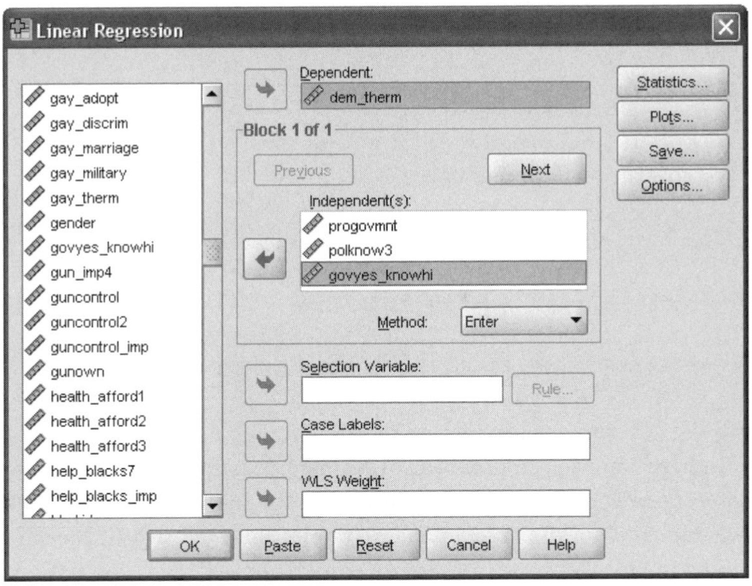

Model Summary

Model	R	R Square	Adjusted R Square	Std. Error of the Estimate
1	.370ª	.137	.135	22.598

a. Predictors: (Constant), govyes_knowhi progovmnt*polknow3 , progovmnt Pro big-government scale, polknow3 Pol know: 3 cats

Coefficientsª

Model		Unstandardized Coefficients		Standardized Coefficients	t	Sig.
		B	Std. Error	Beta		
1	(Constant)	54.099	2.267		23.864	.000
	progovmnt Pro big-government scale	3.507	.985	.168	3.560	.000
	polknow3 Pol know: 3 cats	-8.098	1.679	-.259	-4.822	.000
	govyes_knowhi progovmnt*polknow3	3.423	.791	.252	4.330	.000

a. Dependent Variable: dem_therm Feeling Thermometer: Democratic party

Let's plug the estimates into our model (we'll use the variables' names instead of their lengthy labels):

$$dem\_therm = 54.099 + 3.507*progovmnt - 8.098*polknow3 + 3.423*(govyes\_knowhi).$$

Again use the constant, 54.099, to get oriented. This is the estimated mean of dem_therm for respondents who have values of 0 on all the independent variables: anti–big government individuals (coded 0 on progovmnt) with low political knowledge (a value of 0 on polknow3). So this group averages about 54 on the dependent variable. Now consider the statistically significant coefficient on progovmnt, 3.507. This says that, controlling for political knowledge, each one-unit increase in the pro–big government scale boosts Democratic Party ratings by about 3.5 degrees, on average. For example, the estimated value of the dependent variable for low-knowledge individuals (polknow3 = 0) taking the strongest pro-government stance (progovmnt = 3) would be: 54.099 + 3.507*3 – 8.098*0 + 3.423*(3*0) = 54.099 + 10.521 – 0 + 0 = 64.62. Thus, low-knowledge, anti–big government respondents score about 54 and low-knowledge, pro–big government respondents score about 65—a difference very much in line with the tabular data obtained earlier. What happens as political knowledge increases? According to the regression coefficient on polknow3, –8.098, each one-unit increase in knowledge is associated with more than an 8-point drop on the Democratic rating scale. So as knowledge goes up, mean Democratic ratings go down. But is this what happens for all respondents as polknow3 increases in value? Not according to the large coefficient on the interaction term, 3.423. This coefficient tells us how much to adjust that 8-point decline for each one-unit increase in knowledge as progovmnt shades toward higher values.

Let's use the model to obtain estimates for the staunchest government opponents (progovmnt = 0) and strongest government supporters (progovmnt = 3) for respondents at the highest level of political knowledge (polknow3 = 2). For government opponents we would have

$$54.099 + 3.507*0 - 8.098*2 + 3.423*(0*2) = 54.099 + 0 - 16.196 + 0 = 37.903.$$

For high-knowledge, pro-big government respondents, we have

$$54.099 + 3.507*3 - 8.098*2 + 3.423*(3*2) = 54.099 + 10.521 - 16.196 + 20.538 = 68.962.$$

So high-knowledge, anti-big government people rate the Democratic Party at around 38 degrees, compared with about 69 degrees for high-knowledge, pro–big government individuals—a difference of 30-plus units on the dependent variable. Our regression model has captured the interaction effect quite nicely.

According to their P-values, all of the independent variables achieved significance in the model. But an R-square of .135 is nothing to write home about. Before you save the changes you have made to NES2004, you might want to expand the model and crank out some more regressions. Who knows? An interesting relationship might turn up.

EXERCISES

1. (Dataset: World. Variables: fhrate04_rev, gdp_cap3.) In one of the exercises in Chapter 5, you used World to investigate the relationship between economic development and democracy. In this exercise you will use multiple regression to reanalyze this relationship, using an interval-level measure of democracy (fhrate04_rev) and a set of dummy variables that you will create from gdp_cap3, a three-category ordinal measure of per-capita gross domestic product (GDP).

A. World's gdp_cap3 is coded 1 for countries with "Low" GDP per capita, 2 for countries in the "Middle" category, and 3 for countries with "High" GDP per capita. Use gdp_cap3 to create two dummy variables, one named gdp_mid and labeled "mid-gdp dummy," and the other named gdp_high and labeled "high-gdp dummy." Follow this recoding scheme:

GDP per capita 3 Categories	Old value (gdp_cap3)	New value (gdp_mid)	New value (gdp_high)
Low	1	0	0
Middle	2	1	0
High	3	0	1
	Missing	Missing	Missing

Check your recoding work by running Frequencies on gdp_cap3, gdp_mid, and gdp_high. In the table that follows, write the number of cases (raw frequencies) in the cells that have question marks:

GDP per capita: 3 cats	Frequency	gdp_mid	Frequency	gdp_high	Frequency
Low	59	0	?	0	?
Middle	?	1	?	1	?
High	?				
Valid total	177				

B. Imagine running a regression using gdp_mid and gdp_high to estimate a dependent variable: Dependent variable = Constant + b1*gdp_mid + b2*gdp_high. Complete the matching exercise below by drawing a line connecting the desired estimate on the left to the appropriate coefficient (or combination of coefficients) on the right.

Your estimate of the. . .	Would be provided by (the). . .
mean difference between countries with the lowest gdp and the highest gdp. . .	constant
mean of the dependent variable for the highest gdp countries. . .	b1
mean of the dependent variable for the lowest gdp countries. . .	constant + b2
mean difference between the lowest gdp and the middle gdp countries. . .	b2

World contains fhrate04_rev, a measure of democratic freedoms. The variable fhrate04_rev measures countries on a scale from 1 (least free) to 7 (most free). Run Regression → Linear, using fhrate04_rev as the dependent variable and gdp_mid and gdp_high as independent variables.

C. The regression equation for estimating fhrate04_rev is as follows (fill in the blanks, putting the constant in the first blank):

fhrate04_rev = _____ + _____*gdp_mid + _____*gdp_high.

D. Use the regression coefficients to arrive at estimated mean values of fhrate04_rev for countries at each level of economic development. Write the estimates in the table that follows:

GDP per capita: 3 cats	Estimated mean on democratic freedoms scale
Low	?
Middle	?
High	?

E. Which of the following conclusions are supported by your analysis? (check all that apply)

❏ Countries with middle levels of per-capita GDP do not have significantly higher values on the democratic freedoms scale than do countries with the lowest per-capita GDP.

❏ As per-capita GDP increases, democratic freedoms increase.

❏ Countries with the highest levels of per-capita GDP have significantly higher values on the democratic freedoms scale than do countries with the lowest per-capita GDP.

❏ Per-capita GDP explains less than half of the variation in the democratic freedoms scale.

2. (Dataset: World. Variables: Gini2004, Hi_gdp, democ_regime, rich_democ.) As a country becomes richer, do more of its citizens benefit economically? Or do economic resources become inequitably distributed across society? The answer may depend on the type of regime in power. Democratic regimes, which need

to appeal broadly for votes, may adopt policies that redistribute wealth. Dictatorships, by contrast, are less concerned with popular accountability, and so might hoard economic resources among the ruling elite, creating a less equitable distribution of wealth. This explanation suggests a set of interaction relationships. It suggests that, when we compare poorer democracies with richer democracies, richer democracies will have a more equitable distribution of wealth. However, it also suggests that, when we compare poorer dictatorships with richer dictatorships, richer dictatorships will have a less equitable distribution of wealth. In this exercise you will investigate this set of relationships.

World contains the variable Gini2004, which measures the extent to which wealth is inequitably distributed in society. The variable can take on any value between 0 (equal distribution of wealth) and 100 (unequal distribution of wealth). So, lower values of Gini2004 denote less economic inequality and higher values of Gini2004 denote greater economic inequality. Gini2004 is the dependent variable. World also has a dummy variable, Hi_gdp, that classifies each country as "Low GDP" (coded 0) or "High GDP" (coded 1). Hi_gdp will serve as the measure of the independent variable, level of wealth. Another dummy, democ_regime, which categorizes each country as a democracy (coded 1 and labeled "Yes" on democ_regime) or dictatorship (coded 0 and labeled "No" on democ_regime), is the control variable.

A. Exercise a skill you learned in Chapter 5. To see whether interaction is occurring, obtain a multiple line chart of the Gini2004-Hi_gdp-democ_regime relationships. Click Graphs → Legacy Dialogs → Line → Multiple. Select Other statistic and click Gini2004 into the Variable box in the Lines Represent panel. (SPSS will offer to graph the mean of Gini2004, which suits your purpose.) The independent variable, Hi_gdp, will go in the Category Axis box, and the control variable, democ_regime, will go in the Define Lines by box. Edit the graph for clarity. For example, you will want the line styles to clearly distinguish between dictatorships and democracies. Print the multiple line chart you created.

B. Examine the chart you just created. It would appear that interaction (circle one)

is   is not

occurring in the data.

Explain your reasoning. _____

_____

_____

_____

_____.

C. World contains rich_democ, an interaction variable computed by the expression Hi_gdp*democ_regime. Rich_democ takes on the value of 1 for high-GDP democracies and the value of 0 for all other countries.

Run Regression → Linear, using Gini2004 as the dependent variable and Hi_gdp, democ_regime, and rich_democ as independent variables. The regression equation for estimating Gini2004 is as follows (fill in the blanks, putting the constant in the first blank):

Gini2004 = _____ + _____*Hi_gdp + _____*democ_regime + _____*rich_democ.

D. Use the regression to arrive at estimated mean values of Gini2004 for low-GDP and high-GDP democracies and dictatorships. Write your estimates in the table that follows:

Country GDP and regime	Estimated mean of Gini2004
Low-GDP democracies	?
Low-GDP dictatorships	?
High-GDP democracies	?
High-GDP dictatorships	?

E. Suppose someone claimed that, from the standpoint of statistical significance, low-GDP dictatorships have a significantly more equitable distribution of wealth than do low-GDP democracies. This claim is (circle one)

correct.   incorrect.

Explain your reasoning._____

_____.

_____.

_____.

F. Suppose someone claimed that, as GDP increases, wealth becomes significantly more equitably distributed in democracies and significantly less equitably distributed in dictatorships. This claim is (circle one)

correct.   incorrect.

Explain your reasoning._____

_____.

_____.

_____.

Before proceeding to the next exercise, be sure to save dataset World.

3. (Dataset: GSS2006 [If you are running Student Version, use GSS2006B_Student for Exercises 3 and 5. Exercise 4 is not appropriate for GSS2006B_Student.] Variables: abortion, reliten2, educ.) The abortion issue is a perennially conflictual debate in U.S. politics. What factors divide people on this issue? Because opposition to abortion is often deeply rooted in religious convictions, you could hypothesize that individuals having strong religious ties will be more likely to oppose abortion than will those with weaker affiliations. (If this hypothesis seems too commonplace to test, be patient. It gets more interesting below.) GSS2006 contains the variable abortion, a scale that records the number of conditions under which respondents believe an abortion should be allowed. Scores can range from 0 (abortion should be allowed under none of the conditions) to 6 (abortion should be allowed under all conditions). So higher scale scores denote a stronger pro-abortion stance. GSS2006 also has the dummy variable reliten2, scored 0 for respondents with weak (or no) religious affiliations and 1 for those with strong ties. If the reliten2-abortion hypothesis is correct, then respondents who are coded 0 on reliten2 will have higher scores on the abortion variable than will respondents who are coded 1 on reliten2.

A. Run Regression to test the hypothesis that people with strong religious affiliations will be more likely to oppose abortion than will those with weak religious affiliations. Abortion is the dependent variable, and reliten2 is the independent variable. Examine the output, and fill in the blanks below:

The constant (or intercept) is equal to _____, and reliten2's regression coefficient is equal to _____. Reliten2's regression coefficient has a P-value of _____, and adjusted R-square is equal to _____.

B. Based on your analysis, you can conclude that (check all that apply)

❑ people with weaker religious affiliations score about 4 on the abortion scale.

❑ people with stronger religious affiliations score about 1 on the abortion scale.

❑ reliten2 is significantly related to abortion opinions.

❑ reliten2 explains more than 25 percent of the variation in abortion scores.

C. A critic, upon examining your results, might reasonably ask, "Did you control for education? It could be that individuals with strong religious ties have lower levels of education than do the weakly affiliated. If education is also related to abortion opinions, then you might be confusing the effect of religious attachment with the effect of education."

GSS2006 contains educ, which measures the number of years of formal education for each respondent, from 0 (no formal schooling) to 20 (20 years of formal schooling). Run the regression again, using abortion as the dependent variable and reliten2 and educ as independent variables. Based on your results, you may conclude that (check all that apply)

❑ controlling for education, the relationship between reliten2 and abortion is spurious.

❑ controlling for education, individuals with strong religious ties score more than 1 point lower on the abortion scale than do individuals with weaker religious ties.

❑ according to the regression estimates, strongly affiliated individuals with no formal schooling would score about 1 on the abortion scale.

❑ education is significantly related to abortion opinions.

❑ both independent variables together explain more than 30 percent of the variation in the dependent variable.

4. (Dataset: GSS2006 [GSS2006B_Student does not perform well for this exercise.] Variables: abortion, reliten2, educ.) One of the examples in this chapter discussed the polarization perspective—the idea that political conflict is more pronounced among people who are more knowledgeable about politics than it is among less knowledgeable people. Perhaps the same pattern applies to the relationship between strength of religious attachment and abortion opinions. That is, it could be that religious commitment has a strong effect on abortion attitudes among politically knowledgeable people but that this effect is weaker for people who have lower knowledge about politics. We can use respondents' years of education (educ) as a surrogate for political knowledge because we can reasonably assume that people with more education will be more politically knowledgeable than will less-educated people. In this exercise you will compute an interaction variable. You will then run and interpret a multiple regression that includes the interaction variable you created.

A. Use Transform → Compute to create a new variable, relhi_educhi, by multiplying reliten2 by educ. (The numeric expression will be reliten2*educ.) Give relhi_educhi the label "reliten2*educ." Make sure to restrict the computation of relhi_educhi to respondents having nonmissing values on reliten2 and educ. To do this, you will need to click "If (optional case selection condition)," and type the following: "not missing(reliten2) & not missing(educ)." Think about relhi_educhi, the interaction variable you computed.

A respondent with a weak religious affiliation (coded 0 on reliten2) has what value on relhi_educhi? (circle one)

A value of 0   A value of 1   A value equal to his or her years of education

A respondent with a strong religious affiliation (coded 1 on relint2) has what value on the interaction variable? (circle one)

A value of 0   A value of 1   A value equal to his or her years of education

B. Run a multiple regression, using abortion as the dependent variable and reliten2, educ, and relhi_educhi as the independent variables. According to the Coefficients table, the multiple regression equation for estimating scores on the abortion scale is as follows (fill in the blanks, putting the constant in the first blank):

_____ + _____*reliten2 + _____*educ + _____*relhi_educhi.

D. Consider the regression coefficients. Suppose you were to use this regression to estimate the effect of religious commitment on abortion opinions among people with no formal education, that is, for people with a value of 0 on educ.

First use the regression to estimate the mean abortion score for respondents who have 0 years of education and weak religious affiliations (coded 0 on reliten2). These respondents score (fill in the blank) _____ on the abortion scale.

Now use the regression to estimate the mean abortion score for respondents who have 0 years of education and strong religious affiliations (coded 1 on reliten2). These respondents score (fill in the blank) _____ on the abortion scale.

E. Now suppose you were to use this regression to estimate the effect of religious commitment on abortion opinions among people with 20 years of education, that is, for people with a value of 20 on educ.

First use the regression to estimate the mean abortion score for respondents who have 20 years of education and weak religious affiliations (coded 0 on reliten2). These respondents score (fill in the blank) _____ on the abortion scale.

Now use the regression to estimate the mean abortion score for respondents who have 20 years of education and strong religious affiliations (coded 1 on reliten2). These respondents score (fill in the blank) _____ on the abortion scale.

F. Think about the polarization perspective. Does the analysis support the idea that, as education increases, religious commitment plays a larger role in defining conflict on the abortion issue?

Yes   No

Briefly explain your reasoning. _____

_____.

_____.

_____.

5. (Dataset: GSS2006. Variables: polviews, race_2, homosex2.) If one were trying to predict ideological self-identification on the basis of opinions on social issues, such as homosexuality, one would expect most African Americans to be conservatives. Indeed, blacks are considerably more likely to oppose homosexuality than are whites. According to the GSS2006 data, for example, over 70 percent of blacks say that homosexuality is "always wrong," compared with 50 percent of whites. Yet only about 25 percent of blacks call themselves "conservative," compared with over a third of whites. Why? A commonly accepted view is that social issues lack *salience* for African Americans. Issues such as homosexuality may matter for whites—white opponents of homosexuality are more likely to be self-described conservatives than are white nonopponents—but they have no effect for blacks. According to this argument, blacks who think homosexuality is wrong are no more likely to call themselves conservatives than are blacks who do not think homosexuality is wrong. Or so the familiar argument goes. Is this idea correct? More research needs to be done on this question.[4]

You can model salience with interaction variables. Consider the 7-point ideological scale (polviews) as a dependent variable, ranging from "extremely liberal" at 1 to "extremely conservative" at 7. Now bring in two independent variables: a dummy variable for race (race_2, with blacks scored 1 and whites scored 0) and a dummy variable gauging opposition to homosexuality (homosex2, scored 1 if the respondent said homosexuality is "always wrong" and 0 for "not always wrong"). Finally, think about (but don't compute yet) an interaction variable, black_wrong, created by multiplying race_2 and homosex2. Examine the regression model that follows:

$$polviews = a + b1*race\_2 + b2*homosex2 + b3*black\_wrong.$$

A. The interaction variable, black_wrong, will take on a value of 1 for (check one)

❏ blacks who think that homosexuality is "always wrong."

❏ blacks who think that homosexuality is "not always wrong."

❏ all respondents.

B. To gauge the effect of homosex2 among whites, you would need to compare values of polviews for "not always wrong" whites and "always wrong" whites.

Which of the following will estimate polviews for "not always wrong" whites? (circle one)

a   a + b1   a + b2

Which of the following will estimate polviews for "always wrong" whites? (circle one)

a   a + b1   a + b2

C. Remember that higher scores on polviews denote stronger "conservative" self-identifications.

If the salience argument is correct—the idea that heightened opposition to homosexuality leads to stronger conservative ideological leanings among whites but not blacks—then the sign on the coefficient, b2, will be (circle one)

negative.   positive.   zero.

If the salience argument is correct, then the sign on the coefficient, b3, will be (circle one)

negative.   positive.   zero.

D. Use Compute to create black_wrong. The multiplicative expression is homosex2*race_2. Remember to use the If cases option to restrict computation to respondents having nonmissing values on homo-sex2 and race_2. To accomplish this, click "If (optional case selection condition)" and type: "not missing(homosex2) & not missing(race_2)."

Run Regression → Linear to obtain estimates for the model. The regression equation for estimating polviews is as follows (fill in the blanks, putting the constant in the first blank):

polviews = _____ + _____*race_2 + _____*homosex2 + _____*black_wrong.

E. Which of the variables in the model have statistically significant effects on polviews? (In determining statistical significance, round "Sig." values to the nearest hundredth. Check all that apply.)

❑ race_2

❑ homosex2

❑ black_wrong

F. Use the model to estimate polviews for "not always wrong" whites and "always wrong" whites. For "not always wrong" whites you obtain _____, and for "always wrong" whites you obtain _____.

G. Use the model to estimate polviews for "not always wrong" blacks and "always wrong" blacks. For "not always wrong" blacks you obtain _____, and for "always wrong" blacks you obtain _____.

H. Consider all the evidence you have adduced. Based on the evidence, the salience idea appears to be (circle one)

correct.   incorrect.

Explain your answer._____

_____.

_____.

_____.

That concludes the exercises for this chapter. Before exiting SPSS, be sure to save your output file.

NOTES

1. See John R. Zaller's influential work, *The Nature and Origins of Mass Opinion* (New York: Cambridge University Press, 1992).
2. The pro–big government scale (nes2004 variable progovmnt) was constructed from the following variables in the 2004 National Election Study: Which statement closer to R's views: The main reason government has become bigger over the years is because it has gotten involved in things that people should do for themselves OR Government has become bigger because the problems we face have become bigger (V045150); We need a strong government to handle today's complex economic problems OR The free market can handle these problems without government being

involved (V045151); The less government, the better OR There are more things that government should be doing (V045152). The number of pro-government responses was summed, resulting in a scale ranging from 0 to 3.

3. In calculating a multiplicative product, SPSS will assign a valid code of 0 to any case that has a missing value on one of the variables and a value of 0 on the other variable. For example, a respondent who has a missing value on progovmnt and who has a value of 0 on polknow3 will be assigned a valid, analyzable value on the interaction variable—a value of 0. This respondent should be treated as missing but instead ends up in the analysis. SPSS also returns a valid code of 0 for any expression that divides 0 by a missing value. Avoid the goofy quirk by restricting the Compute procedure to cases having nonmissing values on both variables. See *SPSS 16.0 Command Syntax Reference* (Chicago: SPSS, Inc., 2007), 116.

4. See Quentin Kidd, Herman Diggs, Mehreen Farooq, and Megan Murray, "Black Voters, Black Candidates, and Social Issues: Does Party Identification Matter?" *Social Science Quarterly* 88 (March 2007): 165–176.

10

Logistic Regression*

Procedures Covered

Analyze → Regression → Binary Logistic
Transform → Compute (Predicted probabilities)
Graphs → Legacy Dialogs → Line (Multiple /
Summaries of separate variables)

You now have an array of SPSS skills that enable you to perform the appropriate analysis for just about any situation you will encounter. To analyze the relationship between two categorical variables—variables measured at the nominal or ordinal level—you would enlist Crosstabs. If the dependent variable is an interval-level scale and the independent variable is categorical, then mean comparison analysis would be one way to go. Alternatively, you might create a dummy variable (or variables), specify a linear regression model, and run Regression → Linear to estimate the effects of the categorical variable(s) on the dependent variable. Finally, if both the independent and dependent variables are interval level, then SPSS Correlate or Regression → Linear would be appropriate techniques. There is, however, a common research situation that you are not yet equipped to tackle.

In its most specialized application, logistic regression is designed to analyze the relationship between an interval-level independent variable and a binary dependent variable. A binary variable, as its name suggests, can assume only two values. Binary variables are just like the dummy variables you created and analyzed earlier in this book. Either a case has the attribute or behavior being measured or it does not. Voted/did not vote, married/not married, favor/oppose gay marriage, and South/non-South are examples of binary variables.

Consider a binary dependent variable of keen interest to students of political behavior: whether people voted in an election. This variable, of course, has only two values: Either individuals voted (coded 1 on the binary variable) or they did not vote (coded 0). Now think about an interval-level independent variable often linked to turnout, years of education. As measured by the General Social Survey, this variable ranges from 0 (no formal schooling) to 20 (20 years of education). We would expect a positive relationship between the independent and dependent variables. As years of education increase, the probability of voting should increase as well. So people with fewer years of schooling should have a relatively low probability of voting, and this probability should increase with each additional year of education. Now, we certainly can conceptualize this relationship as positive. However, for statistical and substantive reasons, we cannot assume that it is linear—that is, we cannot assume that a 1-year change in education occasions a consistent increase in the probability of voting. Garden-variety regression, often called ordinary least squares or OLS regression, assumes a linear relationship between the independent and dependent variables.[1] Thus we cannot use Regression → Linear to analyze the relationship between education and the probability of voting. But as luck

---

*For this chapter you will need access to a full-version SPSS installation that includes the SPSS Regression Models module. The full version of SPSS Base, by itself, does not permit the user to perform logistic regression. The SPSS Student Version does not contain the Regression Models module.

and statistics would have it, we can assume a linear relationship between education and the logged odds of voting. Let's put the relationship into logistic regression form and discuss its special properties:

$$\text{Logged odds (voting)} = a + b \text{ (years of education)}.$$

This logistic regression model is quite OLS-like in appearance. Just as in OLS regression, the constant or intercept, a, estimates the dependent variable (in this case, the logged odds of voting) when the independent variable is equal to 0—that is, for people with no formal education. And the logistic regression coefficient, b, will estimate the change in the logged odds of voting for each 1-year increase in education. What is more, the analysis will produce a standard error for b, permitting us to test the null hypothesis that education has no effect on turnout. Finally, SPSS output for logistic regression will provide R-square-type measures, giving us an idea of the strength of the relationship between education and the likelihood of voting. In all of these ways, logistic regression is comfortably akin to linear regression.

However, logistic regression output is more difficult to interpret than are OLS results. In ordinary regression, the coefficients of interest, the constant (a) and the slope (b), are expressed in actual units of the dependent variable. If we were to use OLS to investigate the relationship between years of education (X) and income in dollars (Y), the regression coefficient on education would communicate the dollar-change in income for each 1-year increase in education. With OLS, what you see is what you get. With logistic regression, by contrast, the coefficients of interest are expressed in terms of the logged odds of the dependent variable. The constant (a) will tell us the logged odds of voting when education is 0, and the regression coefficient (b) will estimate the change in the logged odds for each unit change in education. Logged odds, truth be told, have no intuitive appeal. Thus we often must translate logistic regression results into language that makes better intuitive sense.

USING REGRESSION → BINARY LOGISTIC

Let's run the voting-education analysis and clarify these points. GSS2006 contains voted04, coded 0 for respondents who did not vote in the 2004 election and coded 1 for those who voted.[2] GSS2006 also has educ, which records the number of years of schooling for each respondent. Click Analyze → Regression → Binary Logistic, opening the Logistic Regression window (Figure 10-1). Find voted04 in the variable list and click it into the Dependent box. Click educ into the Covariates box. (In logistic regression, independent variables are often called covariates.) For this run, we will do one additional thing. In the Logistic Regression window, click Options. The Logistic Regression: Options window opens (Figure 10-2). Click the box next to "Iteration history." This option will produce output that helps to illustrate how logistic regression works. Click Continue, returning to the main Logistic Regression window. Click OK.

Figure 10–1 The Logistic Regression Window

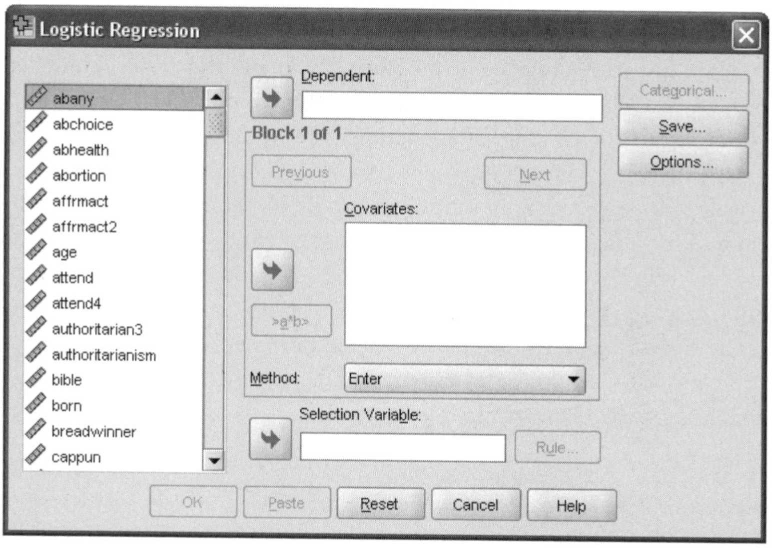

Figure 10–2   Requesting Logistic Regression with Iteration History

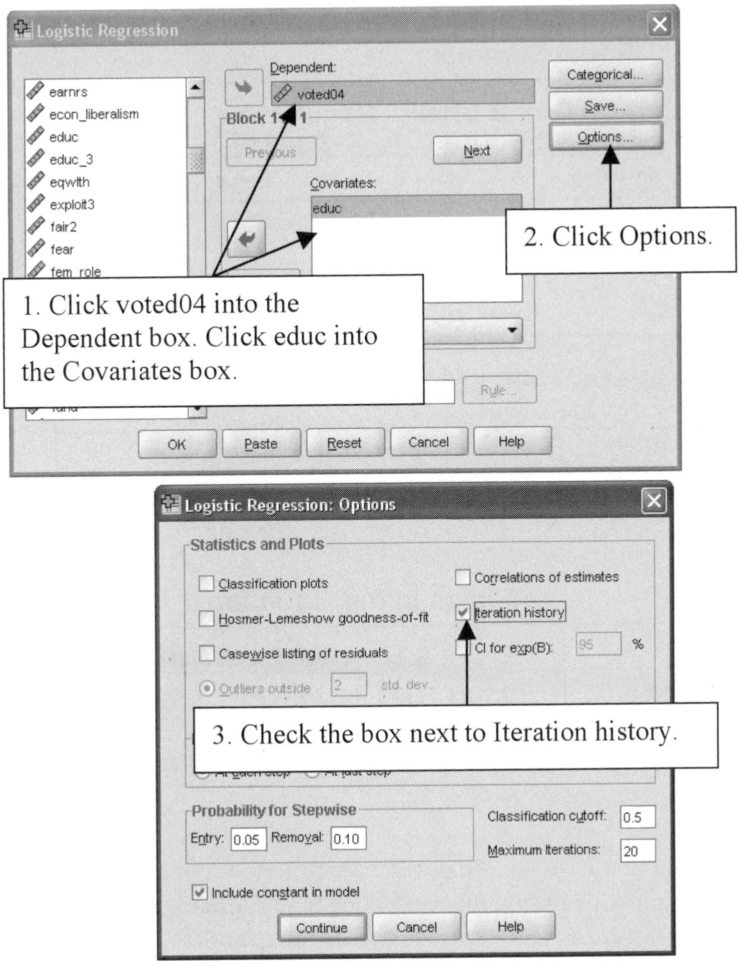

In typical fashion, SPSS has given us a wealth of information. Eleven tables now populate the Viewer. Happily, for the essential purposes of this book, you need to be conversant with only three or four of these tables. Scroll to the bottom of the output, to the table labeled "Variables in the Equation." Here you will find the main results of the voted04-educ analysis (Figure 10-3).

Just as in Regression → Linear, the numbers in the column labeled "B" are the estimates for the constant and the regression coefficient. Plug these estimates into our model:

$$\text{Logged odds (voting)} = -2.215 + .248(\text{educ}).$$

What do these coefficients tell us? Again, the constant says that, for people with no education, the estimated logged odds of voting is equal to –2.215. And the logistic regression coefficient on educ says that the logged odds of voting increases by .248 for each 1-year increase in education. So, as expected, as the independent variable increases, the likelihood of voting increases, too. Does education have a statistically significant effect on the likelihood of voting? In OLS regression, SPSS determines statistical significance by calculating a t-statistic and an accompanying P-value. In logistic regression, SPSS calculates a Wald statistic (which is based on chi-square) and reports a P-value for Wald. Interpretation of this P-value, displayed in the column labeled "Sig.," is directly analogous to ordinary regression. If the P-value is greater than .05, then do not reject the null hypothesis. Conclude that the independent variable does not have a significant effect on the dependent variable. If the P-value is less than or equal to .05, then reject the null hypothesis and infer that the independent variable has a significant relationship with the dependent variable. In our output, the P-value for educ is .000, so we can conclude that, yes, education has a significant effect on voting turnout.

Now let's return to the logistic regression coefficient, .248, and figure out how to make it more meaningful. Consider the right-most column of the Variables in the Equation table, the column labeled "Exp(B)."

Figure 10–3 Logistic Regression Output with One Independent Variable: Variables in the Equation and Model Summary

Model Summary

Step	-2 Log likelihood	Cox & Snell R Square	Nagelkerke R Square
1	4360.527[a]	.091	.134

a. Estimation terminated at iteration number 5 because parameter estimates changed by less than .001.

Classification Table[a]

			Predicted		
			Did R vote in 2004 election?		Percentage Correct
	Observed		0 Did not vote	1 Voted	
Step 1	Did R vote in 2004 election?	0 Did not vote	110	978	10.1
		1 Voted	89	2,941	97.1
	Overall Percentage				74.1

a. The cut value is .500

Variables in the Equation

		B	S.E.	Wald	df	Sig.	Exp(B)
Step 1[a]	educ	.248	.014	321.531	1.000	.000	1.281
	Constant	-2.215	.180	150.702	1	.000	.109

a. Variable(s) entered on step 1: educ.

Here SPSS has reported the value 1.281 for the independent variable, educ. Where did this number originate? SPSS obtained this number by raising the natural log base e (approximately equal to 2.72) to the power of the logistic regression coefficient, .248. This procedure translates the logged odds regression coefficient into an *odds ratio*. An odds ratio tells us by how much the odds of the dependent variable change for each unit change in the independent variable. An odds ratio of less than 1 says that the odds decrease as the independent variable increases (a negative relationship). An odds ratio equal to 1 says that the odds do not change as the independent variable increases (no relationship). And an odds ratio of greater than 1 says that the odds of the dependent variable increase as the independent variable increases (a positive relationship). An odds ratio of 1.281 means that respondents at a given level of education are 1.28 times more likely to have voted than are respondents at the next lower level of education. So people with, say, 10 years of education are 1.28 times more likely to have voted than are people with 9 years of education, people with 14 years are 1.28 times more likely to have voted than people with 13 years, and so on.

The value of Exp(B) is often used to obtain an even more understandable estimate, the *percentage change in the odds* for each unit change in the independent variable. Mercifully, simple arithmetic accomplishes this task. Subtract 1 from Exp(B) and multiply by 100. In our current example: $(1.28 - 1) * 100 = 28$. We can now say that each 1-year increment in education increases the odds of voting by 28 percent. As you can see, when the relationship is positive—that is, when the logistic regression coefficient is greater than 0 and the odds ratio is greater than 1—figuring out the percentage change in the odds requires almost no thought. Just subtract 1 from Exp(B) and move the decimal point two places to the right. But be alert for negative relationships, when the odds ratio is less than 1. (In one of the exercises at the end of this chapter, you will interpret a negative relationship.) Suppose, for example, that Exp(B) were equal to .28, communicating a negative relationship between the independent variable and the probability of the dependent variable. The percentage change in the odds would be equal to $(.28 - 1) * 100 = -72.0$, indicating that a one-unit change in the independent variable decreases the odds of the dependent variable by 72 percent.

How strong is the relationship between years of education and the likelihood of voting? Consider the table labeled "Model Summary," also shown in Figure 10-3. OLS researchers are quite fond of R-square, the overall measure of strength that gauges the amount of variation in the dependent variable that is explained by the independent variable(s). For statistical reasons, however, the notion of "explained variation" has no

Figure 10–4 Logistic Regression Output with One Independent Variable: Iteration History and Omnibus Test of Model Coefficients

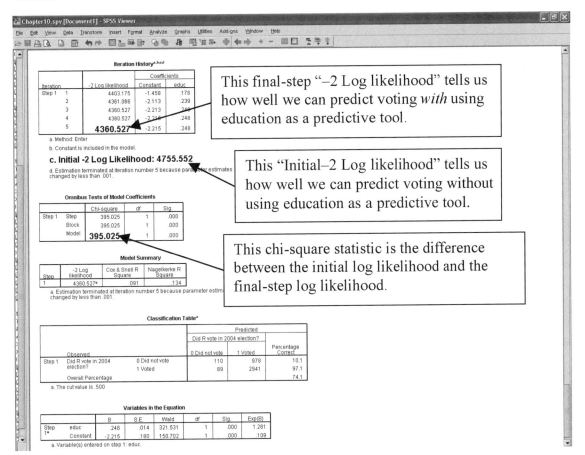

direct analog in logistic regression. Even so, methodologists have proposed various "pseudo R-square" measures that gauge the strength of association between the dependent and independent variables, from 0 (no relationship) to 1 (perfect relationship). SPSS reports two of these: the Cox and Snell R-square and the Nagelkerke R-square. Cox-Snell is the more conservative measure—that is, its maximum achievable value is less than 1. The Nagelkerke measure adjusts for this, and so it generally reports a higher pseudo R-square than does Cox-Snell.[3] These two measures are never wildly different, and they do give the researcher a ballpark feel for the strength of the relationship. With values in the range of .091 to .134, you could conclude that education, though related to voting, by itself provides a less-than-complete explanation of it.

One other measure is reported in the Model Summary table, "–2 Log likelihood," equal to 4360.527. In some ways this is the most important measure of strength produced by logistic regression. By itself, however, the magnitude of –2 log likelihood doesn't mean very much. But scroll up a bit, so that you can view the tables labeled "Omnibus Tests of Model Coefficients" and "Iteration History" together on your screen (Figure 10-4).[4]

In figuring out the most accurate estimates for the model's coefficients, logistic regression uses a technique called maximum likelihood estimation (MLE). When it begins the analysis, MLE finds out how well it can predict the observed values of the dependent variable without using the independent variable as a predictive tool. So MLE first determined how accurately it could predict whether individuals voted by not knowing how much education they have. The number labeled "Initial –2 Log Likelihood" (equal to 4755.552 and found beneath the Iteration History table) summarizes this "know-nothing" prediction. MLE then brings the independent variable into its calculations, running the analysis again—and again and again—to find the best possible predictive fit between years of education and the likelihood of voting.

According to the Iteration History table, SPSS ran through five iterations, finally deciding that it had maximized its ability to predict voting by using education as a predictive instrument. This final-step log likelihood, 4360.527, is recorded in the Iteration History table and it appears, as well, in the Model Summary table. The amount of explanatory leverage gained by including education as a predictor is determined by subtracting the final-step –2 log likelihood (4360.527) from the initial –2 log likelihood (4755.552). If you performed this calculation by hand, you would end up with 395.025, which appears in the Omnibus Tests of Model Coefficients table next to "Model." This number, which could be more accurately labeled "Change in –2 log likelihood," is a chi-square test statistic. In the "Sig." column of the Omnibus Tests of Model Coefficients table, SPSS has reported a P-value of .000 for this chi-square statistic. Conclusion: Compared with how well we can predict voting without knowing education, including education as a predictor significantly enhances the performance of the model.

By now you are aware of the interpretive challenges presented by logistic regression analysis. In running good old Regression → Linear, you had a mere handful of statistics to report and discuss: the constant, the regression coefficient(s) and accompanying P-value(s), and adjusted R-square. That's about it. With Regression → Binary Logistic, there are more statistics to record and interpret. Below is a tabular summary of the results of the voted04-educ analysis. You could use this tabular format to report the results of any logistic regressions you perform:

Model estimates and model summary: Logged odds (voting) = a + b (educ)

Model estimates	Coefficient	Significance	Exp(B)*	Percentage change in odds
Constant	–2.215			
Education	.248	.000	1.281	28.1

Model summary	Value	Significance
Chi-square**	395.025	.000
Cox-Snell R-square	.091	
Nagelkerke R-square	.134	

 * Alternatively, this column could be labeled "Odds ratio."
** Alternatively, this row could be labeled "Change in –2 Log likelihood."

LOGISTIC REGRESSION WITH MULTIPLE INDEPENDENT VARIABLES

The act of voting might seem simple, but we know that it isn't. Certainly, education is not the only characteristic that shapes the individual's decision whether to vote or to stay home. Indeed, we have just seen that years of schooling, although clearly an important predictor of turnout, returned so-so pseudo-R square statistics, indicating that other factors might also contribute to the explanation. Age, race, marital status, strength of partisanship, political efficacy—all of these variables are known predictors of turnout. What is more, education might itself be related to other independent variables of interest, such as age or race. Thus you might reasonably want to know the partial effect of education on turnout, controlling for the effects of these other independent variables. When performing OLS regression, you can enter multiple independent variables into the model and estimate the partial effects of each one on the dependent variable. Logistic regression, like OLS regression, can accommodate multiple predictors of a binary dependent variable. Consider this logistic regression model:

$$\text{Logged odds (voting)} = a + b_1(\text{educ}) + b_2(\text{age}).$$

Again we are in an OLS-like environment. As before, educ measures number of years of formal education. The variable age measures each respondent's age in years, from 18 to 89. From a substantive standpoint, we would again expect b_1, the coefficient on educ, to be positive: As education increases, so too should the logged odds of voting. We also know that older people are more likely to vote than are younger people. Thus we should find a positive sign on b_2, the coefficient on age. Just as in OLS, b_1 will estimate the effect of education on voting, controlling for age, and b_2 will estimate the effect of age on the dependent variable, controlling for the effect of education. Finally, the various measures of strength—Cox-Snell, Nagelkerke, –2 log likelihood—will give us an idea of how well both independent variables explain turnout.

Let's see what happens when we add age to our model. Click Analyze → Regression → Binary Logistic. Everything is still in place from our previous run: voted04 is in the Dependent box and educ is in the Covariates box. Good. Now locate age in the variable list and click it into the Covariates box. Click OK to run the analysis. Now scroll to the bottom of the output and view the results displayed in the Variables in the Equation and Model Summary tables.

Model Summary

Step	-2 Log likelihood	Cox & Snell R Square	Nagelkerke R Square
1	4101.451[a]	.144	.210

a. Estimation terminated at iteration number 5 because parameter estimates changed by less than .001.

Variables in the Equation

		B	S.E.	Wald	df	Sig.	Exp(B)
Step 1[a]	educ	.298	.015	383.823	1	.000	1.347
	age	.036	.002	212.986	1	.000	1.037
	Constant	-4.555	.252	326.059	1	.000	.011

a. Variable(s) entered on step 1: educ, age.

Plug these estimates into our model:

$$\text{Logged odds (voting)} = -4.555 + .298(\text{educ}) + .036(\text{age}).$$

Interpretation of these coefficients follows a straightforward multiple regression protocol. The coefficient on educ, .298, tells us that, controlling for age, each additional year of education increases the logged odds of voting by .298. And notice that, controlling for education, age is positively related to the likelihood of voting. Each 1-year increase in age produces an increase of .036 in the logged odds of voting. According to Wald and accompanying P-values, each independent variable is significantly related to the dependent variable.

Now consider SPSS's helpful translations of the coefficients, from logged odds to odds ratios, which are displayed in the "Exp(B)" column. Interestingly, after controlling for age, the effect of education is somewhat stronger than its uncontrolled effect, which we analyzed earlier. Taking respondents' age differences into account, we find that each additional year of schooling increases the odds ratio by 1.347 and boosts the odds of voting by 34.7 percent: $(1.347 - 1) * 100 = 34.7$.[5] For age, too, the value of Exp(B), 1.037, is greater than 1, again communicating the positive relationship between age and the likelihood of voting. If you were to compare two individuals having the same number of years of education but who differed by 1 year in age, the older person would be 1.037 times more likely to vote than the younger person. Translating 1.037 into a percentage change in the odds: $(1.037 - 1) * 100 = 3.7$. Conclusion: Each additional year in age increases the odds of voting by about 4 percent.[6]

According to Cox-Snell (.144) and Nagelkerke (.210), adding age to the model increased its explanatory power, at least when compared with the simple analysis using education as the sole predictor. The value of -2 log likelihood, 4101.451, is best viewed through the lens of the chi-square test, which you will find by scrolling up to the tables labeled "Omnibus Tests of Model Coefficients" and "Iteration History."

Iteration History[a,b,c,d]

			Coefficients		
Iteration		-2 Log likelihood	Constant	educ	age
Step 1	1	4190.425	-2.794	.194	.023
	2	4104.162	-4.214	.278	.034
	3	4101.455	-4.541	.297	.036
	4	4101.451	-4.555	.298	.036
	5	4101.451	-4.555	.298	.036

a. Method: Enter

b. Constant is included in the model.

c. Initial -2 Log Likelihood 4738.154

d. Estimation terminated at iteration number 5 because parameter estimates changed by less than .001.

Omnibus Tests of Model Coefficients

		Chi-square	df	Sig.
Step 1	Step	636.703	2	.000
	Block	636.703	2	.000
	Model	636.703	2	.000

MLE's initial know-nothing model—estimating the likelihood of voting without using education or age as predictors—returned a –2 log likelihood of 4738.154. After bringing the independent variables into play and running through five iterations, MLE settled on a –2 log likelihood of 4101.451, an improvement of 636.703. This value, which is a chi-square test statistic, is statistically significant ("Sig." = .000). This tells us that, compared with the know-nothing model, both independent variables significantly improve our ability to predict the likelihood of voting.

WORKING WITH PREDICTED PROBABILITIES: MODELS WITH ONE INDEPENDENT VARIABLE

You now know how to perform basic logistic regression analysis, and you know how to interpret the logistic regression coefficient in terms of an odds ratio and in terms of a percentage change in the odds. No doubt, odds ratios are easier to comprehend than are logged odds. And percentage change in the odds seems more understandable still. Not surprisingly, most researchers prefer to think in terms of probabilities. One might reasonably ask, "What is the effect of a 1-year increase in education on the probability of voting?" Inconveniently, with logistic regression the answer is always, "It depends."

In the first analysis we ran, which examined the voting-education relationship, logistic regression assumed that a linear relationship exists between years of education and the logged odds of voting. This linearity assumption permitted us to arrive at an estimated effect that best fits the data. However, the technique also assumed a nonlinear relationship between years of education and the probability of voting. That is, it assumed that for people who lie near the extremes of the independent variable—respondents with either low or high levels of education—a 1-year increase in education will have a weaker effect on the probability of voting than will a 1-year increase for respondents in the middle range of the independent variable. Because people with low education are unlikely to vote, a 1-year change should not have a huge effect on this likelihood. Ditto for people with many years of schooling. They are already quite likely to vote, and a one-unit increase should not greatly enhance this probability. By contrast, in the middle range of the independent variable, education should have its most potent marginal impact, pushing individuals over the decision threshold from "do not vote" to "vote." So the effect of a 1-year change in education is either weaker or stronger, depending on where respondents "are" on the education variable.

In logistic regression models having more than one independent variable, such as the voted04-educ-age analysis, working with probabilities becomes even more problematic. The technique assumes that the independent variables have additive effects on the logged odds of the dependent variable. Thus for any combination of values of the independent variables, we arrive at an estimated value of the logged odds of the dependent variable by adding up the partial effects the predictor variables. However, logistic regression also assumes that the independent variables have interactive effects on the probability of the dependent variable. For example, in the case of younger respondents (who have a lower probability of voting), the technique might estimate a large effect of education on the probability of voting. For older respondents (who have a higher probability of voting), logistic regression might find a weaker effect of education on the probability of voting. So the effect of each independent variable on the probability of the dependent variable will depend on the values of the other predictors in the model.

Let's explore these issues one at a time, beginning with the simple model that used education alone to predict voting. Even though we cannot identify a single coefficient that summarizes the effect of education on the probability of voting, we can use SPSS to calculate a predicted probability of voting for respondents at each level of education. How does this work? Recall the logistic regression equation SPSS estimated in our first analysis:

$$\text{Logged odds (voting)} = -2.215 + .248(\text{educ}).$$

SPSS would use this logistic regression model to obtain an estimated logged odds of voting for each respondent. It would plug in each respondent's education level, do the math, and calculate an estimated value of the dependent variable, the logged odds of voting. SPSS would then use the following formula to convert the estimated logged odds of voting into a predicted probability of voting:

$$\text{Probability of voting} = \text{Exp(Logged odds of voting)}/(1 + \text{Exp(Logged odds of voting)}).$$

Figure 10–5 Requesting Predicted Probabilities

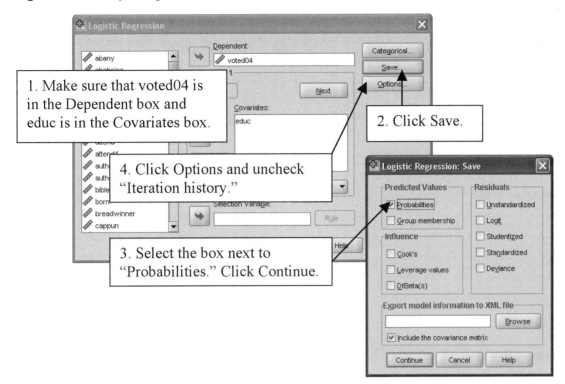

According to this formula, we retrieve the probability of voting by first raising the natural log base *e* to the power of the logged odds of voting. We then divide this number by the quantity one plus *e* raised to the power of the logged odds of voting.[7] Clear as mud.

To get an idea of how SPSS calculates predicted probabilities, let's work through an example. Consider respondents who have a high school education: 12 years of schooling. Using the logistic regression equation obtained in the first guided example, we find the logged odds of voting for this group to be $-2.215 + .248(12) = -2.215 + 2.976 = .761$. What is the predicted probability of voting for people with 12 years of education? It would be $\text{Exp}(.761)/(1 + \text{Exp}(.761)) = 2.140/3.140 \sim .68$. So for respondents with a high school education, the estimated probability of voting is .68. At the user's request, SPSS will follow this procedure to calculate predicted probabilities for individuals at all values of education, and it will save these predicted probabilities as a new variable in the dataset.

Let's run the voted04-educ analysis again and request that SPSS calculate and save the predicted probability of voting for each respondent. Click Analyze → Regression → Binary Logistic. All of the variables are still in place from our previous run. This time, however, we want to use only one independent variable, educ. Select age with the mouse and click it back into the variable list, leaving educ in the Covariates box and voted04 in the Dependent box. Now click the Save button in the Logistic Regression window. This opens the Logistic Regression: Save window (Figure 10-5). In the Predicted Values panel, click the Probabilities box. Click Continue, which returns you to the Logistic Regression window. One more thing. We won't be discussing iteration history on this run, so click Options and uncheck the Iteration history box. Click Continue. Ready to go. Click OK.

SPSS generates output that is identical (except for the iteration history) to our earlier run. So where are the predicted probabilities that we requested? Because we just ran the analysis, SPSS has taken us to the Viewer. Return to the Variable View of the Data Editor. Scroll to the bottom of the Variable View. As you know, this is where SPSS puts the new variables that you create using Recode or Compute. There you will find a new variable bearing the innocuous name "PRE_1" and the equally bland label "Predicted probability" (Figure 10-6). SPSS has performed just as requested. It ran the analysis, generated the logistic regression output, and discreetly saved a new variable, the predicted probability of voting for each case in the dataset. We will want to have a look at PRE_1. But first we need to give it a more descriptive label. Click in the Label cell and type a more informative variable label, such as "Pred prob: educ-voted04."

Figure 10–6 Predicted Probability Saved as a New Variable in the Data Editor

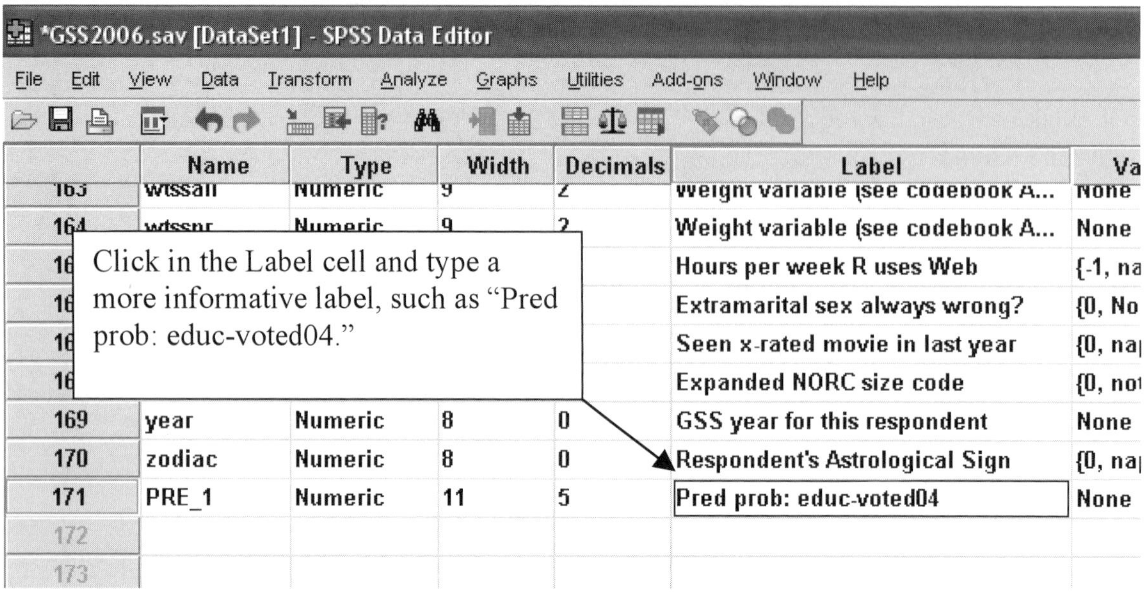

In what ways can this new variable, PRE_1, help us to describe changes in the estimated probability of voting as education increases? Remember, SPSS now has a predicted probability of voting for respondents at each value of the education variable, from 0 years to 20 years. So there are two complementary ways to describe the relationship between education and PRE_1, the predicted probability of voting. First, we can perform Analyze → Compare Means → Means, asking SPSS to calculate the mean values of PRE_1 (dependent variable) for each value of educ (independent variable). This would show us by how much the estimated probability of voting increases between groups of respondents having different numbers of years of schooling. Second, we can obtain a line chart of the same information. To obtain a line chart, click Graphs → Legacy Dialogs → Line → Simple and click educ into the "Category Axis" box. Then select "Other statistic," and click PRE_1 into the Line Represents panel. This allows us to visualize the nonlinear relationship between education and the predicted probability of voting.

To you, both of these modes of analysis are old hat, so go ahead and perform the analyses. In the mean comparison results (Figure 10-7), the values of educ appear in ascending order down the left-hand column, and mean predicted probabilities (somewhat distractingly, to 7-decimal point precision) are reported in the column labeled "Mean." The line chart (Figure 10-8) adds clarity and elegance to the relationship. To get a feel for what is going on, scroll back and forth between the tabular analysis and the graphic output. What happens to the predicted probability of voting as education increases? Notice that, in the lower range of the independent variable, between 0 years and about 6 years, the predicted probabilities are quite low (between .10 and about .33) and these probabilities increase on the order of .02 to .04 for each increment in education. Now shift your focus to the upper reaches of education and note much the same thing. Beginning at about 13 years of schooling, the estimated probability of voting is at or above about .73—a high likelihood of turning out—and so increments in this range have weaker effects on the probability of voting. In the middle range, from 7 to 12 years, the probabilities increase at a "faster" marginal rate, and within this range the graphic curve shows its steepest slope.

Although most political researchers like to get a handle on predicted probabilities, as we have just done, there is no agreed-upon format for succinctly summarizing logistic regression results in terms of probabilities. One commonly used approach is to report the so-called full effect of the independent variable on the probability of the dependent variable. The full effect is calculated by subtracting the probability associated with the lowest value of the independent variable from the probability associated with the highest value of the independent variable. According to our Compare Means analysis, the predicted probability of voting for people with no formal schooling is about .10, and the predicted probability for those with 20 years of education is .94. The full effect would be .94 – .10 = .84. So, measured across its full range of observed values, education boosts the probability of voting by a healthy .84.

Figure 10–7 Mean Comparison Table for Predicted Probabilities

PRE 1 Pred prob: educ-voted04

educ ...	Mean	N
0	.0984381	22
1	.1227243	4
2	.1519921	28
3	.1867540	13
4	.2273348	11
5	.2737653	23
6	.3256816	69
7	.3822610	32
8	.4422235	85
9	.5039191	127
10	.5654956	152
11	.6251150	215
12	.6811661	1204
13	.7324240	422
14	.7781254	628
15	.8179612	212
16	.8520051	687
17	.8806116	167
18	.9043097	208
19	.9237113	78
20	.9394425	112
Total	.7254533	4499

Notice that, between 8 and 9 years of education, the predicted probability switches from less than .5 to greater than .5.

Figure 10–8 Line Chart for Predicted Probabilities

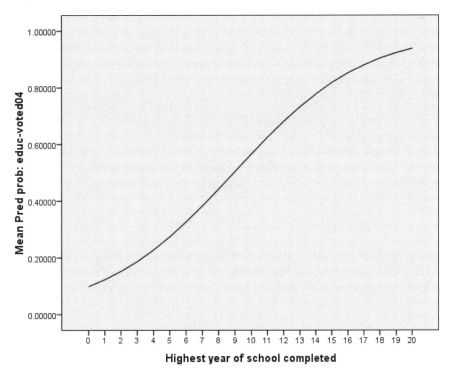

Another way of summarizing a relationship in terms of probabilities is to report the interval of the independent variable that has the biggest impact on the probability of the dependent variable. Suppose that you had to pick the 1-year increment in education that has the largest impact on the probability of voting. What would that increment be? Study the output and think about the phenomenon you are analyzing. Remember that voting is an up or down decision. A person either decides to vote or decides not to vote. But between which two values of education does a "vote" decision become more likely than a "do not vote" decision? You may have noticed that, between 8 years and 9 years, the predicted probabilities increase from .442 to .504, a difference of .062 and the largest marginal increase in the data. And it is between these two values of education that, according to the analysis, the binary decision shifts in favor of voting—from a probability of less than .50 to a probability of greater than .50. So the interval between 8 years and 9 years is the "sweet spot"—the interval with the largest impact on the probability of voting, and the interval in which the predicted probability switches from less than .50 to more than .50.[8]

WORKING WITH PREDICTED PROBABILITIES: MODELS WITH MULTIPLE INDEPENDENT VARIABLES

Saving predicted probabilities using the Logistic Regression: Save option works fine for simple models with one independent variable. By examining these predicted probabilities, you are able to summarize the full effect of the independent variable on the dependent variable. Furthermore, you can describe the interval of the independent variable having the largest impact on the probability of the dependent variable. Of course, SPSS also will gladly save predicted probabilities for logistic regression models having more than one independent variable. With some specialized exceptions, however, these predicted probabilities are not very useful for summarizing the effect of each independent variable on the probability of the dependent variable, controlling for the other independent variables in the model. As noted earlier, although logistic regression assumes that the independent variables have an additive effect on the logged odds of the dependent variable, the technique also assumes that the independent variables have an interactive effect on the probability of the dependent variable. Thus the effect of, say, education on the probability of voting will be different for younger people than for older people. And the effect of age will vary, depending on the level of education being analyzed. How can we summarize these interaction effects? In dealing with logistic regression models with multiple independent variables, many researchers use the *sample averages method* for presenting and interpreting probabilities. Another approach, the *probability profile method,* permits the researcher to compare probabilities across groups. Let's take a look at both of these methods.

The Sample Averages Method

In the sample averages approach, the analyst examines the effect of each independent variable while holding the other independent variables constant at their sample means. For example, we would ask and answer these questions: "For people of 'average' age, what effect does education have on the probability of voting?" and "For respondents with 'average' levels of education, what effect does age have on the probability of voting?" In this way, we can get an idea of the effect of each variable on individuals who are "average" on all the other variables being studied. Unfortunately, Regression → Binary Logistic will not calculate the predicted probabilities associated with each value of an independent variable while holding the other variables constant at their sample means.[9] That's the bad news. The good news is that the desired probabilities can be obtained using Transform → Compute, and they are readily analyzed using Compare Means.

Here is the logistic regression model that SPSS estimated for the voted04-educ-age relationships:

$$\text{Logged odds (voting)} = -4.555 + .298(\text{educ}) + .036(\text{age}).$$

We can enlist this equation for two tasks. First, we can plug in the sample mean of age and calculate the full effect of educ on the probability of voting. Second, we can plug in the sample mean of educ and calculate the full effect of age on the probability of voting. Here we will work through the first task only—figuring out the full effect of education on the probability of voting for people of average age. Before proceeding, of course, we need to obtain the sample mean of age. A quick Descriptives run (which you are invited to replicate) reveals that age has a mean value of 47.14 years. The following equation would permit us to estimate the logged odds of voting at any value of educ, holding age constant at its mean:

$$\text{Logged odds (voting)} = -4.555 + .298(\text{educ}) + .036(47.14).$$

We have already seen that probabilities may be retrieved from logged odds via this conversion:

$$\text{Probability of voting} = \text{Exp(Logged odds of voting)}/(1 + \text{Exp(Logged odds of voting)}).$$

The following equation, therefore, would convert a logged odds of voting into a predicted probability of voting for any plugged-in value of educ, holding age constant at its mean of 47.14:

$$\text{Probability of voting} = \text{Exp}(-4.555 + .298^*\text{educ} + .036^*47.14) / $$
$$(1 + \text{Exp}(-4.555 + .298^*\text{educ} + .036^*47.14)).$$

Of course, we could figure all this out using a hand calculator—first finding the predicted probability of voting for individuals having no formal schooling (educ = 0) and then calculating the predicted probability of people with 20 years of education (educ = 20). By subtracting the first probability (when educ = 0) from the second probability (when educ = 20), we would arrive at the full effect of educ at the mean value of age. But let's ask SPSS to do the work for us. Click Transform → Compute. What do we want SPSS to do? We want it to calculate the predicted probability of voting for respondents at each level of education, holding age constant at its sample mean. Because we are holding age constant but allowing educ to vary, we will name this variable "pre_educ." Type "pre_educ" in the Target Variable box. In the Numeric Expression box, type this expression: "Exp(−4.555 + .298*educ + .036*47.14)/(1 + Exp(−4.555 + .298*educ + .036*47.14))," as shown in Figure 10-9.[10] Click Type & Label. In the Label box, give pre_educ the descriptive label "Pred prob: educ-voted04, mean age." Click OK. SPSS computes a new variable, pre_educ, and enters this variable into the dataset.

Figure 10–9   Computing a Predicted Probability for Different Values of an Independent Variable at the Mean Value of Another Independent Variable

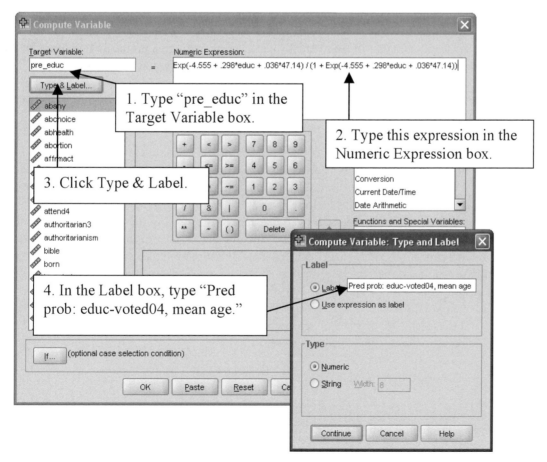

Finally we have the estimates that permit us to examine the effect of education on the probability of voting for respondents of average age. Click Analyze → Compare Means → Means. Click our newly computed variable, pre_educ, into the Dependent List, and click educ into the Independent List. Click OK.

pre_educ Pred prob: educ-voted04, mean age

edu...	Mean	N
0	.0543	22
1	.0718	4
2	.0943	28
3	.1230	13
4	.1590	11
5	.2029	23
6	.2554	69
7	.3161	32
8	.3837	85
9	.4561	127
10	.5305	152
11	.6035	215
12	.6722	1204
13	.7342	422
14	.7882	628
15	.8337	212
16	.8710	687
17	.9010	167
18	.9246	208
19	.9429	78
20	.9570	112
Total	.7232	4499

What is the full effect of education? Notice that people with 0 years of education have a probability of voting equal to about .05 (5 chances in 100 that the individual voted), compared with a probability of about .96 for individuals with 20 years of education (96 chances in 100 that the individual voted). Thus holding age constant at its sample mean, we find that the full effect of education is equal to .96 – .05 = .91. Note that the largest marginal effect of education, a boost of more than .07 in the probability of voting, occurs between 9 and 10 years of schooling.

The Probability Profile Method

As you can see, the sample averages method is a convenient way of summarizing the effect of an independent variable on the probability of a dependent variable. A more sophisticated approach, the *probability profile method,* affords a closer and more revealing look at multivariate relationships. In the probability profile method, we compare the effects of one independent variable at two (or more) values of another independent variable. Such comparisons can provide a clearer picture of the interaction relationships between the independent variables and the probability of the dependent variable. In estimating the voted04-educ-age model, for example, logistic regression assumed that the effect of education on the probability of voting is not the same for people of all ages. But just how much does the effect of education vary across age groups? If we were to estimate and compare the effects of education among two groups of respondents—those who are, say, 26 years of age, and those who are 65 years of age—what would the comparison reveal? Once again, here are the logistic regression estimates for the voted04-educ-age model:

$$\text{Logged odds (voting)} = -4.555 + .298(\text{educ}) + .036(\text{age}).$$

In the sample averages method, we used these estimates to compute the probability of voting at each value of education while holding age constant at its sample mean. In the probability profile method, we can

Figure 10–10 Computing a Predicted Probability for Different Values of an Independent Variable at a Fixed Value of Another Independent Variable

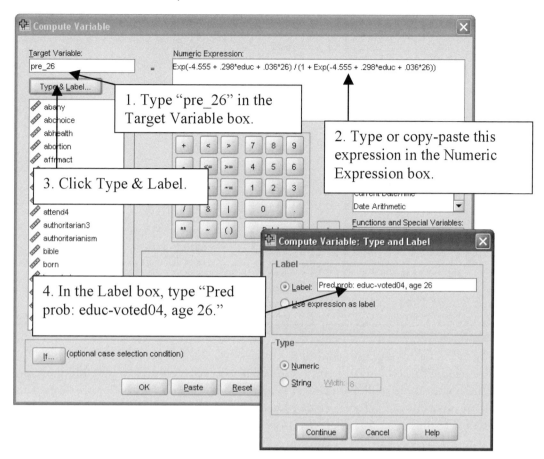

ask SPSS to compute two estimated probabilities for the effect of education on voting—one while holding age constant at 26 years, and one while holding age constant at 65 years. Consider the two numeric expressions that follow:

$$\text{Probability of voting, age 26:}$$
$$\text{pre\_26} = \text{Exp}(-4.555 + .298*\text{educ} + .036*26)/(1 + \text{Exp}(-4.555 + .298*\text{educ} + .036*26)).$$

$$\text{Probability of voting, age 65:}$$
$$\text{pre\_65} = \text{Exp}(-4.555 + .298*\text{educ} + .036*65)/(1 + \text{Exp}(-4.555 + .298*\text{educ} + .036*65)).$$

The first command will estimate the probability of voting at each value of education, while holding age constant at 26 years. And it asks SPSS to save these predicted probabilities in a new variable, pre_26. The second statement estimates the probability of the dependent variable at each value of education, holding age constant at 65 years, and it too will save a new variable, pre_65. To obtain these estimates, we'll need to make two circuits through Transform → Compute, each requiring a fair amount of typing in the Numeric Expression box—although you may have already figured out a useful shortcut that greatly reduces the keyboard drudgery.[11] Figure 10-10 shows the Compute Variable window for obtaining pre_26. To get pre_65, modify the relevant details of the window: Change the Target Variable name to "pre_65," replace "26" with "65" in the Numeric Expression box, and alter Label to read "Pred prob: educ-voted04, age 65." Go ahead and run the Computes.

Run Means → Compare Means, putting educ in the Independent List and both pre_26 and pre_65 in the Dependent List. To enhance readability, in Options click Standard Deviation and Number of Cases back into the Statistics list. SPSS will return a bare-bones mean comparison table:

Mean		
educ Highest year of school completed	pre_26 Pred prob: educ-voted04, age 26	pre_65 Pred prob: educ-voted04, age 65
0	.0261	.0984
1	.0349	.1282
2	.0464	.1653
3	.0615	.2107
4	.0811	.2644
5	.1063	.3263
6	.1381	.3948
7	.1776	.4678
8	.2253	.5421
9	.2815	.6147
10	.3455	.6824
11	.4156	.7433
12	.4893	.7959
13	.5634	.8401
14	.6348	.8762
15	.7008	.9051
16	.7593	.9278
17	.8095	.9454
18	.8513	.9589
19	.8852	.9691
20	.9122	.9769
Total	.5773	.8191

Consider the dramatically different effects of education for these two age groups. To be sure, the full effect of education is the same for 26 year olds (.91 − .03 = .88) and 65 year olds (.98 − .10 = .88). But the patterns of marginal effects are not the same at all. For younger people with low levels of education (between 0 and 8 years of schooling), the probability of voting is extraordinarily low, in the range of .03–.23. Indeed, the educational increment with the largest marginal effect—the increment in which the probability of voting switches from less than .50 to more than .50—occurs at a fairly high level of educational attainment, between 12 and 13 years of schooling. So 26 year olds with a year of junior college are somewhat more likely to vote than not to vote. Compare the probability profile of younger respondents—sluggish marginal effects in the lower range of education, a high "switchover" threshold—with the probability profile of older respondents. Does education work the same way as we read down the column labeled "pre_65"? Here the probabilities start at a higher level (about .10) and build quite rapidly, in increments of .02 to .05, crossing the .50 threshold at a fairly low level of education, between 7 and 8 years of schooling. So 65 year olds with a junior high school education are almost as likely to vote as are 26 year olds with a year of junior college.[12]

When you use the probability profile method to explore complex relationships, you will want to complement your analyses with appropriate graphic support. Consider Figure 10-11, a multiple line chart that has

Figure 10–11   Edited Multiple Line Chart of Two Logistic Regression Curves

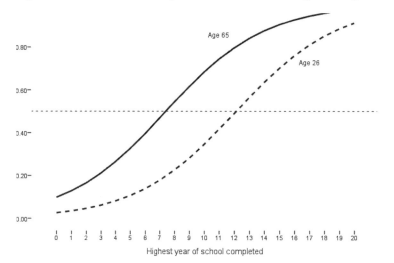

spent some "erasing" time in the Chart Editor. This chart instantly communicates the remarkably different ways in which education affects turnout for 26 year olds (dashed line) and 65 year olds (solid line). With one minor exception, the skills you developed earlier in this book will allow you to obtain an unedited version of this graphic. By using the editing skills you already have—and acquiring additional skills through practice and experimentation—you can create the edited version. Click Graphs → Legacy Dialogs → Line (Figure 10-12). In the Line Chart window, select Multiple. Here is something new: Instead of the default setting, "Summaries for groups of cases," select the radio button next to "Summaries of separate variables," as shown in Figure 10-12. Click Define. The Summaries of Separate Variables window is a reasonably familiar-looking sight (Figure 10-13). The variable whose effects we want to display, educ, goes in the Category Axis box. The

Figure 10–12   Changing the Defaults in the Line Chart Windows

Figure 10–13   The Define Multiple Line: Summaries of Separate Variables Window (modified)

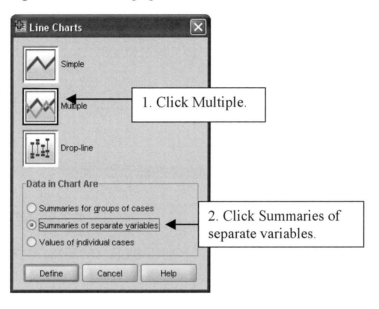

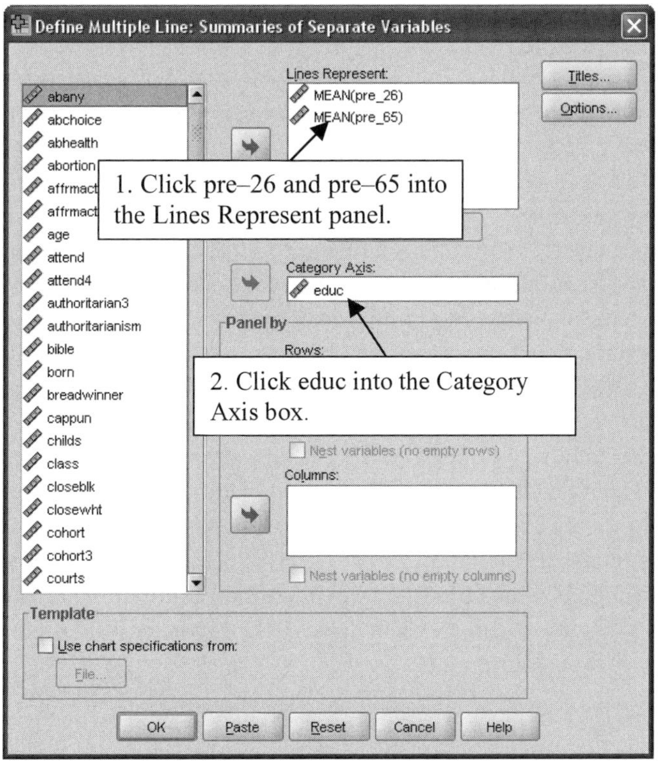

Figure 10-14 Multiple Line Chart of Two Logistic Regression Curves

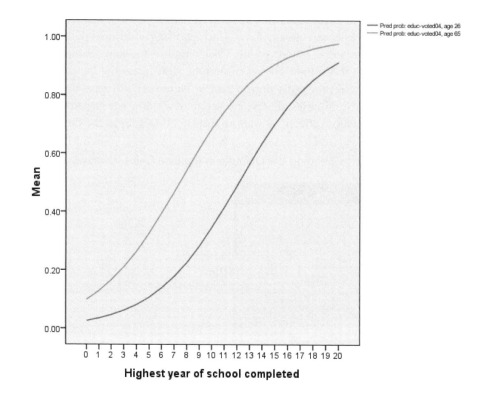

two predicted-probability variables, pre_26 and pre_65, go in the Lines Represent panel. SPSS offers to graph mean values of pre_26 and pre_65, which fits our purpose. Nothing more to it. Clicking OK produces the requested graphic result (Figure 10-14). This is a line chart with a lot of potential. Before moving on to the exercises, see what improvements you can make.[13]

EXERCISES

1. (Dataset: States. Variables: gb_win04, attend.) As you know, presidential elections in the United States take place within an unusual institutional context. Naturally, candidates seek as many votes as they can get, but the real electoral prizes come in winner-take-all, state-sized chunks: The plurality-vote winner in each state receives all the electoral college votes from that state. Cast in logistic regression terms, each state represents a binary outcome—it goes either Republican or Democratic. What variables shape this outcome? What factors make states more likely to end up in the Republican column than in the Democratic column? Given the Republican Party's reputation for conservative positions on social issues, the following hypothesis seems plausible: In a comparison of states, those with higher percentages of residents who frequently attend religious services will be more likely to be won by the Republican candidate than will states having lower percentages of residents who frequently attend religious services.

   States contains gb_win04, a binary variable coded 1 if the state's electoral vote went to Republican George W. Bush in 2004, and coded 0 if the state went to Democrat John Kerry. This is the dependent variable. States also has attend, the percentage of state residents who frequently attend religious services. Attend, which displays interesting variation across states, is the independent variable.[14] Run Regression → Binary Logistic, clicking gb_win04 into the Dependent box and clicking attend into the Covariates box. In Options, request Iteration history. Click Save. In the Predicted Values panel of the Logistic Regression: Save window, select Probabilities.

A. The following table contains seven question marks. Fill in the correct value next to each question mark.

Model estimates	Coefficient	Significance
Constant	?	
attend	?	?
Model summary	Value	Significance
Chi-square	?	?
Cox-Snell R-square	?	
Nagelkerke R-square	?	

B. For the variable attend, Exp(B) is equal to (fill in the blank) _____ . After converting this number to a percentage change in the odds of a Bush win, you could say that a one-unit increase in attend increases the odds of a Bush win by (fill in the blank) _____ percent.

C. Run Analyze → Compare Means → Means, obtaining mean values of the predicted probability of gb_win04 (dependent variable, which SPSS saved as PRE_1) for each value of attend (independent variable). Print the mean comparison table. Use the results to answer parts D, E, and F.

D. The full effect of attend on the probability of gb_win04 is equal to (fill in the blank) _____ .

E. A Republican strategist must decide in which states to concentrate her limited campaign resources. She calculates that, with an all-out effort, her team can increase the probability of a Republican win by about .03 in one state. To achieve maximum effect, this strategist should concentrate her campaign on (check one)

❏ a state in which about 24 percent of its residents frequently attend religious services.

❏ a state in which about 34 percent of its residents frequently attend religious services.

❏ a state in which about 44 percent of its residents frequently attend religious services.

F. Briefly explain your reasoning in part E. _____

_____

_____

_____

_____ .

2. (Dataset: World. Variables: democ_regime, frac_eth, gdp_10_thou.) In Chapter 5 you tested this hypothesis: In a comparison of countries, those having lower levels of ethnic heterogeneity will be more likely to be democracies than will those having higher levels of ethnic heterogeneity. This hypothesis says that, as heterogeneity goes up, the probability of democracy goes down. You then reran the analysis, controlling for a measure of countries' economic development, per-capita gross domestic product (GDP). For this independent variable, the relationship is thought to be positive: As economic development increases, so does the likelihood that a country will be democratic. In the current exercise, you will reexamine this set of relationships, using interval-level independent variables and a more powerful method of analysis, logistic regression.

World contains these three variables: democ_regime, frac_eth, and gdp_10_thou. Democ_regime is coded 1 if the country is a democracy and coded 0 if it is not a democracy. This is the dependent variable. One of the independent variables, frac_eth, can vary between 0 (denoting low heterogeneity) and 1 (high heterogeneity). The other independent variable, gdp_10_thou, measures per-capita GDP in units of $10,000.

A. Run Regression → Binary Logistic, clicking democ_regime into the Dependent box and clicking frac_eth and gdp_10_thou into the Covariates box. In Options, request Iteration history. (For this exercise, you will not be saving predicted probabilities.) Click OK to run the analysis. The following table contains eight question marks. Fill in the correct value next to each question mark.

Model estimates	Coefficient	Significance	Exp(B)
Constant	.842		
frac_eth	−1.592	?	?
gdp_10_thou	.713	?	?

Model summary	Value	Significance
Chi-square	?	?
Cox-Snell R-square	?	
Nagelkerke R-square	?	

B. Use each value of Exp(B) to calculate a percentage change in the odds. Controlling for gdp_10_thou, a one-unit change in frac_eth, from low heterogeneity to high heterogeneity, (check one)

❏ increases the odds of democracy by about 20 percent.

❏ decreases the odds of democracy by about 20 percent.

❏ decreases the odds of democracy by about 80 percent.

Controlling for frac_eth, each $10,000 increase in per-capita GDP (check one)

❏ increases the odds of democracy by about 104 percent.

❏ increases the odds of democracy by about 204 percent.

❏ increases the odds of democracy by about 40 percent.

To respond to parts C, D, E, and F, you will need to use Compute to calculate a new variable, which you will name "pre_frac" and label "Pred prob: frac_eth-democ, mean gdp." Pre_frac will estimate the probability of democracy for each value of frac_eth, holding gdp_10_thou constant at its mean. *Useful fact:* The mean of gdp_10_thou is equal to .602. *Helpful hint:* The numeric expression for computing the predicted probability of democracy for each value of frac_eth is "Exp(.842 − 1.592*frac_eth + .713*.602)/(1 + Exp(.842 − 1.592*frac_eth + .713*.602))." After computing pre_frac, run Compare Means → Means, entering pre_frac as the dependent variable and frac_eth as the independent variable. (*Note:* Because frac_eth has many different values, you may not wish to print the means table. However, you will need to make reference to it for parts C, D, and E.)

C. As an empirical matter, the most homogeneous country in World has a value of 0 on frac_eth, and the most heterogeneous country has a value of .93 on frac_eth. The predicted probability of democracy for a highly homogeneous country (frac_eth = 0) with an average level of gdp_10_thou is equal to (fill in the blank) _____. The predicted probability of democracy for a highly heterogeneous country (frac_eth = .93) with an average level of gdp_10_thou is equal to (fill in the blank) _____.

D. As frac_eth increases, from low heterogeneity to high heterogeneity, the predicted probability of democracy (circle one)

decreases.   does not change.   increases.

E. At mean levels of gdp_10_thou, the full effect of frac_eth (from 0 to .93) on the probability of democracy is equal to (fill in the blank) _____.

F. Imagine a country that has average per-capita GDP and also has an average level of ethnic fractionalization: .44 on the frac_eth scale. This country (circle one)

is probably not a democracy.   is probably a democracy.

3. (Dataset: NES2004 Variables: gay_marriage, egalitarianism, gay_therm.) What factors affect individuals' attitudes toward laws that would permit homosexuals to marry? One possibility: the general principle of egalitarianism. People holding more egalitarian views will be more likely to support gay marriage than will people holding less egalitarian views. Another idea has less to do with principle and more to do with feelings. It may be that, regardless of their professed levels of egalitarianism, people who have negative feelings toward gays will oppose gay marriage, and people with positive feelings toward gays will support gay marriage. NES2004 contains gay_marriage, coded 0 for respondents who oppose and coded 1 for respondents who support. The dataset also has egalitarianism, a scale that ranges from 0 (low egalitarianism) to 24 (high egalitarianism). Affect toward gays is measured by gay_therm, a 100-point feeling thermometer.[15]

A. Run the analysis. Write the correct values next the question marks in the following table:

Model estimates	Coefficient	Significance	Exp(B)
Constant	?		
egalitarianism	?	?	?
gay_therm	?	?	?
Model summary	Value	Significance	
Chi-square	?	?	
Cox-Snell R-square	?		
Nagelkerke R-square	?		

B. Based on your analysis in part A, you may conclude that (check all that apply)

❏ A model that uses egalitarianism and gay_therm to predict the likelihood of supporting gay marriage performs significantly better than the know-nothing model.

❏ Each one-unit increase in the egalitarianism scale increases the odds of supporting gay marriage by about 8 percent.

❏ Each one-unit increase in the gay feeling thermometer increases the odds of supporting gay marriage by about 5 percent.

❏ Both independent variables are significantly related to the dependent variable.

Next you are going to see how egalitarianism affects the probability of supporting gay marriage for people having different levels of affect toward gays: People who give gays fairly low ratings (a rating of 30 on gay_therm), people falling at the sample-wide mean of gay_therm (a rating of 50 on gay_therm), and people who give gays fairly high ratings (a rating of 70 on gay_therm).[16] You will need to process three Compute expressions and create three new variables. Name the new variables "gay_low," "gay_mean," and "gay_high." *Hint:* The following expression will create gay_low, the predicted probability of supporting gay marriage for each value of egalitarianism, holding gay_therm constant at 30: "Exp(−4.315 + .08*(egalitarianism) + .048*30)/(1 + Exp(−4.315 +.08*(egalitarianism) + .048*30))."

C. Run a means analysis, using egalitarianism as the independent variable and gay_low, gay_mean, and gay_high as the dependent variables. In Options, click Standard Deviation and Number of Cases back into the Statistics list. Examine the effects of egalitarianism for each level of affect toward gays. For example, according to the means table, for respondents who give gays a rating of 30 (the gay_low column) and who score 0 on the egalitarianism scale, the probability of supporting gay marriage is equal to .053. For respondents who give gays a rating of 30 and who score 24 on the egalitarianism scale, the probability of supporting gay marriage is equal to .278. For these respondents, the full effect of egalitarianism increases the probability of supporting gay marriage by .225. Fill in the missing values in the following table:

	Probability at 0 on the egalitarianism scale	Probability at 24 on the egalitarianism scale	Full effect of egalitarianism on probability of supporting gay marriage
gay_low	.053	.278	.225
gay_mean	?	?	?
gay_high	?	?	?

D. Based on your analysis in part C, you may infer that (check two)

❑ The effect of egalitarianism on support for gay marriage is about the same for people with low affect toward gays (gay_low) as for people with high affect toward gays (gay_high).

❑ Egalitarianism has a weaker effect on support for gay marriage for people with low affect toward gays (gay_low) than for people with high affect toward gays (gay_high).

❑ Among people with average affect toward gays (gay_mean), strong egalitarians—individuals who score 24 on the egalitarian scale—are more likely not to support gay marriage than to support gay marriage.

❑ Among people with average affect toward gays (gay_mean), strong egalitarians—individuals who score 24 on the egalitarian scale—are more likely to support gay marriage than not to support gay marriage.

E. Use Graphs → Legacy Dialogs → Line to create a multiple line chart showing the relationship between egalitarianism and predicted probabilities of supporting gay marriage for individuals with low affect toward gays (gay_low) and for individuals with high affect toward gays (gay_high). In the Line Charts window, don't forget to request Summaries of separate variables. Edit the graph for clarity. Print the line graph you have created.

That concludes the exercises for this chapter. Before exiting SPSS, be sure to save your output file.

NOTES

1. In arriving at the estimated effect of the independent variable on the dependent variable, linear regression finds the line that minimizes the square of the distance between the observed values of the dependent variable and the predicted values of the dependent variable—predicted, that is, on the basis of the independent variable. The regression line is often referred to as the "least squares" line or "ordinary least squares" line.

2. For all guided examples and exercises in this chapter, the binary dependent variables are naturally coded 0 or 1. To get logistic regression to work, SPSS must have 0/1 binaries. However, here is a bit of SPSS trivia. In running Regression → Binary Logistic, SPSS will check to make sure that the dependent variable has only two values. The values could be 0 and 1, 3 and 5, 2 and 6, or any two (but only two) unique values. If the two values are not 0 and 1, then SPSS will temporarily recode the variable for the immediate purposes of the analysis, encoding one value of the dependent as 0 and the other as 1. SPSS output informs you which natural code it changed to 0 and which it changed to 1. The encoding does not alter your permanent dataset codes.

3. Cox-Snell's maximum achievable value depends on the analysis at hand, but it can never exactly equal 1. For a binary dependent in which the probabilities of 0 and 1 are equal (probability of 0 = .5 and probability of 1 = .5), Cox-Snell reaches a maximum of only .75 for a model in which all cases are predicted perfectly. Nagelkerke's adjustment divides the calculated value of Cox-Snell by the maximum achievable value of Cox-Snell, returning a coefficient that varies between 0 and 1. See D. R. Cox and E. J. Snell, *The Analysis of Binary Data* (London: Chapman and Hall, 1989); N. J. D. Nagelkerke, "A Note on a General Definition of the Coefficient of Determination," *Biometrika* 78 (September 1991): 691–692.

4. When you request Iteration history, SPSS will by default produce two histories—one appearing near the beginning of the output beneath the label "Block 0: Beginning Block" and one appearing later beneath the label "Block 1: Method = Enter." In most situations, all of the information you will need can be found under the Block 1 entry. Figure 10-4 portrays the information contained in the Block 1 entry.

5. Why is the controlled effect of education (a 34.7-percent increase in the odds of voting) somewhat greater than its uncontrolled effect (a 28.1-percent increase in the odds of voting)? Running Analyze → Correlate → Bivariate for educ and age provides an important clue: educ and age are negatively correlated ($r = -.06$). Thus in the earlier analysis, in which we compared respondents having less education with respondents having more education (but in which we did not control for age), we were also comparing older respondents (who, on average, have fewer years of schooling) with younger respondents (who, on average, have more years of schooling). Because younger people are less likely to vote than are older people, the uncontrolled effect of age weakens the zero-order relationship between educ and voted04. In a situation like this, age is said to be a *suppressor variable*, because it suppresses or attenuates the true effect of education on turnout.

6. When using interval-level independent variables with many values, you will often obtain logistic regression coefficients and odds ratios that appear to be quite close to null hypothesis territory (coefficients close to 0 and odds ratios close to 1) but that nonetheless trump the null hypothesis. Remember that logistic regression, like OLS, estimates the marginal effect of a one-unit increment on the logged odds of the dependent variable. In the current example, logistic regression estimated the effect of a 1-year change in age (from, say, an age of 20 years to 21 years) on the logged odds of voting. The researcher may describe the relationship in terms of larger increments. Thus, if a 1-year increase in age (from 20 years to 21 years) increases the odds of voting by an estimated 3.7 percent, then a 10-year increase in age (from 20 years to 30 years) would produce a 37-percent increase in the odds of voting.

7. The expression "Exp(Logged odds of voting)" translates logged odds into odds: Exp(Logged odds of voting) = Odds of voting. You get from an odds to a probability by dividing the odds by the quantity one plus the odds: Probability of voting = Odds of voting/(1 + Odds of voting). Thus the formula for the probability of voting, "Exp(Logged odds of voting)/(1 + Exp(Logged odds of voting))," is equivalent to the formula "Odds of voting/(1 + Odds of voting)."

8. The largest marginal effect of the independent variable on the probability of the dependent variable is sometimes called the *instantaneous effect*. In our example, the instantaneous effect is equal to .062, and this effect occurs between 8 years and 9 years of education. The effect of a one-unit change in the independent variable on the probability of the dependent variable is always greatest for the interval containing a probability equal to .5. The instantaneous effect, calculated by hand, is equal to $b*.5*(1 - .5)$, in which b is the value of the logistic regression coefficient. For a discussion of the instantaneous effect, see Fred C. Pampel, *Logistic Regression: A Primer,* Sage University Papers Series on Quantitative Applications in the Social Sciences, series no. 07-132 (Thousand Oaks, Calif.: Sage Publications, 2000), 24–26.

9. In calculating predicted probabilities for multivariate logistic regression models, SPSS returns estimated probabilities for subjects having each combination of values on the independent variables. It does not calculate the probabilities associated with each value of a given independent variable while holding the other predictors constant.

10. SPSS has a large repertoire of canned statistical functions. The function Exp(numerical expression) returns the natural log base *e* raised to the power of the numerical expression. This is precisely what we want here, because the estimated probability of voting is equal to Exp(Logged odds of voting)/(1 + Exp(Logged odds of voting)).

11. The Compute Variable window's micro-font Numeric Expression box is a tedious and error-prone place to work, especially if you wish to create long, nested expressions. On the plus side, the box accepts cut-and-paste editing, and it's not choosy about word processing software. In a word processor, you can type an expression just as you want it to appear in the Numeric Expression box, and then copy and paste it into the box. Because the current example's two expressions differ in only one detail—the pre_26 expression requires a "26" and the pre_65 expression requires a "65"—we could type and copy-paste the pre_26 statement, and then compute pre_26. We would then return to Transform → Compute, change the Target Variable to pre_65, replace "26" with "65" in the Numeric Expression box, and compute pre_65.

12. See Raymond E. Wolfinger and Steven J. Rosenstone's classic study of turnount, *Who Votes?* (New Haven: Yale University Press, 1980). Using probit analysis, a technique that is very similar to logistic regression, Wolfinger and Rosenstone explored the effects of a range of demographic characteristics on the likelihood of voting.

13. You may want to experiment with a few choices in the Chart Editor's Options menu: Y Axis Reference Line, Annotation, and Hide Legend. SPSS sometimes produces charts that are too "square." A chart with a width of between 1.3 and 1.5 times its height may be more pleasing to the eye. In any event, you can alter the aspect ratio (the ratio of width to height) in the Chart Size tab of the Properties window. After unchecking the box next to "Maintain aspect ratio," click in the Width box and type a number that is between 1.3 and 1.5 times the value appearing in the Height box.

14. Attend varies between 24.28 percent (Washington) and 60.95 percent (Louisiana). A total of 13 states lack sufficient information on attend and have been set to missing on this variable. The remaining 37 states are analyzed in this exercise.

15. Gay_therm is based on NES variable V045074, which asks respondents to rate "gay men and lesbians, that is, homosexuals." You analyzed the egalitarianism scale in Chapter 6. Scores range from 0 (low egalitarianism) to 24 (high egalitarianism).

16. The true mean of gay_therm is 48.52, but use the simpler value of 50.

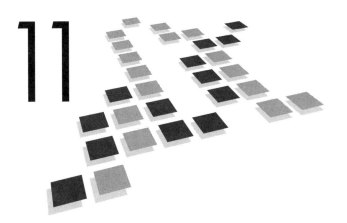

11

Doing Your Own Political Analysis

Procedures Covered

File → Read Text Data

In working through the guided examples in this book, and in performing the exercises, you have developed some solid analytic skills. The datasets you have analyzed throughout this book could, of course, become the raw material for your own research. You would not be disappointed, however, if you were to look elsewhere for excellent data. High-quality social science data on a variety of phenomena and units of analysis—individuals, census tracts, states, countries—are widely available via the Internet and might serve as the centerpiece for your own research. Your school, for example, may be a member of the Inter-university Consortium for Political and Social Research (ICPSR), the premier organizational clearinghouse for datasets of all kinds.[1] You will find that the expertise you have gained can be productively applied to any number of research questions that interest you.

Even so, there is much to be said for striking out on your own—observing an interesting fact or behavior, developing an explanation and hypothesis, and collecting and analyzing your own original dataset. Doing original research from the ground up can yield large intellectual dividends. (Besides, your instructor may require it.) In this chapter we explore the do-it-yourself route. We begin by laying out the stages of the research process and by offering some manageable ideas for original analysis. We illustrate how the raw dataset included with this book, senate2003.txt, was created from information collected from an Internet site. By following the steps described in this chapter, you will learn how to get SPSS to read the raw data. Finally, we describe a serviceable format for an organized and presentable research paper.

FIVE DOABLE IDEAS

Let's begin by describing the stages involved in an ideal research procedure and then discuss some practical considerations and constraints. In an ideal world you would

1. observe an interesting behavior or relationship and frame a research question about it;
2. develop a causal explanation for what you have observed and construct a hypothesis;
3. read and learn from the work of other researchers who have tackled similar questions;
4. collect and analyze the data that will address the hypothesis; and
5. write a research paper or article in which you present and interpret your findings.

In this scenario the phenomenon that you observe in stage 1 drives the whole process. First think up a question and then research it and obtain the data that will address it. As a practical matter, the process is almost never this clear-cut. Often someone else's idea or assertion may pique your interest. For example, you might read articles or attend lectures on a variety of topics—democratization in developing countries, global environmental issues, ideological change in the Democratic or Republican Party, the effect of election laws on turnout and party competition, and so on—that suggest hypotheses you would like to examine. So you may begin the process at stage 3, and then return to stage 1 and refine your own ideas. Furthermore, the availability of relevant data, considered in stage 4, almost always plays a role in the sorts of questions we address. Suppose, for example, that you want to assess the organizational efforts to mobilize African Americans in your state in the last presidential election. You want precinct-level registration data, and you need to compare these numbers with the figures from previous elections. You would soon become an expert in the bureaucratic hassles and expense involved in dealing with county governments, and you might have to revise your research agenda. Indeed, for professional researchers and academics, data collection in itself can be a full-time job. For students who wish to produce a competent and manageable project, the so-called law of available data can be a source of frustration and discouragement.

A doable project often requires a compromise between stage 1 and stage 4. What interesting question can you ask, given the available data? Fortunately, this compromise need not be as restrictive as it sounds. Consider five possibilities: political knowledge, economic performance and election outcomes, judicial selection and public opinion, electoral turnout in comparative perspective, and the U.S. Congress.

Political Knowledge

As you may have learned in other political science courses, scholars continue to debate the levels of knowledge and political awareness among ordinary citizens. Do citizens know the length of a U.S. senator's term of office? Do they know what constitutional protections are guaranteed by the First Amendment? Do people tend to know more about some things—Internet privacy or abortion policy, for example—and less about other things, such as foreign policy or international politics? Political knowledge is a promising variable because you are likely to find some people who know a lot about politics, some who know a fair amount, and others who know very little. You could ask, "What causes this variation?" Imagine constructing a brief questionnaire that asks 8 or 10 multiple-choice questions about basic facts and is tailored to the aspects of political knowledge you find most thought provoking. After including questions that gauge some potentially important independent variables (partisanship, gender, liberalism/conservatism, college major, class standing), you could conduct an exploratory survey among perhaps 50 or 100 of your fellow students. (Your sample would not be random, but you would still get some compelling results that could guide you toward a more ambitious study.)[2]

Economic Performance and Election Outcomes

Here is one of the most widely discussed ideas in political science: The state of the economy before an election has a big effect on the election result. If the economy is strong, then the candidate of the incumbent party does well, probably winning. If the economy is performing poorly, then the incumbent party's nominee pays the price, probably losing. This idea has a couple of intriguing aspects. For one thing, it works well—but not perfectly. (The 2000 presidential election is a case in point.) Moreover, the economy-election relationship has several researchable layers. Focusing on presidential elections, you can imagine a simple two-category measure of the dependent variable—the incumbent party wins or the incumbent party loses. Now consider several stints in the reference section of the library, collecting information on some potential independent variables for each presidential election year: inflation rates, unemployment, economic growth, and so on. Alternatively, you could look at congressional or state-level elections, or elections in several different countries. Or you could modify and refine the basic idea, as many scholars have done, by adding additional noneconomic variables you believe to be important. Scandal? Foreign policy crises? With some hands-on data collection and guidance from your instructor, you can produce a well-crafted project.[3]

Judicial Selection and Public Opinion

To what extent does a justice's partisanship or ideology affect his or her ruling in a case? Are judges somehow different from ordinary public figures, or is their behavior shaped by the same political forces—partisanship, ideology, and the need for public approval? Again, the 2000 election comes to mind. The U.S. Supreme Court based its pivotal decision on judicial principles, but the Court split along partisan lines. And, given the level of partisan acrimony that accompanies the nominations of would-be federal judges, members of the U.S. Senate behave as if political ideology plays a role in judicial decision making. Consider also state and county judicial systems. In many states judges are elected—some in partisan contests, others in nonpartisan elections—and in other states they are appointed. Imagine classifying a number of county judgeships as elected or appointed. Further imagine comparing these two groups on a number of attributes and behaviors. For example, perhaps appointed judges have more legal experience than do elected judges. Perhaps elected judges make decisions more in line with local interests than do appointed judges. The questions are interesting, and appropriate data are available.[4]

Electoral Turnout in Comparative Perspective

The record of voter turnout in U.S. presidential elections, while showing an encouraging reversal in 2004, generally is pretty anemic. Despite the down-to-the-wire closeness of the 2000 contest, for example, just slightly more than half of the voting-age population went to the polls on the day of the election. The situation in other democratic countries is strikingly different. Turnouts in some Western European countries average well above 70 percent. Why? More generally, what causes turnout to vary between countries? Some scholars have focused on legal factors. Unlike the United States, some countries may not require their citizens to register beforehand, or they may penalize citizens for not voting. Other scholars look at institutional differences in electoral systems. Many countries, for example, have systems of proportional representation in which narrowly focused parties with relatively few supporters nonetheless can gain representation in the legislature. Are citizens more likely to be mobilized to vote under such institutional arrangements? Using sources available in the reference section of the library, you could gather information on 15 or 20 nations. You could then look to see if different legal requirements and institutional arrangements are associated with differences in turnout. This area of research might also open the door for some informed speculation on your part. What sort of electoral reforms, if instituted in the United States, might enhance electoral turnout? What other (perhaps unintended) consequences might such reforms have?[5]

Congress

Political scholars have long taken considerable interest in questions about the U.S. Congress. Some researchers focus on internal dynamics: the role of leadership, the power of party ties versus the pull of constituency. Others pay attention to demographics: Has the number of women and minorities who serve in Congress increased in the recent past? Still others look at ideology: Are Republicans, on average, becoming more conservative and Democrats more liberal in their congressional voting? The great thing about Congress is the rich data that are available. The U.S. House and the U.S. Senate are among the most-studied institutions in the world. Several annual or biannual publications chronicle and report a large number of attributes of members of the House and the Senate.[6] And the Internet is rife with information about current and past Congresses. Liberal groups, such as Americans for Democratic Action (ADA), and conservative groups, such as the American Conservative Union, regularly rate the voting records of elected officials and post these ratings on their Internet sites.

Each of these five possibilities represents a practical compromise between posing an interesting question and obtaining the data to address it. In the next section we take a closer look at one of these examples—research on the U.S. Senate. Using for illustration a dataset included with this book, we consider the nuts and bolts of the research: where to find worthwhile data, how to code the data, and how to get SPSS to read the data.

DOING RESEARCH ON THE U.S. SENATE

The Senate is an intriguing institution. Recall a seemingly insignificant historical event that illustrates the institution's complexity. Following the 2000 elections, the Senate was split right down the middle on partisanship: There were 50 Republicans and 50 Democrats. Republican vice president Dick Cheney, acting in his constitutional capacity, was the tie-breaker, giving the Republican Party a nominal majority. However, a few months after the elections, in May 2001, one of the Republicans, Sen. James Jeffords of Vermont, formally abandoned his party affiliation and became an Independent, making the Democrats the majority party in the Senate.

This was a controversial event. But what *general* questions did it raise about Congress? For one thing, it underscored the fact that each house of Congress is formally organized on the basis of party. The party that holds a majority gains control of the leadership posts, and to be chosen to chair a committee or subcommittee, a member of Congress must be from the majority party. Thus, following the partisan shift of a solitary member, all of the leadership and committee chair positions in the Senate went to members of the Democratic Party—a situation that remained in place until after the 2002 elections, when Republicans reclaimed a majority. Yet Jeffords's shift also illustrated that individual members of Congress have a measure of autonomy and that the parties do not always behave as cohesive partisan blocs. Why?

Congress is, of course, charged with making laws that affect the entire country. Democratic and Republican leaders put forward agendas that often reflect philosophically different approaches to lawmaking and governance. The success or failure of these agendas often hinges on whether Democrats and Republicans vote together along party lines. So partisan affiliation exerts an influence on elected representatives. But Congress is also designed as a representative institution—a forum in which the disparate interests of districts and states are brought to bear on public policy. Because representatives and senators must be careful not to support policies that may displease their constituents, they must often weigh the views of their constituencies against the aims of their party leaders.

Consider the hypothetical example of an antiabortion measure supported by the Republican leadership and opposed by the Democratic leadership. Would a Democrat from a traditionally conservative region, such as the South, follow the Democratic leadership and oppose such a measure? What about a Republican from a liberal region, such as the Northeast? More generally, one could ask whether partisan and regional interests are more likely to clash on certain types of issues, such as abortion or military spending, than on other issues, such as energy policy or tax cuts. Or do region and party reinforce one another? Is the United States becoming, as some analysts have said, "two political countries," with the interior South and mountain West aligned on the conservative side of issues and the Northeast, upper Midwest, and coastal West aligned on the liberal side? These are general questions. And they are questions worth asking. Where do we find the data?

Finding Raw Data

In this example we show how one of this book's datasets, senate2003.txt, was created from information that appears on an Internet site. This site, maintained by ADA, provides ratings of each member of Congress across a range of congressional votes.[7] Specifically, for each session of Congress ADA selects a number of key votes, which reflect either a liberal position, supported by ADA, or a conservative position, opposed by ADA. The organization selects these votes from different policy areas: abortion, taxes, health care, defense spending, and so on. Thus we can look for variations in voting patterns, depending on the type of policy being considered. Make no mistake—ADA has an agenda to promote. Its members are down-the-line liberal on every issue. But this consistency works to our advantage in measuring congressional ideology because we can be reasonably certain that a "more favorable" ADA rating is a valid gauge of liberalism and that a "less favorable" rating is a valid measure of conservatism.

Figure 11-1 displays a partial Web page from ADA's site, which reports ratings of the 2003 Senate. The leftmost column shows the state's name and the names of the senators from that state, and the next column displays party affiliation. The next 20 columns show how each senator voted on the 20 key votes, with a plus sign (+) denoting a liberal vote and a minus sign (−) denoting a conservative vote. If the senator did not vote, ADA records this absence with a question mark (?). The substantive policy of each vote, appearing as unhelpful numbers 1 through 20 on the Web page shown in Figure 11-1, can be obtained elsewhere at the ADA site. For example, a little browsing reveals that vote number 1 dealt with President George W. Bush's judicial nomination of Miguel A. Estrada, which ADA opposed. Senators who voted no on this measure got a plus and those supporting it received a minus. Vote number 2, which ADA supported, was an abortion measure expressing Senate support for the Supreme Court's decision in *Roe v. Wade.1*. Senators voting in favor of this measure were given a plus, and those voting against got a minus. On the right side of the page we find ADA's overall liberalism score for 2003, under the heading "LQ," which stands for Liberal Quotient. ADA arrives at the Liberal Quotient by determining the percentage of votes on which the senator supported ADA's position. Scores can range from 0% (most conservative) to 100% (most liberal). How would you take the information from this site and put it into a form that SPSS can analyze?

How to Code Raw Data

Coding raw data is the sweat equity phase of original research, the stage of tedious labor well invested. You must start with numbers or symbols that appear as text on a printed page—or pixels on a computer screen—and record them as numbers or codes that SPSS can interpret. This coding can be done in several ways. A popular approach is to type the data into a spreadsheet, such as Microsoft Excel. Excel, of course,

Figure 11–1   Internet Site of Americans for Democratic Action: Ratings of the U.S. Senate (2003)

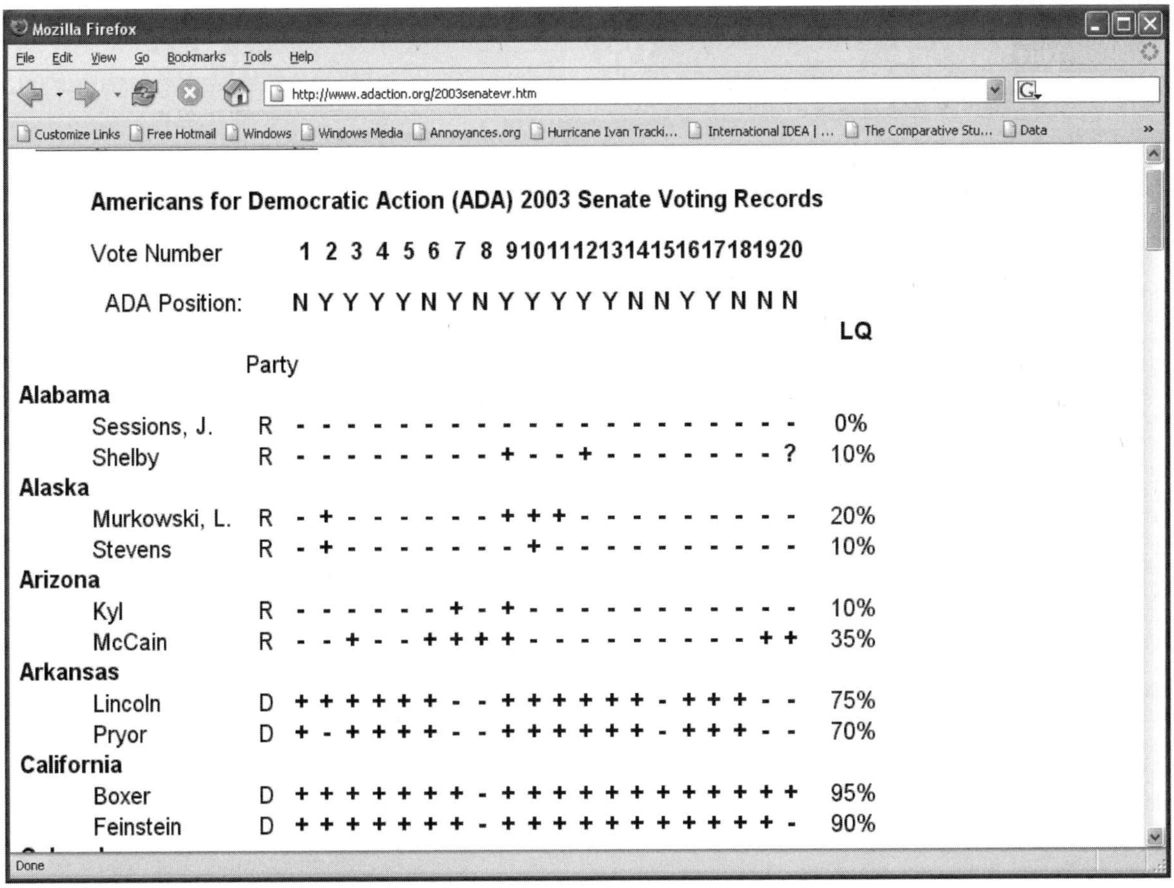

has the advantages of being widely available and easy to use. Alternatively the data could simply be typed into a plain text file editor (like Wordpad) or a favorite word processor and converted to plain text. Another approach is to enter the information directly into the SPSS Data Editor and then proceed with the analysis. Fortunately, the choice of method doesn't really matter. SPSS will read a large variety of data formats, including Excel spreadsheets and plain text. What *does* matter is that the coded information be in a form that SPSS understands. So whether you are using a spreadsheet, a text editor, or the SPSS Data Editor, some coding rules are in order. To make sure that all contingencies are covered, we use the most accessible approach—typing data into a word processor or text file—to illustrate how to code the ADA data. We then describe some coding shortcuts.

To learn some coding essentials, consider the second line of information from the ADA page, data on Senator Shelby of Alabama:

Shelby   R   – – – – – – – – + – – + – – – – – – – ?   10%

Although SPSS would read this data line, it would have trouble analyzing it. SPSS recognizes two basic forms of data, string and numeric. String data are words or symbols, such as Senator Shelby's name, his party designation, the question mark denoting a missed vote, and the plus and minus signs. In fact, in the preceding data line, none of variables are numeric data. SPSS would encounter the percentage sign in "10%" and interpret the entire value as a string. It is a good practice to use numeric codes whenever possible. To convert this line of information into an SPSS-friendly form, we will code the pluses and minuses as 1s and 0s, with 1 representing a liberal vote and 0 representing a conservative vote. Because "?" represents a missed vote, we will replace it with a period (.), recognized as missing data by SPSS. Also, we will delete the "%," leaving the numeric value 10 as Senator Shelby's overall LQ score:

Shelby   R   0 0 0 0 0 0 0 0 1 0 0 1 0 0 0 0 0 0 0 . 10

Figure 11–2 Coding Data in a Text Editor

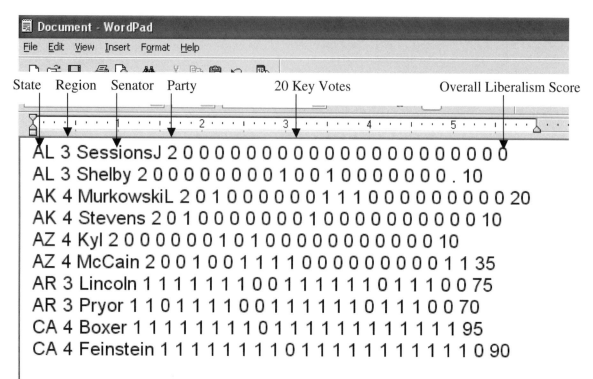

Now each liberal vote is coded 1, and each conservative vote is coded 0. Only the senator's name and his party affiliation ("R" for Republican) remain in nonnumeric form. Notice, too, that the codes are separated by spaces. SPSS interprets spaces as delimiters, boundaries that separate the values of different variables. Alternatively, we can use commas, semicolons, or tabs, because SPSS also recognizes them as delimiters.[8]

We will want to add the state name and census region to each data line. And we also will replace the "D" and "R" party codes with a numeric code for the party affiliation of each senator. State names will be represented by two-character strings, but for region and party we will use numeric codes. For region, each senator from the Northeast will be coded 1; from the Midwest, 2; from the South, 3; and from the West, 4. For party, we'll code Democrats 1 and Republicans 2. (We will code Vermont's Senator Jeffords, the institution's lone Independent, 3 on the party variable.) Here is Senator Shelby's complete typed data line:

$$\text{AL} \quad 3 \quad \text{Shelby} \quad 2\ 0\ 0\ 0\ 0\ 0\ 0\ 0\ 0\ 1\ 0\ 0\ 1\ 0\ 0\ 0\ 0\ 0\ 0\ 0\ .\ 10$$

Senator Shelby represents Alabama (clearly enough), which is in the South census region (coded 3), and he is a Republican (coded 2). We would follow the same coding scheme for each case in the ADA data, consistently typing the state, region code, senator's name, party code, codes for each of the 20 votes, and the overall ADA score. The first 10 cases, typed into a text editor, would look like Figure 11-2. (Notice that, to avoid potential confusion, spaces and commas have been removed from senators' names. So, "Sessions, J." becomes "SessionsJ," and "Murkowski, L." becomes "MurkowskiL.")

Thus far, all of our coding efforts have been aimed at making the data SPSS-friendly—using numeric codes when possible, leaving spaces between the values of different variables, and using periods for missing data. Before we read the data into SPSS, however, we will want to invent a brief but descriptive name for each variable—a word or abbreviation that communicates what the variable is measuring.

Keep in mind two goals when you invent variable names. The first goal is to follow SPSS's variable-naming rules. If you are running SPSS 11.5 or earlier, these rules are rather restrictive. An SPSS variable name must be no longer than eight characters, must begin with a letter, and must not contain any blanks or special characters, such as dashes or commas. If you are running SPSS 12.0 or later, life is easier. Although variable names must begin with a letter and cannot contain special characters, they can be up to 64 characters in length. But to keep the presentation general, here we will use 11.5-compatible variable names. The variable-naming rules are easily followed for the first four variables, which we'll call state, region, senator, and party.

Table 11-1   Key Votes in the U.S. Senate (2003), as Identified by Americans for Democratic Action (ADA), with Corresponding Variable Names and Variable Labels

Key Vote	Vote Description	Outcome of a yes vote	ADA's position	Winning position	Variable name	Variable label
1	Estrada judicial nomination	Limit debate and vote on nomination	No	No	estrada	Cloture on Estrada
2	Amendment to abortion bill	Express support for *Roe v. Wade*	Yes	Yes	abort1	Support Roe
3	ANWR oil drilling	Kill language authorizing oil drilling	Yes	Yes	energy1	ANWR drilling
4	Budget resolution on childcare	Increase mandatory childcare spending	Yes	No	daycare	Childcare spending
5	Unemployment insurance extension	Waive Budget Act and extend unemployment benefits	Yes	No	employ1	Unemployment insurance
6	Tax reduction	Reduce taxes by $350 billion over 11 years	No	Yes	tax	Tax cuts
7	Nuclear power plants	Kill loan guarantees for seven new nuclear power plants	Yes	No	energy2	Nuclear power
8	Prescription drug benefit	Authorize creation of prescription drug benefit	No	Yes	rxplan1	Rx benefit
9	State Department reauthorization, HIV/AIDS	Support full funding of global HIV/AIDS bill	Yes	Yes	aids	AIDS funding
10	Overtime pay regulations	Prohibit regulations taking away overtime pay	Yes	Yes	employ2	Overtime pay
11	Student financial aid eligibility	Prohibit change in student loan formulas	Yes	Yes	govloan	Student loans
12	FCC media ownership rule	Disapprove allowing media companies to own more television stations	Yes	Yes	fcc_rule	FCC media rule
13	Iraq reconstruction funding	Eliminate money allocated for reconstruction of Iraq	Yes	No	iraq1	Iraq reconstruction
14	Supplemental funds for Iraq and Afghanistan	Kill above amendment (see key vote number 13)	No	Yes	iraq2	Iraq/Afghanistan
15	Late term abortion ban	Ban medical procedure known as partial birth abortion	No	Yes	abort2	Partial birth abortion
16	Competitive sourcing, government/private	Prohibit competition between government and private sources	Yes	No	sourcing	Competitive sourcing
17	Election systems overhaul	Add grants to states for improving election technology	Yes	Yes	vot_tech	Voting technology
18	Pickering judicial nomination	Limit debate and vote on nomination	No	No	pickerin	Cloture on Pickering
19	Bush energy policy	End debate on energy bill	No	No	energy3	Energy policy
20	Prescription drug benefit conference report	End debate on bill that would create a prescription drug benefit	No	Yes	rxplan2	Rx benefit, conference report

Source: For key vote descriptions, ADA positions, and winning positions, see the Internet site of Americans for Democratic Action, www.adaction.org.

Note: SPSS variable names and variable labels are author's suggestions.

The second goal is to give similar names to variables that measure similar characteristics. According to the ADA site, for example, key votes 2 and 15 dealt with abortion, and key votes 3, 7, and 19 had to do with energy policy. We would keep such similarities in mind when devising names for the key votes. As noted earlier, a measure involving the Estrada nomination was the first key vote, which we can name estrada. We'll call the second and third votes, abort1 and energy1, respectively. (See Table 11-1 for a description of every

Figure 11–3 Typing Variable Names in a Text Editor

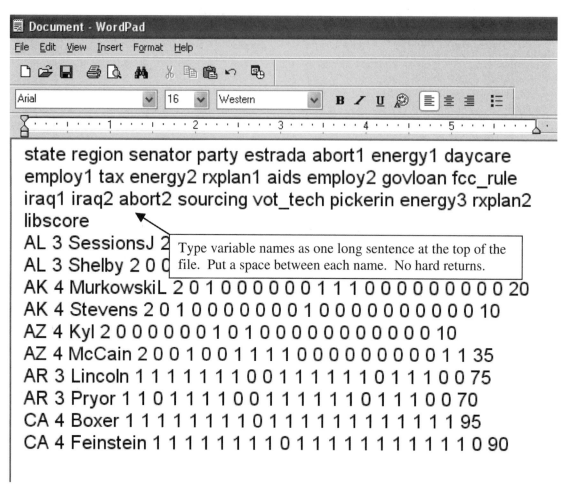

key vote.) We type these names at the top of the text data file in precisely the order in which they appear for each case. We type them as if they were one long sentence, with a space between each name and no hard returns. So the variable names—including state, region, senator, party, the 20 key votes, and the overall liberalism score (named "libscore")—and the first 10 cases in the Senate data would look like Figure 11-3.

Two Possible Coding Shortcuts

Typing a raw data file, line by line, can be tedious work. If you want to code data that you find in printed sources, there is no other option. You've got to hand-code the numbers. However, if you want to retrieve data from an Internet site, there are two ways in which you might cut your coding time. First, most browsers allow you to save a Web page in a format that can be accessed by other applications. Mozilla Firefox, for example, permits you to save the ADA page as a file with the *.htm extension, which Microsoft Excel will gladly open. To be sure, once the data are in Excel, much editing and labeling are required, but at least the data are in the spreadsheet and in workable form. The second method is somewhat more mundane: Select the desired text on the Web page, copy it to the clipboard, and then paste it into your word processor. Once it is in the word processor, you can make abundant use of the Find and Replace editing functions to alter the form of the data—replacing all the plus signs with 1s and all the minus signs with 0s, for example. The Senate dataset used in this illustration was created in just this way, although a fair amount of hand-typing—the state abbreviations and the region codes—was still required. The data were edited using Microsoft Word, and then the file was saved as a plain text file, with the *.txt extension.

You have seen how senate2003.txt was created. Now you will read the data file into SPSS.

Using the SPSS Text Import Wizard

Open SPSS. In the Data View of the Data Editor, click File → Read Text Data. Browse the file directory and locate the text data file, senate2003.txt. Select the data file and click Open. Step 1 of the Text Import Wizard

Figure 11-4 Step 1 of the Text Import Wizard

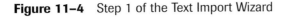

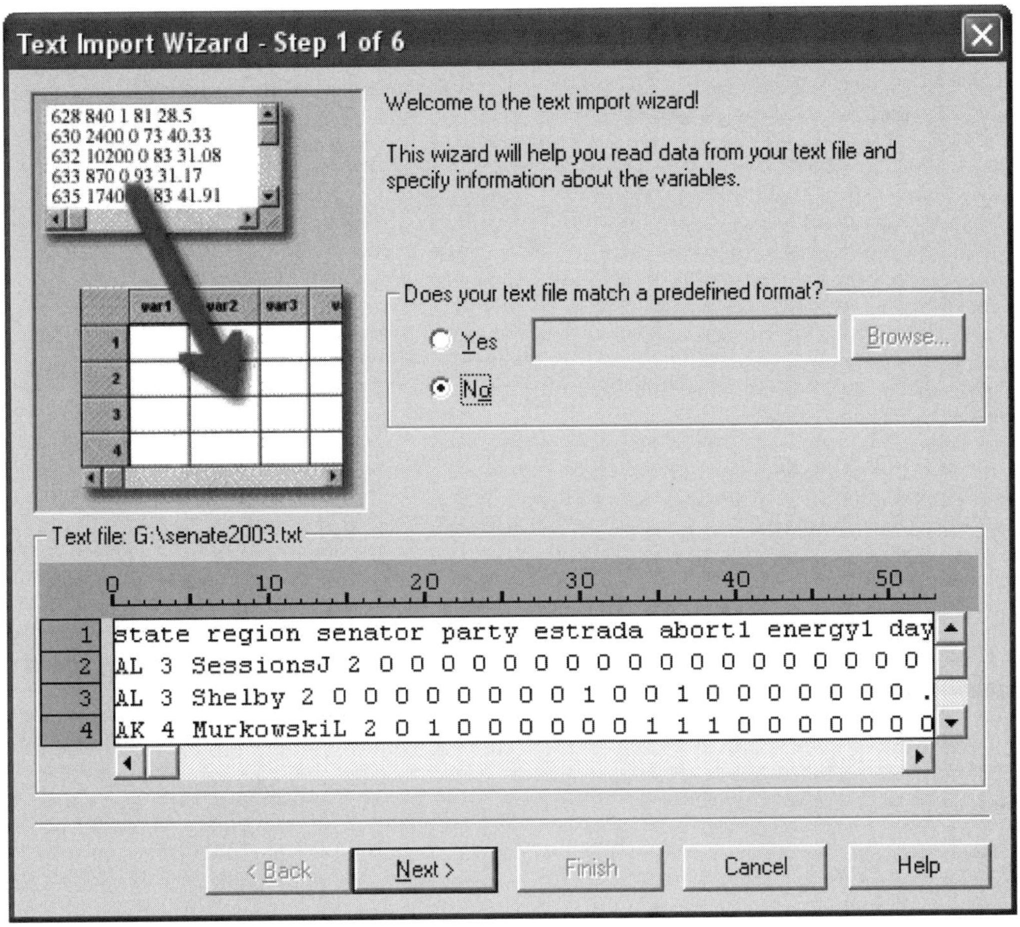

appears (Figure 11-4). SPSS asks one question about the data: "Does your text file match a predetermined format?" Make sure that the radio button next to "No" is clicked. At the bottom of the Step 1 window, SPSS offers a preview window. You can scroll through the preview and see if everything checks out. If all is well, click Next. Step 2 opens (Figure 11-5). Because the ADA variables are separated by spaces, in Step 2 you will answer "Delimited" to the question "How are your variables arranged?" and "Yes" to "Are variable names included at the top of your file?" Each succeeding step in the Text Import Wizard follows in this vein as SPSS elicits more specific information about the text data.

In Step 3 (Figure 11-6) you will tell SPSS that the first case of data begins on line 2 (because line 1 of the dataset contains the variable names), that each line represents a case, and that you want SPSS to import all the cases. In Step 4 (Figure 11-7) you will make sure that "Space" is selected because spaces—not tabs, commas, or semicolons—define the boundaries between variables. In Step 5 (Figure 11-8) you will make a minor change to the SPSS defaults. With the sort of text data you are importing in this example, SPSS sets a default of 10 characters for string variables. This limit works fine for the state abbreviation, but it won't accommodate the full text of all senators' names. To remedy this problem, click on the variable name "senator." In the Characters box, change the value to 12. In Step 6 (Figure 11-9), answer no to both questions, and click Finish. The dataset appears in the SPSS Data Editor (Figure 11-10). Data for the by-now familiar Senator Sessions appear along the first data row, data for his fellow Alabaman, Senator Shelby, in the second row, and so on. Naturally, you can scroll through the data to ensure that you imported the text file correctly. If so, you will want to save your data file in SPSS format. Click File → Save and choose a name and location. Save the Senate data as Senate2003.sav.

This is an eminently analyzable dataset. You may want to explore it. You can create new variables using Recode or Compute.[9] You can perform mean comparison analyses, comparing scores on the liberalism scale for Democrats and Republicans from different regions of the country. You can create liberalism scales for different policy areas. For example, you could sum the abortion votes (abort1 and abort2) and the drug

Figure 11–5   Step 2 of the Text Import Wizard

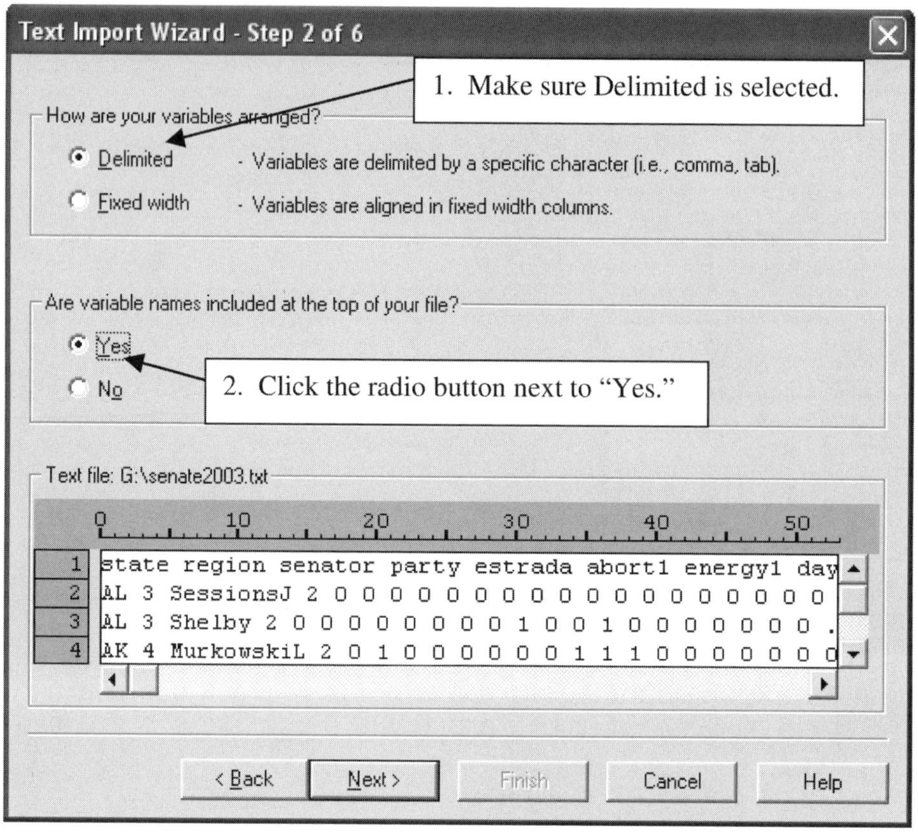

Figure 11–6   Step 3 of the Text Import Wizard

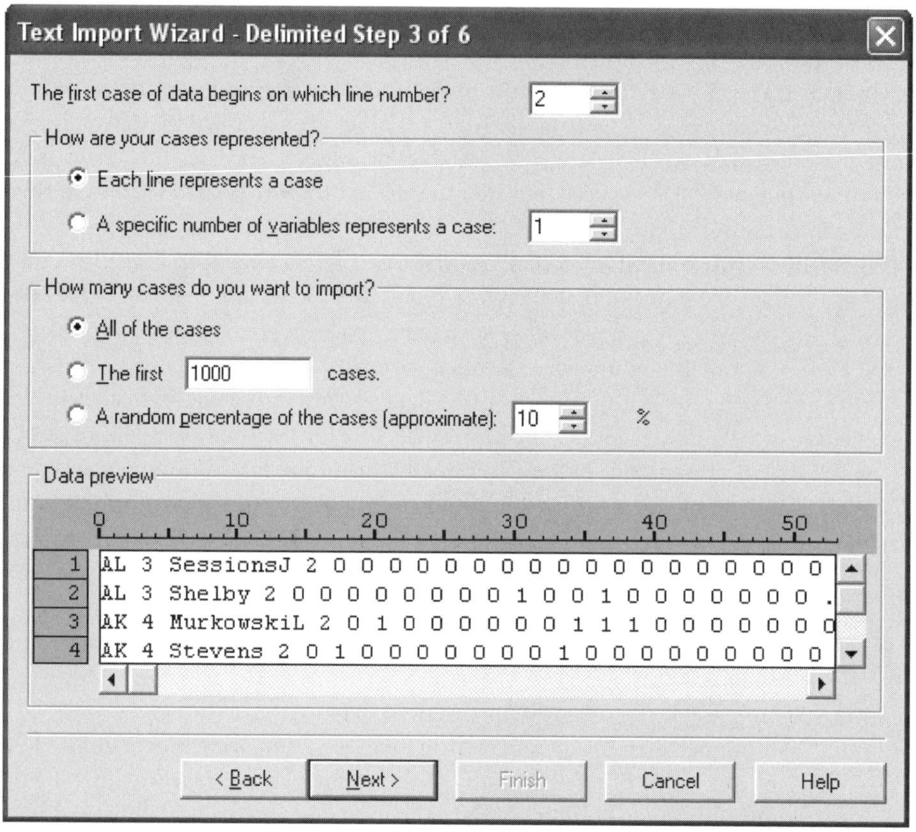

Figure 11–7   Step 4 of the Text Import Wizard

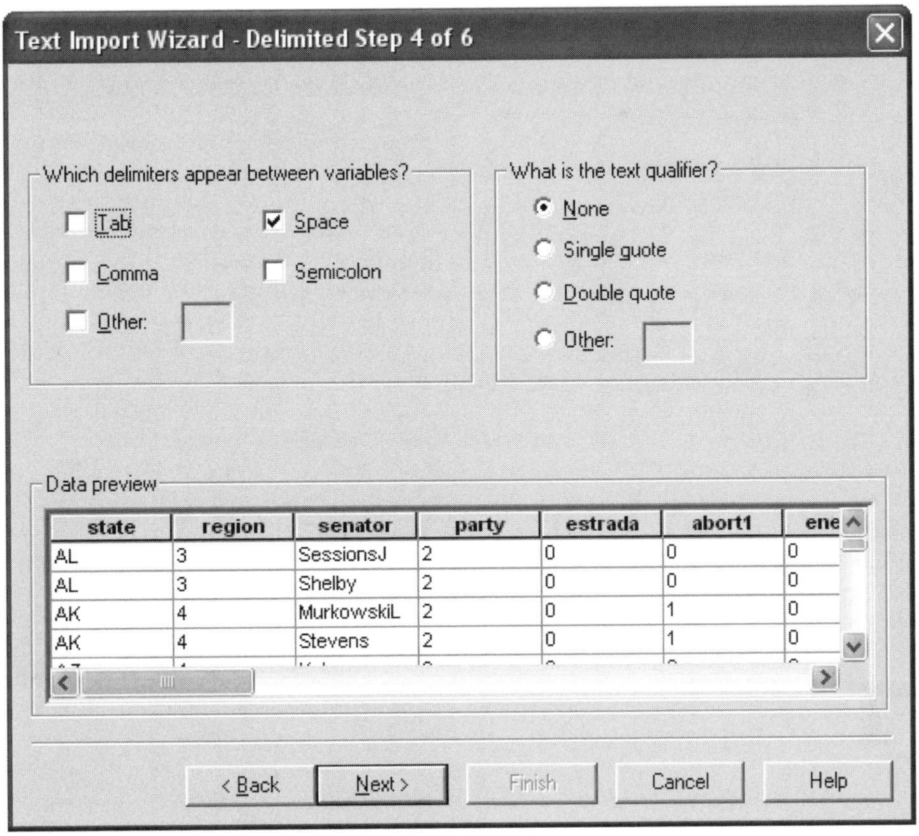

Figure 11–8   Step 5 of the Text Import Wizard

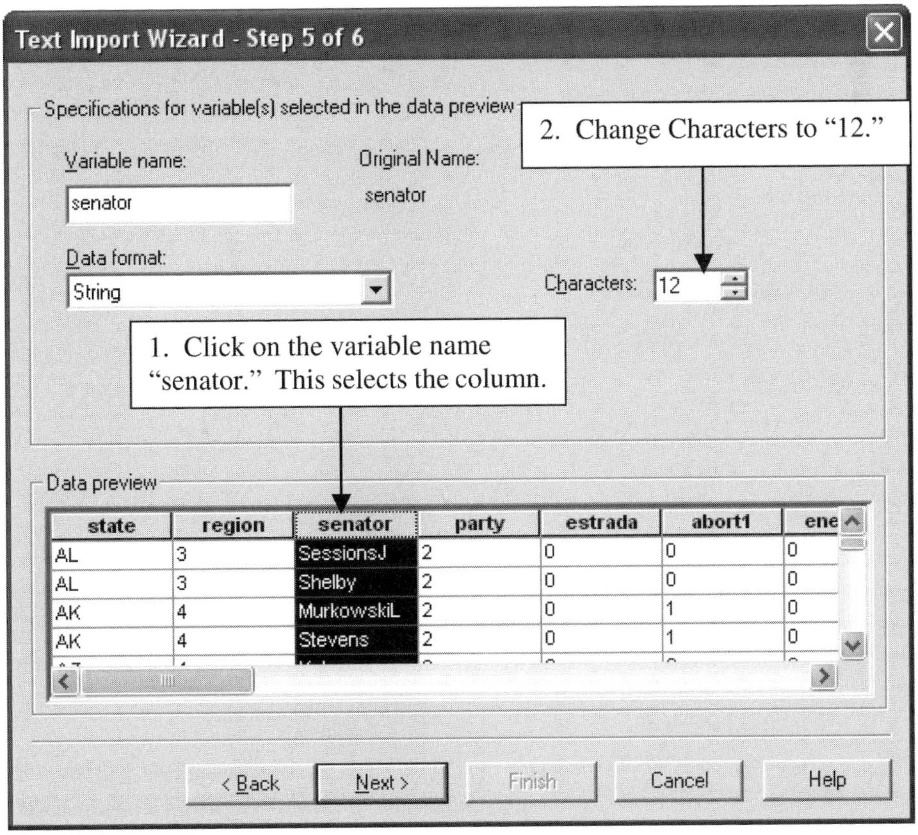

Figure 11–9 Step 6 of the Text Import Wizard

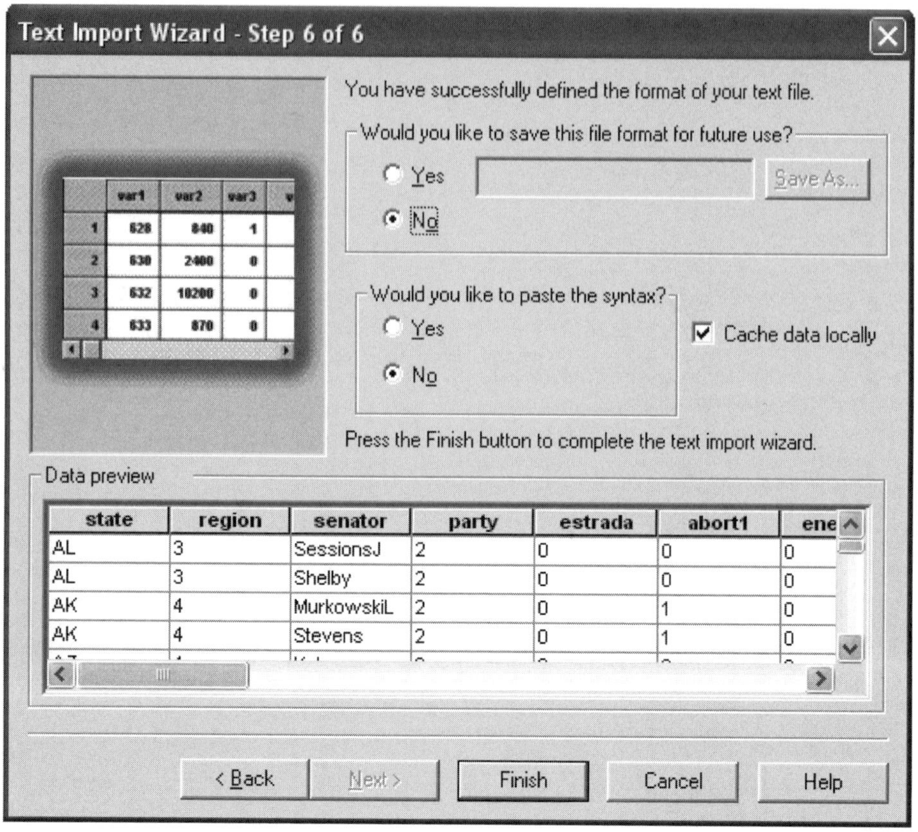

Figure 11–10 Senate Data in the Data Editor

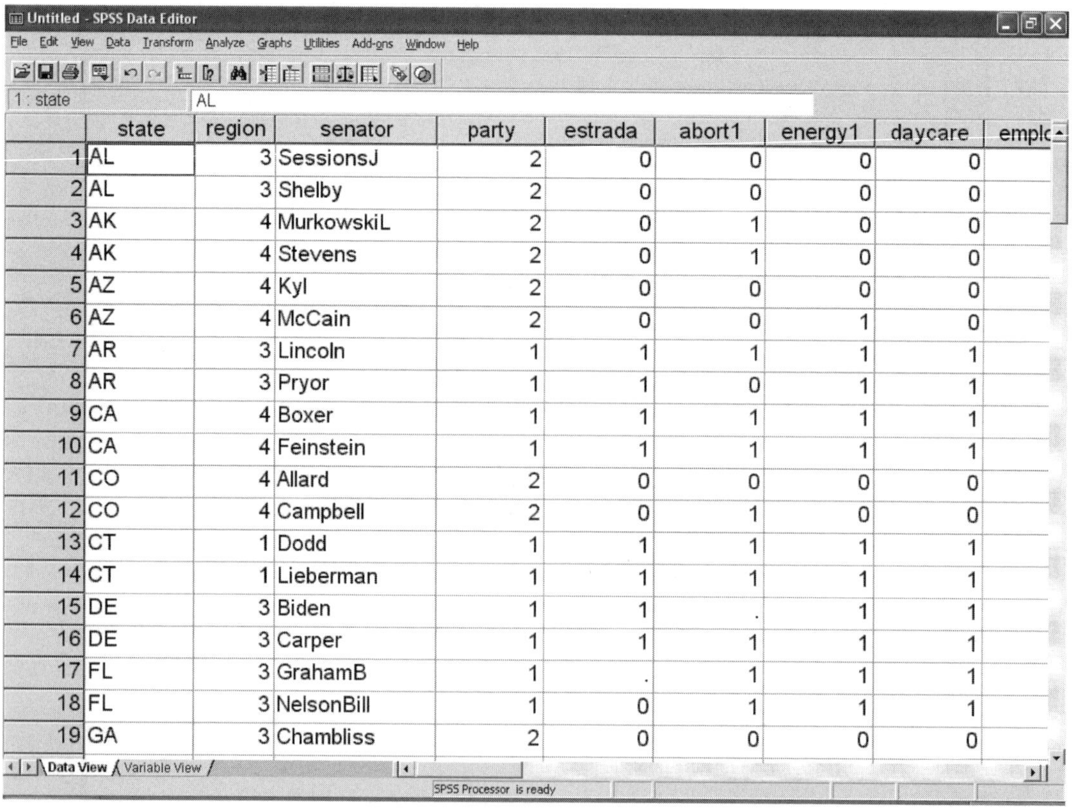

benefit votes (rxplan1 and rxplan2) and then use Correlate to see how strongly these two new variables are related. You can find additional information on each senator—the percentage of the vote they received in their most recent elections, the amount of money they spent on their campaigns, the number of years they have served in the Senate—and augment the dataset with this new information.[10] If you wanted to write a research paper on party, regionalism, and ideology in the Senate, Senate2003.sav would give you a good place to start.[11] Rewarding findings are guaranteed.

WRITING IT UP

At some point the analysis ends and the writing must begin. At this point, two contradictory considerations often collide. On one hand, you have an embarrassment of riches. You have worked on your research for several weeks, and you know the topic well—better, perhaps, than does anyone who will read the paper. You know the ins and outs of the data analysis. You know the inconsistencies in others' research on the topic. You know which variables and relationships yield notable measurements and outcomes and which do not. You want to include all of these things in your paper. On the other hand, you want to get it written, and you do not want to write a book. Viewed from an instructor's perspective, the two questions most frequently asked by students are "How should my paper be organized?" and "How long should it be?" (The questions are not necessarily asked in this order.)

Of course, different projects and instructors might call for different requirements. But here is a rough outline for a well-organized paper of 16 to 24 double-spaced pages (in a 12-point font).

I. The research question (3–4 pages)
    A. Introduction to the problem (1 page)
    B. Theory and process (1–2 pages)
    C. Propositions (1 page)
II. Previous research (2–4 pages)
    A. Descriptive review (1–2 pages)
    B. Critical review (1–2 pages)
III. Data and hypotheses (3–4 pages)
    A. Data and variables (1–2 pages)
    B. Measurement (1 page)
    C. Hypotheses (1 page)
IV. Analysis (5–8 pages, including tables)
    A. Descriptive statistics (1–2 pages)
    B. Bivariate comparisons (2–3 pages)
    C. Controlled comparisons (2–3 pages)
V. Conclusions and implications (3–4 pages)
    A. Summary of findings (1 page)
    B. Implications for theory (1–2 pages)
    C. New issues or questions (1 page)

The Research Question

Because of its rhetorical challenges, the opening section of a paper is often the most difficult to write. In this section you must both engage the reader's interest and describe the purpose of the research. Here is a heuristic device that may be useful: In the first page of the write-up, place the specific research problem (which the reader may or may not find fascinating) in the context of larger, clearly important issues or questions. Consider the example of Senator Jeffords, described earlier. You might begin a research paper on the Senate by recounting the Jeffords case, reminding the reader of the close partisan split in the chamber following the 2000 elections and describing Jeffords's high-profile shift in party alliances. A narrowly focused example? Yes. A dry topic? Not at all. The opening page of this paper could frame larger questions about the role of the Senate in U.S. government. To what extent does party serve as an organizing principle in Congress? How important are constituency or regional factors in shaping the behavior of elected representatives? Your analysis may appear "small," but it will advance our knowledge by illuminating one facet of a larger, more complex question.

Following the introduction, you begin to zero in on the problem at hand. The "theory and process" section describes the logic of the relationships you are studying. Many political phenomena, as you have learned, have competing or alternative explanations. Describe these alternatives, and the tension between them, in this section. Would you expect senators from the South to be more conservative than senators from the Northeast? If so, why? How important is party in explaining congressional voting behavior? Although a complete description of previous research does not appear in this section, you should give appropriate attribution to the most prominent work. These references tie your work to the scholarly community, and they raise the points you will cover in a more detailed review.

Round out the introductory section of your paper with a brief statement of purpose or intent. Think about it from the reader's perspective. Thus far you have made the reader aware of the larger context of the analysis, and you have described the process that may explain the relationships of interest. If this process has merit, then it should submit to an empirical test of some kind. What test do you propose? The "Propositions" page serves this role. Here you set the parameters of the research—informing the reader about the units of analysis, the concepts to be measured, and the type of analysis to be performed.

Previous Research

In this section you provide an intellectual history of the research problem, a description and critique of the published research on which the analysis is based. First you describe these previous analyses in some detail. What data and variables were used? What were the main findings? Did different researchers arrive at different conclusions? Political scientists who share a research interest often agree on many things. Yet knowledge is nourished through criticism, and in reviewing previous work you will notice key points of disagreement—about how concepts should be measured, what are the best data to use, or which variables need to be controlled. In the latter part of this section of the paper, you review these points and perhaps contribute to the debate. Previous studies of congressional voting, for example, have approached the measurement of ideology from different angles. Might not the use of separate measures—a measure of ideology on economic policy and a separate measure on social policy—be superior in some ways? A final, practical point: The frequently asked question "How many articles and books should be reviewed?" has no set answer. It depends on the project. However, here is an estimate: A well-grounded yet manageable review should discuss at least four references.

Data, Hypotheses, and Analysis

Together, the sections "Data and Hypotheses" and "Analysis" form the heart of the project, and they have been the primary concerns of this book. By now you are well versed in how to describe your data and variables and how to frame hypotheses. You also know how to set up a cross-tabulation or mean comparison table, and you can make controlled comparisons and interpret your findings.

In writing these sections, however, bear in mind a few reader-centered considerations. First, assume that the reader might want to replicate your study—collect the data you gathered, define and measure the concepts as you have defined and measured them, manipulate the variables just as you have computed and recoded them, and produce the tables you have reported. By explaining precisely what you did, your write-up should provide a clear guide for such a replication. Second, devote some space to a statistical description of the variables. Often you can add depth and interest to your analysis by briefly presenting the frequency distributions of the variables, particularly the dependent variable. A frequency distribution (or bar chart) of the ADA liberalism scale, for example, might show a distinct bimodal shape, with one group of senators at the liberal end and another at the conservative end. This distribution would make an effective prelude to the region-ideology analysis: Do regional differences help account for this pattern? Finally, exercise care in constructing readable tables. You can select, copy, and paste the tables generated by SPSS directly into a word processor, but they almost always require further editing for readability.

Conclusions and Implications

No section of a research paper can write itself. But the final section comes closest to realizing this optimistic hope. Here you discuss the analysis on three levels. First, you provide a condensed recapitulation. What are the main findings? Are the hypotheses borne out? Were any findings unexpected? Second, you describe where the results fit in the larger fabric of scholarly research on the topic. In what ways are the findings consistent with the work of previous researchers? Does your analysis lend support to one scholarly perspective as opposed to another? Third, research papers often include obligatory "suggestions for further research." Indeed, you might have encountered some methodological problems that still must be worked out, or you might have unearthed a noteworthy substantive relationship that could bear future scrutiny. Describe these new issues or questions. Here, too, you are allowed some room to speculate—to venture beyond the edge of the data and engage in a little "What if?" thinking. After all, the truth is still out there.

NOTES

1. You can browse ICPSR's holdings at www.icpsr.umich.edu.
2. For excellent guidance on the meaning and measurement of political knowledge, see Michael X. Delli Carpini and Scott Keeter, "Measuring Political Knowledge: Putting First Things First," *American Journal of Political Science* 37 (November 1993): 1179–1206.
3. The literature on retrospective voting in national elections is voluminous. For a general overview (and a specific accounting of the 2000 election), see Morris Fiorina, Samuel Abrams and Jeremy Pope, "The 2000 US Presidential Election: Can Retrospective Voting Be Saved?" *British Journal of Political Science* 33 (2003): 163–187. For an interesting analysis of retrospective voting at the local level, see Christopher R. Berry and William G. Howell, "Accountability and Local Elections: Rethinking Retrospective Voting," *Journal of Politics* 69 (August 2007): 844–858.
4. Data resources include State Court Processing Statistics, available through ICPSR. For an excellent example of how to merge county-level data with judicial statistics, see Noelle E. Fearn, "A Multilevel Analysis of Community Effects on Criminal Sentencing," *Justice Quarterly* 22 (December 2005): 452–487.
5. For a classic introduction to the study of turnout in comparative perspective, see the contributions to Richard Rose (ed.), *Electoral Participation: A Comparative Analysis* (Beverly Hills, Calif.: Sage, 1980).
6. Examples include three books published by CQ Press: *Who's Who in Congress,* offered twice a year through 2001; *CQ's Politics in America*, ed. Brian Nutting and H. Amy Stern, published every 2 years; and *Vital Statistics on American Politics*, by Harold W. Stanley and Richard G. Niemi, which also appears every 2 years. *Vital Statistics* is an excellent single-volume general reference on U.S. politics.
7. ADA's homepage is www.adaction.org.
8. SPSS will read data that are crunched together, with no delimiters between values. Fixed-format data, however, can present some complications worth avoiding. If you are entering your own data in a text file, the best practice is to put a space, comma, or tab between the values you are coding.
9. Because party is a potentially important independent variable, you probably want to treat Senator Jeffords (currently coded 3 on party) as a Democrat. Using Recode, create a new party variable for which old value 1 equals new value 1, old value 2 equals new value 2, and old value 3 equals new value 1.
10. For purposes of data augmentation, you can treat the Data Editor's Data View as a spreadsheet. Go to the first blank data column and type the data for each case directly into the Data Editor.
11. If you decide to use Senate2003.sav, you will want to type descriptive variable labels and value labels in the Variable View of the Data Editor. Table 11-1 suggests variable labels that you could use.

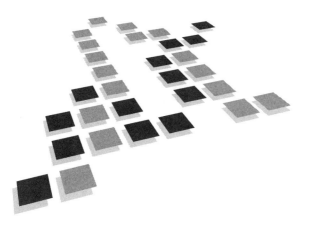

Appendix

Table A-1 Descriptions of Constructed Variables in GSS2006

Variable name and label	Source variables, GSS 1972–2006 cumulative data file[a]	Notes on construction of variable
authoritarianism Authoritarianism scale	thnkself obey	thnkself minus obey (thnkself–obey), rescaled 0 (low authoritarianism) to 7 (high authoritarianism). Measure based on Karen Stenner, *The Authoritarian Dynamic* (Cambridge: Cambridge University Press, 2005).
breadwinner Percent of total family $ earned by R	rincom06 income06	(rincom06 / income06) × 100.
dogmatism Religious dogmatism scale	punsin blkwhite rotapple permoral	Variables coded in "more dogmatic" direction, summed and rescaled, 0 (low dogmatism) to 12 (high dogmatism). Based on Ted G. Jelen and Clyde Wilcox, "Religious Dogmatism among White Christians: Causes and Effects," *Review of Religious Research* 33 (September 1991): 32–46.
econ_liberalism Economic liberalism scale	cutgovt lessreg	Variables summed, rescaled, 0 (most conservative) to 8 (most liberal).
fem_role Female role: Children, home, politics	fechld fepresch fefam fepol	Variables coded in "liberal" direction, summed, and rescaled, 0 (women domestic) to 12 (women in work/politics).

Continued

Table A-1 Descriptions of Constructed Variables in GSS2006 *Continued*

Variable name and label	Source variables, GSS 1972–2006 cumulative data file[a]	Notes on construction of variable
gun_control Gun control scale	gunsales gunsdrug semiguns guns911 rifles50 gunlaw	Number of pro-control responses summed, 0 (oppose control) to 6 (favor control).
job_autonomy How much autonomy does R have in job?	wkfreedm lotofsay	Variables summed, recoded in "more autonomy" direction, 1 (low autonomy) to 7 (high autonomy).
policy_know R's self-assessed knowledge: foreign / economic policy	knwforgn knwecon	Variables summed, recoded, and rescaled, 0 (low knowledge) to 8 (high knowledge).
pro_secure Civil liberties or security?	Based on GSS2006 variable civlibs	civlibs recoded (0–4 = 1) (5–9 = 0).
racial_liberal Racial liberalism scale	affrmact wrkwayup natrace	Variables collapsed, "most liberal" responses summed, 0 (least liberal) to 3 (most liberal).
science_quiz 10-item science test	hotcore boyorgrl electron bigbang condrift evolved earthsun radioact lasers viruses	Number of correct responses summed.
social_connect Social connectedness	socrel socommun socfrend socbar	Variables recoded and summed. Final scale collapsed, 0 (low connectedness) to 2 (high connectedness).
social_cons Social conservatism scale	abany grass cappun homosex pornlaw	Variables recoded, "least permissive" responses summed, 0 (most permissive) to 5 (least permissive).

Table A-1 *Continued*

Variable name and label	Source variables, GSS 1972–2006 cumulative data file[a]	Notes on construction of variable
spend6 Government increase $ on how many programs?	natcity natdrug nateduc natenvir natfare natheal	Number of "increase spending" responses summed.
suicide Incurable patient / incurable disease: suicide permissible?	letdie1 suicide1	Number of "yes" responses summed.
sz_place Size of place where R resides	xnorcsiz	Source variable recoded using this protocol: (1–2 = 1; 5–7 = 1; 3–4 = 2; 8–10 = 3). sz_place categories: 1 = city; 2 = suburb; 3 = rural.
tol_communist Tolerance twrd communists tol_racist Tolerance twrd racists	 libcom spkcom librac spkrac	Variables coded 0/1 in "more tolerant" direction and summed, 0 (least tolerant) to 2 (most tolerant).
trust_gov Level of trust in government	confed conjudge conlegis	Variables summed and recoded, 1 (low trust) to 7 (high trust).
trust_media Level of trust in media	conpress contv	Variables summed and recoded, 1 (low trust) to 4 (high trust).
trust_social Level of social trust	helpful trust fair	Number of "trusting" responses summed, 0 (low trust) to 3 (high trust).

[a]To find source variable question wording, go to the University of California Berkeley's Survey Documentation and Analysis (SDA) web site (http://sda.berkeley.edu/archive.htm).

Table A-2 Descriptions of Variables in States

Variable name and label	Description	Source and notes
abortion Abortions per 1000 women	Number of legal abortions per 1,000 women aged 15–44, 2000.	Alan Guttmacher Institute (www.guttmacher.org). Lawrence B. Finer and Stanley K. Henshaw, "Abortion Incidence and Services in the United States in 2000," *Perspectives on Sexual and Reproductive Health* 35 (2003): 6–15.
abortlaw Number of restrictions on abortion	Five variables coded separately and summed: (i) parental involvement = 2, none = 0; (ii) partial birth banned = 2, not = 0; (iii) counseling and waiting period = 2, counseling or waiting period = 1, neither = 0; (iv) public funding limited to life-endangerment/rape/incest = 2, exceptions for health reasons = 1, funding for all or most medically necessary = 0; (v) private insurance coverage limited to life endangerment = 2, providers may refuse to participate = 1, no restrictions = 0. Final variable: 0 (least restrictive) to 10 (most restrictive).	Guttmacher Institute, "State Policies in Brief: An Overview of Abortion Laws" (as of November 1, 2007; www.guttmacher.org/statecenter/spibs/spib_OAL.pdf). *Note:* All provisions enacted into law, including legally enjoined provisions, are counted as restrictions.
attend Percent frequent church attenders	Percentage attending religious services "every week" or "almost every week."	Pooled National Election Study, 1988–2002. Percentage coded 1 ("every week") or 2 ("almost every week") on question #vcf0130. *Note:* Thirteen states have fewer than 30 cases and were set to missing values on attend.
battle04 Battleground state	Coded 1 for 2004 battleground states, coded 0 for non–battleground states.	U.S. Department of State (http://usinfo.state.gov/dhr/Archive/2004/Jul/12-250886.html).
blkleg Percent of state legislators who are black	Percent of state legislators who are black, 2003.	National Conference of State Legislatures (www.ncsl.org/programs/legismgt/about/afrAmer.htm).
blkpct Percent black	Percent black or African American, 2004.	U.S. Census Bureau (www.census.gov/compendia/smadb/SMADBstate.html).

Table A-2   *Continued*

Variable name and label	Description	Source and notes
bush00 Percent voting for Bush 2000 gore00 Percent voting for Gore 2000 nader00 Percent voting for Nader 2000	Number of votes cast for Bush [Gore, Nader] as a percentage of all votes cast.	Federal Election Commission (www.fec .gov/pubrec/fe2000/2000presge.htm).
bush04 Percent voting for Bush 2004 kerry04 Percent voting for Kerry 2004	Number of votes cast for Bush [Kerry] as a percentage of all votes cast.	Federal Election Commission (www.fec .gov/pubrec/fe2004/federalelections2004 .shtml).
carfatal Motor vehicle fatalities (per 100,000 pop)	Motor vehicle fatalities per 100,000 population, 2002.	U.S. Census Bureau (www.census.gov/ compendia/smadb/TableA-13.pdf).
christad Percent of pop who are Christian adherents	Percentage of population who are Christian adherents, 2000.	U.S. Census Bureau (www.census.gov/ compendia/statab/tables/07s0075.xls). Dale E. Jones, Sherri Doty, Clifford Grammich, James E. Horsch, Richard Houseal, John P. Marcum, Kenneth M. Sanchagrin, and Richard H. Taylor. 2002. *Religious Congregations and Membership in the United States: 2000* (Nashville: Glenmary Research Center; www.glenmary.org/grc).
cigarettes Packs bimonthly per adult pop	Bimonthly pack sales per adult pop, 2003.	University of California, San Diego, Social Sciences Data Collection (http://ssdc.ucsd.edu/tobacco/sales/).
cig_tax cig_tax_3 Cigarette tax per pack	Cigarette tax per pack, 2007.	Campaign for Tobacco-Free Kids (http://tobaccofreekids.org/).

Continued

Table A-2 Descriptions of Variables in States   *Continued*

Variable name and label	Description	Source and notes
college Percent of pop w/college or higher	Percentage of population 25 years and over who have college degrees or higher.	U.S. Census Bureau (www.census.gov/compendia/smadb/TableA-22.pdf).
cons_hr Conservatism score, US House delegation	Mean American Conservative Union rating of state's delegation to House of Representatives, 2006.	American Conservative Union (http://acuratings.com).
defexpen Federal defense expenditures per capita	Federal defense expenditures per capita, 2003.	U.S. Census Bureau (www.census.gov/compendia/smadb/TableA-84.pdf).
demnat % US House and Senate Democratic	Number of Democratic House and Senate members as a percentage of total House and Senate delegations, 2006.	American Conservative Union (http://acuratings.com)
dempct_m Percent mass public Democratic reppct_m Percent mass public Republican indpct_m Percent mass public Independent libpct_m Percent mass public Liberal modpct_m Percent mass public Moderate conpct_m Percent mass public Conservative	Pooled CBS News/*New York Times* poll party identification and ideology estimates, 1977–1999.	Gerald C. Wright, Indiana University. Data used with permission. Robert S. Erikson, Gerald C. Wright, and John P. McIver, *Statehouse Democracy* (Cambridge, U.K.: Cambridge University Press, 1993).

Table A-2 *Continued*

Variable name and label	Description	Source and notes
demstate Percent state legislators Democratic	Number of Democratic state house and senate members as a percentage of total house and senate members, 2006 post-election/2007 pre-election.	National Conference of State Legislatures (www.ncsl.org/statevote/partycomptable2007.htm).
density Population per square mile	Population per square mile, 2005.	U.S. Census Bureau (www.census.gov/compendia/smadb/TableA-01.pdf).
gb_win00 Did Bush win electoral vote, 2000?	0 = Gore win / 1 = Bush win.	Federal Election Commission (www.fec.gov/).
gb_win04 Did Bush win electoral vote, 2004?	0 = Kerry win / 1 = Bush win.	Federal Election Commission (www.fec.gov/).
hispanic Percent hispanic	Percentage of population Hispanic, 2004.	U.S. Census Bureau (www.census.gov/compendia/smadb/TableA-05.pdf).
HS_or_more Percent of pop w/HS or higher	Percentage of population 25 years and over with high school degree or higher.	U.S. Census Bureau (www.census.gov/compendia/smadb/TableA-22.pdf).
over64 Percent of pop age 65 or older	Percentage of population aged 65 or older, 2004.	U.S. Census Bureau (www.census.gov/compendia/smadb/TableA-04.pdf).
permit Percent public "Always allow" abortion	Percentage responding that abortion should "always" be permitted.	Pooled National Election Study, 1988–2002. Percentage coded 4 on question #vcf0838. *Note:* Ten states have fewer than 30 cases and were set to missing values on permit.
pop_18_24 Percent age 18-24	Percentage of population aged 18 to 24, 2004.	U.S. Census Bureau (www.census.gov/compendia/smadb/TableA-04.pdf).
prcapinc Percapita income	Income per capita, 2004.	U.S. Census Bureau (www.census.gov/compendia/smadb/TableA-43.pdf).

Continued

Table A-2 Descriptions of Variables in States   *Continued*

Variable name and label	Description	Source and notes
relig_import Percent religion "A great deal of guidance"	Percentage responding that religion provides "a great deal of guidance" in life.	Pooled National Election Study, 1984–2004. Percentage coded 3 on question #vcf0847. *Note:* Twelve states have fewer than 30 cases and were set to missing values on relig_import.
turnout00 turnout04	Number of voters as a percentage of 2000 [2004] voting age population.	Committee for the Study of the American Electorate, press release, November 4, 2004 (http://american.edu/ia/cdem/csae/pdfs/csae041104.pdf).
unemploy Unemployment rate	Percent unemployed of the civilian labor force, 2004.	U.S. Census Bureau (www.census.gov/compendia/smadb/TableA-29.pdf).
union Percent workers who are union members	Percentage of workers who are members of a labor union, 2004.	U.S. Census Bureau (www.census.gov/compendia/smadb/TableA-33.pdf).
urban Percent urban pop	Percentage of population in urban areas, 2000.	U.S. Census Bureau (www.census.gov/compendia/smadb/TableA-02.pdf).
womleg Percent of state legislators who are women	Percentage of state legislators who are women, 2007.	Center for American Women and Politics (www.cawp.rutgers.edu/Facts/Officeholders/stleg.pdf).

Table A-3 Descriptions of Variables in World

Variable name	Variable label	Source and notes
colony	Colony of what country?	*CIA World Factbook.*
confidence	Confidence in institutions scale	Based on Global Indicators variable v52 (Confidence in state institutions early-mid 1990s; WVS).
decentralization decentralization4	Decentralization scale	Based on the sum of Global Indicators variables Political, Fiscal, and Admin. See A. Schneider, "Who Gets What from Whom? The Impact of Decentralisation on Tax Capacity and Pro-Poor Policy," *Institute of Development Studies Working Paper*, No. 179 (Brighton, U.K.: University of Sussex, 2003).
dem_other dem_other5	Percentage of other democracies in region	Calculated by author from region and democ_regime.
district_size3	Average # of members per district	Based on Global Indicators variable dm, mean district magnitude.
durable	Number of years since the last regime transition	Polity IV, 2000.
effectiveness	Government effectiveness scale	Kaufmann, 2002.
enpp_3	Effective number of parliamentary parties, june 2000 (Agora) (Banded)	Based on Global Indicators variable enpp (2000).
enpp3_democ	Effective number of parliamentary parties	Variable enpp3, restricted to democracies.
eu	EU member state (yes/no)	
fhrate04_rev	Freedom House rating of democracy (reversed)	Freedom House.
frac_eth frac_eth3	Ethnic fractionalization (combined linguistic and racial)	Alesina, 2003.
gdp_10_thou	GDP per capita in 10K US$	UNDP, 2004. Income data are from 2002.
gdp_cap2	GDP per capita (US$): 2 cats	UNDP.
gdp_cap3	GDP per capita (US$): 3 cats	UNDP.
gender_equal3	Gender empowerment measure	Based on Global Indicators variable GEMValue2004 (UNDP, 2004). http://hdr.undp.org/en/statistics/faq/question,81,en.html
Gini2004 gini04_4	GINI coefficient	UNDP.

Continued

Table A-3   Descriptions of Variables in World   *Continued*

Variable name	Variable label	Source and notes
indy	Year of independence	*CIA World Factbook.*
old2003	Population ages 65 and above (% of total) 2003	WB, 2004.
pmat12_3	Post-materialism	Based on Global Indicators variable pmat12 (postmaterialism 12 cat; WVS, 1995).
Pop2003	Population 2003	World Bank.
Popcat3	Size of country by population (3-categories)	UNDP.
pr_sys	PR system?	Based on Global Indicators variable elecpr (IDEA).
protact3	Protest activity	Based on Global Indicators variable protact (summary mean protest activity; WVS, 1995).
regime_type3	Regime types	Based on Global Indicators variable Cheibub4Type (2000). Author set mixed systems to missing.
typerel	Predominant religion	*CIA World Factbook.*
unions	Union density	ILO, 1995.
Urban2003	Urban population (% of total) 2003	WB, 2004.
vi_rel3	Percent saying religion "very" important	Based on Global Indicators variable vi_rel (Percent religion 'very' important; WVS, 1995).
votevap	Vote/vap during 1990s	IDEA.
women05	Percent women in lower house of parliament, 2005	IPU, 2005. Author restricted this measure to democracies only.
womyear	Year women first enfranchised	IPU.
yng2003	Population ages 0-14 (% of total) 2003	WB, 2004.

Source: All variables used in World were compiled and made available by Pippa Norris, John F. Kennedy School of Government, Harvard University, www.pippanorris.com. Professor Norris's dataset, Global Indicators Shared Dataset V2.0 (Updated Fall 2005 [September 13, 2005]), may be accessed at http://ksghome.harvard.edu/~pnorris/Data/Data.htm. Alesina = Alberto Alesina, Arnaud Devleeschauwer, William Easterly, Sergio Kurlat, and Romain Wacziarg, "Fractionalization," *Journal of Economic Growth* 8 (June 2003): 155–194 (retrieved from www.stanford.edu/~wacziarg/papersum.html). *CIA World Factbook* (retrieved from www.cia.gov). Freedom House = Freedom House, Gastil index, Annual to 2004 (retrieved from www.freedomhouse.org). IDEA = IDEA, voter turnout since 1945 (retrieved from www.idea.int). IPU = Inter-Parliamentary Union (retrieved from www.ipu.org). Kaufmann = Kaufmann Governance indicators (retrieved from www.worldbank.org/wbi/governance/govdata2002/). Polity IV = Polity IV Project (retrieved from www.cidcm.umd.edu/polity/data/). UNDP = Human Development Report OUP 2003 (retrieved from www.undp.org). WB = World Bank Development Indicators (retrieved from www.worldbank.org). World Bank = DPI2000 Database of Political Institutions (retrieved from http://econ.worldbank.org/). WVS = World Values Survey (1995–2000 waves; retrieved from wvs.isr.umich.edu).

Table A-4   Descriptions of Variables and Constructed Scales in NES2004

Variable name	Variable label	ANES source variables[a]	Notes on construction of variable
abort_funding	Favor/oppose govt funds to pay for abortion	V043179	
abort_imp	Importance of abortion issue to R	V045133	
abort_partial	Favor/oppose ban on partial-birth abortions	V043181	
abort_rights	Abortion position: self-placement	V045132	
afghan_cost	War in Afghanistan worth the cost	V043131	
age	Respondent age	V043250	
ashcroft_therm	Feeling Thermometer: John Ashcroft	V043047	
asian_therm	Feeling Thermometer: Asian Americans	V045075	
asian_traits asian_traits3	R's stereotypes of Asians: lazy/hrdwrking, unintel/intel, untrstwrthy/trstwrthy	V045225, V045229, V045233	Recoded (1=3, 2=2, 3=1, 4=0, 5=−1, 6=−2, 7=−3), summed; range = −9 (negative traits) to 9 (positive traits)
attend attend3	Attendance at religious services	V043223, V043224	
attent	Interested in following campaigns?	V043001	
bible	Bible is word of God or men	V043222	
bigbus_therm	Feeling Thermometer: Big Business	V045067	
black_affirmact	R favors/opposes preference for blacks in jobs?	V045207a	
black_jobs	Should government see to fair employment for blacks?	V045109a	
black_link	Scale: Strength of link between R and blacks in society	V045177, V045177a, V045178, V045179	Recoded in "stronger link" direction; V045177 and V045177a combined, summed with V045178 and V045179; rescaled 0 (weak link) to 9 (strong link); observed range = 1–9; black Rs only
black_therm	Feeling Thermometer: Blacks	V045077	

Continued

Table A-4 Descriptions of Variables and Constructed Scales in NES2004 *Continued*

Variable name	Variable label	ANES source variables[a]	Notes on construction of variable
black_traits black_traits3	R's stereotypes of blacks: lazy/hrdwrking, unintel/intel, untrstwrthy/trstwrthy	V045223, V045227, V045231	Recoded (1=3, 2=2, 3=1, 4=0, 5=−1, 6=−2, 7=−3), summed; range = −9 (negative traits) to 9 (positive traits)
bush_budget_dfct	Pres approval: handling budget deficit	V043031	
bush_econ	Pres approval: handling of econ	V043027	
bush_foreign_rel	Pres approval: handling foreign relations	V043029	
bush_job	Pres approval: general job handling	V043025	
bush_likes	Bush likes minus dislikes	V043007, V043009	V043007 (likes) minus V043009 (dislikes)
bush_terror	Pres approval: handling war on terror	V043033	
bush_therm	Feeling Thermometer: GW Bush	V043038	Pre-election wave
bush_therm_pst	Feeling Thermometer: GW Bush	V045043	Post-election wave
buspeople	Feeling Thermometer – business people	V045084	
camp_contrib	Summary: $ to cands / parties	V045014, V045014a, V045015, V045015a	Variables combined; range −2 (R $ to Rep cand and Rep pty) to 2 (R $ to Dem cand and Dem pty)
cand_contrib	Give $ to candidate?	V045014, V045014a	
care_house	Care who wins House election?	V043035	
care_pres	Care who wins presidential election?	V043092	
cathchurch_therm	Feeling Thermometer - the Catholic Church	V045085	
catholic_therm	Feeling Thermometer: Catholics	V045058	
cheney_therm	Feeling Thermometer: Cheney	V043041	
clinton_therm	Feeling Thermometer: Bill Clinton	V043045	
congress_job	Approve/disapprove Congress handling its job?	V043037	

Table A-4 *Continued*

Variable name	Variable label	ANES source variables[a]	Notes on construction of variable
congress_therm	Feeling Thermometer: Congress	V045076	
conservative_therm	Feeling Thermometer: Conservatives	V045069	
deathpen deathpen_strong	Strength R favors/opposes death penalty	V043187	
defense7	Defense spending—7-point scale self-placement	V043142	
defense_imp	Importance of defense spending issue to R	V043143	
dem_likes	Dem party likes minus dislikes	V043053, V043055	V043053 (likes) minus V043055 (dislikes)
dem_therm	Feeling Thermometer: Democratic party	V043049	
democrat	Is R Democrat?		See partyid3
denom	Summary: R's major religious group	V043247	
divided_govt	Better when one party controls presidency and congress?	V043060	
early_voter	Did R vote early?	V045023	
educ educ2 educ3 educ7	Highest grade of school or year of college R completed	V043252	
educ7_spouse	Summary: Spouse/partner education level	V043257	
educ_spouse	Spouse: highest grade or year of college	V043255	
edwards_therm	Feeling Thermometer: John Edwards	V043042	
egalitarianism egalit3	Egalitarianism scale	V045212, V045213, V045214, V045215, V045216, V045217	All coded in "more egalitarian" direction, summed, then rescaled from 0 (low egalitarianism) to 24 (high egalitarianism)
elderly_therm	Feeling Thermometer: Older people (the elderly)	V045071	

Continued

Table A-4 Descriptions of Variables and Constructed Scales in NES2004 *Continued*

Variable name	Variable label	ANES source variables[a]	Notes on construction of variable
employstat	R employment status	V043260b	
enviro_therm	Feeling Thermometer: environmentalists	V045072	
envirojobs	Environment vs. jobs tradeoff scale—self-placement	V043182	
envirojobs_imp	Importance of environment/jobs issue to R	V043183	
external_poleff	External political efficacy	V045201, V045202	Summed and collapsed (2–4 = 1; 5–7 = 2; 8–10 = 3)
fair_2000	Was 2000 Pres election fair/unfair	V043005	
fedgov_therm	Feeling Thermometer: Federal Government in Washington	V045060	
fedspend_borders	Fed Budget Spending: border sec to prevent illeg imm	V043173	
fedspend_child	Federal Budget Spending: child care	V043170	
fedspend_crime	Federal Budget Spending: dealing with crime	V043168	
fedspend_foreign	Federal Budget Spending: foreign aid	V043171	
fedspend_poor	Federal Budget Spending: aid to the poor	V043172	
fedspend_roads	Federal Budget Spending: building/repairing highways	V043164	
fedspend_school	Federal Budget Spending: public schools	V043166	
fedspend_science	Federal Budget Spending: science and technology	V043167	
fedspend_socsec	Federal Budget Spending: Social Security	V043165	
fedspend_terror	Federal Budget Spending: war on terrorism	V043174	
fedspend_welfare	Federal Budget Spending: welfare programs	V043169	

Table A-4   *Continued*

Variable name	Variable label	ANES source variables[a]	Notes on construction of variable
feminist_therm	Feeling Thermometer: Feminists	V045059	
foreign_expand	Foreign pol goals: Expansionist scale	V045100, V045104, V045106	Number of "Very important" responses summed
foreign_protect	Foreign pol goals: Protectionist scale	V045103, V045105, V045107	Number of "Very important" responses summed
funda_therm	Feeling Thermometer: Christian Fundamentalists	V045057	
gay_adopt	Gay adoptions OK?	V045158	
gay_discrim	Favor laws protecting gays?	V045156a	
gay_marriage	Allow gay marriage?	V043210	
gay_military	Allow gays in military?	V045157a	
gay_therm	Feeling Thermometer: Gay Men and Lesbians	V045074	
gender	R gender	V041109a	
gun_imp4	Importance of gun access issue to R	V043190	
guncontrol guncontrol2	How much easier/harder to buy gun—self-placement	V043189	
guncontrol_imp	Importance of gun access issue to R	V043190	
gunown	Does R have a gun in his or her home or garage?	V043195	
health_afford1	Did R put off medical treatment R could not afford?	V043066	
health_afford2	Can R afford needed health care?	V043067	
health_afford3	Does R have health insurance?	V043068	
help_blacks7	Government assistance to blacks-7 point scale self-placement	V043158	
help_blacks_imp	Importance of aid to blacks issue to R	V043159	
hh_kids	R have children in HH?	V041109c	Collapsed (0 = 0; 1–6 = 1)

Continued

Table A-4 Descriptions of Variables and Constructed Scales in NES2004 *Continued*

Variable name	Variable label	ANES source variables[a]	Notes on construction of variable
hhold_comp	Household composition	V041110	
hillary_therm	Feeling Thermometer: Hillary Clinton	V043044	
hispanic7	R position on aid to Hispanic-Americans scale	V045140	
hispanic_link	Scale: Strength of link between R and Hispanics in society	V045180, V045180a, V045181, V045182	Recoded in "stronger link" direction; V045180 and V045180a combined, summed with V045181 and V045182; rescaled 0 (weak link) to 9 (strong link); Hispanic Rs only
hispanic_therm	Feeling Thermometer: Hispanics (Hispanic-Americans)	V045056	
hispanic_traits hispanic_traits3	R's stereotypes of hispanics: lazy/hrdwrking, unintel/intel, untrstwrthy/trstwrt	V045224, V045228, V045232	Recoded (1=3, 2=2, 3=1, 4=0, 5=−1, 6=−2, 7=−3), summed; range = −9 (negative traits) to 9 (positive traits); observed range = −8 to 9
illegals_therm illegals_therm3	Feeling Thermometer—illegal immigrants	V045081	
immigration	Should immigration be increased, decreased, stay same?	V045115	
imports	Favor or oppose limits on foreign imports	V045114	
income_hh income_hh3	Household income	V043293x	
income_r	Respondent income	V043294	
inflation_future	Inflation in next year?	V043106	
inflation_past	Inflation much or somewhat better/worse in last year?	V043105	
interventionism	Interventionism by diplomacy/military: self-placement	V043107	
interventionism_imp	Importance of diplomacy issue to R	V043108	
interventionism_imp_pst	Interventionism by diplomacy/military: importance	V045125	

Table A-4   *Continued*

Variable name	Variable label	ANES source variables[a]	Notes on construction of variable
interventionism_pst	Interventionism by diplomacy/military: self-placement	V045124	
iraq_approve	How much approve/disapprove Bush handling Iraq war	V043133	
iraq_cost	Was Iraq war worth the cost	V043134	
iraq_terror	Iraq war increased or decreased threat of terrorism	V043135	
israel_therm	Feeling Thermometer: Israel	V045087	
jew_therm	Feeling Thermometer: Jews	V045061	
jobs7	Job and Good Standard of Living - scale self-placement	V043152	
jobs_imp	Importance of guaranteed jobs/standard living issue	V043153	
kerry_likes	Kerry likes minus dislikes	V043011, V043013	V043011 (likes) minus V043013 (dislikes)
kerry_therm	Feeling Thermometer: John Kerry	V043039	Pre-election wave
kerry_therm_pst	Feeling Thermometer: John Kerry	V045044	Post-election wave
laura_bush_therm	Feeling Thermometer: Laura Bush	V043043	
libcon7_bush	Liberal/conservative placement—GW Bush	V043087	
libcon7_dem	Liberal/conservative placement—Dem Party	V043090	
libcon7_kerry	Liberal/conservative placement—Kerry	V043088	
libcon7_nader	Liberal/conservative placement—Nader	V043089	
libcon7_r libcon3_r	Liberal/conservative self-placement—7-point scale	V043085	
libcon7_rep	Liberal/conservative placement—Rep party	V043091	
liberal_therm	Feeling Thermometer: Liberals	V045062	

Continued

Table A-4 Descriptions of Variables and Constructed Scales in NES2004 *Continued*

Variable name	Variable label	ANES source variables[a]	Notes on construction of variable
marital	Marital status	V043251	
mccain_therm	Feeling Thermometer: John McCain	V043048	
medinsur7	Govt/private medical insurance scale: self-placement	V043150	
medinsur7_imp	Importance of govt health insurance issue to R	V043151	
memnum memnum2	Number of group memberships	V045170, V045170a	
men_therm	Feeling Thermometer—men	V045079	
midclass_therm	Feeling Thermometer—middle class people	V045063	
military_imp	How important is it for US to have strong military?	V045108	
military_therm	Feeling Thermometer: The Military	V045066	
moralism moralism3	Moralism scale	V045189, V045190, V045191, V045192	Recoded in "more moralistic" direction, summed; rescaled 0 (low moralism) to 16 (high moralism)
muslim_therm	Feeling Thermometer: Muslims	V045088	
nader_therm	Feeling Thermometer: Nader	V043040	Pre-election wave
nader_therm_pst	Feeling Thermometer: Ralph Nader	V045045	Post-election wave
online_news_days	How many days past week read a daily online newspaper	V043020	
paper_days	How many days past week read a daily newspaper?	V043019	
parents_usa	Both parents born in U.S.?	V043300	
party_better_econ	Which party better?: handling nation's economy	V043109	
party_better_nowar	Which party better?: handle keeping out of war	V043111	
party_better_terror	Which party better?: handle war on terrorism	V043110	

Table A-4 *Continued*

Variable name	Variable label	ANES source variables[a]	Notes on construction of variable
party_contrib	Give $ to party?	V045015, V045015a	
partyid7 partyid3	Summary: R party ID	V043116	
patriotism patriotism_nonH	Patriotism scale	V043205, V043208, V043209	Number of "extremely" responses summed; range = 0–3
polinfo_iw	Interviewer observation: R level of political information	V043403	
polknow polknow3	Wh/party controls Hse, Sen?; Who are Hastert, Rehnquist, Blair, Cheney?	V045089, V045090, V045162, V045163, V045164, V045165	Number of correct responses summed
poor_therm	Feeling Thermometer: Poor people	V045065	
powell_therm	Feeling Thermometer: Colin Powell	V043046	
pray	How often does R pray	V043221	
predict_close_state	Will pres race be close in state?	V043096	
predict_close_win	Will pres race be close or will (winner) win by a lot?	V043094	
predict_win	Who does R think will be elected president?	V043093	
predict_win_state	Which pres cand will carry state?	V043095	
prochoice	Prochoice abortion scale		Constructed from created variables: abort_rights, abort_funding (recoded 1=4, 2=3, 4=2, 5=1), abort_partial (recoded 1=1, 2=2, 4=3, 5=4); variables summed, rescaled; range = 0 (low prochoice) to 9 (high prochoice)
progovmnt	Pro big-government scale	V045150, V045151, V045152	Number of pro-government responses summed; range = 0–3

Continued

Table A-4 Descriptions of Variables and Constructed Scales in NES2004 *Continued*

Variable name	Variable label	ANES source variables[a]	Notes on construction of variable
r_econ_future	R better/worse off 1 year from now?	V043064	
r_econ_past	R better/worse off than 1 year ago?	V043062	
race race_2 race_3	Race of Respondent	V043299	
racial_cons4	Racial conserv: 4 cats		
racial_conservatism	Racial conservatism scale	V045193, V045194, V045195, V045196	All variables coded in "more conservative" direction, summed, rescaled; range 0 (low racial conservatism) to 16 (high racial conservatism)
reagan_therm	Feeling Thermometer: Ronald Reagan	V043051	
region	Region	V041205	
relimp	Religion: guidance in day-to-day living?	V043220	
rep_likes	Rep party likes minus dislikes	V043057, V043059	V043057 (likes) minus V043059 (dislikes)
rep_therm	Feeling Thermometer: Republican party	V043050	
retired	Is R retired?		Based on employstat (5 = 1; 1–4 = 0; 6–8 = 0)
rich_therm	Feeling Thermometer—rich people	V045082	
right_track	Are things in the country on right track?	V043034x	
servspend7	Spending and Services—7-point scale self-placement	V043136	
servspend_imp	Importance of spending/services issue to R	V043137	
social_trust	Social trust	V045186, V045187, V045188	Number of "more trusting" responses summed; range = 0–3
socsec_reform	Summary: favor/oppose investing Soc Sec funds	V045143b	

Table A-4   *Continued*

Variable name	Variable label	ANES source variables[a]	Notes on construction of variable
south	Non-South/South		Based on region (3 = 1; else 0)
southern_therm	Feeling Thermometer—Southerners	V045078	
spend11	IncrFed$, 11 prgms	V043164, V043165, V043166, V043167, V043168, V043169, V043170, V043171, V043172, V043173, V043174	Number of "spending should be increased" responses summed
state	FIPS state code	V041202	
stock_investor	Does R/spouse have any money invested in stock market?	V043065	
supcourt_therm	Feeling Thermometer: U.S. Supreme Court	V045073	
taxcut	Does R favor/oppose tax cuts Pres. Bush initiated	V043148	
timing timing2	Timing of R's vote choice	V045027	
tv_days	Days past week watch natl news on TV	V043014	
unemploy_future	R think there more or less unemployment in next year?	V043103	
unemploy_past	Unemployment better or worse in last year?	V043102	
union_hh	Anyone in HH belong to labor union?	V043290	
union_therm	Feeling Thermometer: Labor Unions	V045064	
us_econ_bush	National economy better/worse since GW Bush took ofc	V043213	
us_econ_future	Will national economy be better or worse in next 12 mos?	V043100	

Continued

Table A-4 Descriptions of Variables and Constructed Scales in NES2004 *Continued*

Variable name	Variable label	ANES source variables[a]	Notes on construction of variable
us_econ_past us_econ_past3	National economy better/worse in last year?	V043098	
us_isolate	Country would be better off if we just stayed home	V043113	
us_moral_bush	Has US moral climate gotten better/worse since 2000	V043217	
us_secure_bush	Has current admin made U.S. more/less secure	V043215	
us_world	During last year, U.S. position in world weaker/stronger?	V043112	
veteran	Ever served or currently serving in US military?	V043258	
vote_2000	Did R vote in 2000?	V043002	
vote_2004	Did R vote in 2004?	V045018x	
vouchers	Favor/oppose school vouchers	V045144a	
welfare3	Increase fed welfare$?		Based on fedspend_welfare (1=1)(5=2)(3=3)(7=.)
welfare_therm	Feeling Thermometer: People on welfare	V045068	
white_therm	Feeling Thermometer: Whites	V045086	
white_traits white_traits3	R's stereotypes of whites: lazy/hrdwrking, unintel/intel, untrstwrthy/trstwrthy	V045222, V045226, V045230	Recoded (1=3, 2=2, 3=1, 4=0, 5=–1, 6=–2, 7=–3), summed; range = –9 (negative traits) to 9 (positive traits)
who04_2	Kerry or Bush?	V045026	
who_2000	2000 vote choice	V043003	
who_2004	R's vote for president	V045026	
women_jobs	Equal treatment for women in jobs	V045161a	
women_link	Scale: Strength of link between R and women in society	V045174, V045174a, V045175, V045176	Recoded in "stronger link" direction; V045174 and V045174a combined, summed with V045175 and V045176; rescaled 0 (weak link) to 9 (strong link); female Rs only

Table A-4 _Continued_

Variable name	Variable label	ANES source variables[a]	Notes on construction of variable
women_therm	Feeling Thermometer—women	V045083	
women_work	Women cause problems or face discrimination?	V045183, V045184, V045185	Coded in "pro-female" direction, summed; rescaled 0 (women not discriminated against) to 12 (women discriminated against)
womrole7	Women's role—7-point scale self-placement	V043196	
womrole_imp	How important is the issue of women's equal role	V043197	
workclass_therm	Feeling Thermometer: working class people	V045070	
yob	Year of birth	V043249a	
young_therm	Feeling Thermometer—young people	V045080	

[a]To find source variable question wording, go to the University of California Berkeley's Survey Documentation and Analysis (SDA) web site (http://sda.berkeley.edu/archive.htm).